Integrated Circuits and Systems

Series editor

Anantha P. Chandrakasan, Massachusetts Institute of Technology
Cambridge, MA, USA

More information about this series at http://www.springer.com/series/7236

Gaudenzio Meneghesso • Matteo Meneghini
Enrico Zanoni
Editors

Gallium Nitride-enabled High Frequency and High Efficiency Power Conversion

Springer

Editors
Gaudenzio Meneghesso
Department of Information Engineering
University of Padova - DEI
Padova, Padova, Italy

Matteo Meneghini
Department of Information Engineering
University of Padova - DEI
Padova, Padova, Italy

Enrico Zanoni
Department of Information Engineering
University of Padova - DEI
Padova, Padova, Italy

ISSN 1558-9412
Integrated Circuits and Systems
ISBN 978-3-030-08594-0 ISBN 978-3-319-77994-2 (eBook)
https://doi.org/10.1007/978-3-319-77994-2

Printed on acid-free paper

This Springer imprint is published by the registered company Springer International Publishing AG part of Springer Nature.
The registered company address is: Gewerbestrasse 11, 6330 Cham, Switzerland

Preface

Gallium nitride represents today a semiconductor that may provide significant improvements in many applications, spanning from the solid-state lighting to very-high-frequency transmissions to power-switching converters. In this book, we will focus on the electrical properties of electronic devices (transistors) for very-high-power and very-high-frequency applications.

Radio-frequency (rf) power is important to a wide range of applications, including radio transmitters, plasma generation, medical imagers (e.g. MRI), power converters and wireless power transfer (WPT) among myriad other applications. If we consider rf applications, the material properties of GaN offer clear advantages over the other more commonly used semiconductors, such as silicon (Si), gallium arsenide (GaAs) and indium phosphide (InP). RF transistors offer orders of magnitude more times the theoretical maximum output power density of GaAs or silicon transistors. Additional key characteristics of GaN transistors include high cut-off frequency and good thermal conductivity. GaN devices offer the best solution for simultaneous high-power, high-frequency and high-temperature operation.

Power electronics is the technology associated with the efficient conversion, control and conditioning of electric energy from the source to the load. It is the enabling technology for the generation, distribution and efficient use of electrical energy. It is a cross-functional technology covering the very high Giga-Watt (GW) power (e.g. in energy transmission lines) down to the very low milli-Watt (mW) power needed to operate a mobile phone. Many market segments, such as domestic and office appliances, computer and communication, ventilation, air conditioning and lighting, factory automation and drives, traction, automotive and renewable energy, can potentially benefit from the application of power electronics technology.

Power devices based on wide-bandgap semiconductors (WBS) like gallium nitride (GaN) and silicon carbide (SiC) are poised to play an important role in future power electronics systems. WBS has a high breakdown strength and, in the case of GaN, allows for fabrication of lateral and vertical devices: (i) lateral high-electron mobility lateral transistors, for which the electron mobility is not degraded as would be the case for traditional silicon MOSFETs; (ii) vertical devices allowing large current densities and very high breakdown voltage. Together, these facts allow the fabrication of devices, which have orders of magnitude better performances than the other more known semiconductors (Si, GaAs, InP).

After being the semiconductor of the endless promises without ever coming to the reality today, things are changing: (i) the cost of GaN materials is largely decreasing, making these devices economically competitive; (ii) devices stability and robustness have been already proven among several GaN devices providers; and (iii) system employing GaN devices start to appear in the market. These facts pushed us to write this book reporting the latest results on this technology spanning from available material to the final applications.

Padova, Italy

Gaudenzio Meneghesso
Matteo Meneghini
Enrico Zanoni

Introduction

It is well established that today semiconductor devices based on silicon (Si) have reached their physical limits either in terms of scaling or in terms of their physical properties. To further improve systems performance and reliability, new materials must be explored. To this aim, gallium nitride (GaN) (a wide-bandgap semiconductor material) is the semiconductor of choice: its basic physical material properties in terms of high-conductivity, high-temperature, high-frequency and high-power operation make this semiconductor the unique materials of choice to address the requirements of the future advance systems in the field of radio frequency and power conversion applications.

The technological maturity of GaN has been largely improved in the past years; however, work still has to be carried out to exploit the full capability of GaN devices by keeping the cost at the same level of the silicon devices and guaranteeing the high demand of reliability and robustness that are required in the fields of application (satellite, automotive, industrial, etc.).

This book reports an overview of the GaN-based technology starting from the material availability, to the device architecture, analysing reliability and parasitic aspects, and finally looking at the application of these devices in the power conversion field as well as in the radio-frequency amplifier applications.

Chapter 1 (by Joff Derluyn – *Epigan nv*, Marianne Germain – *Epigan nv* and Elke Meissner – *Fraunhofer IISB*) discusses all the issues related to the growth of GaN substrates. GaN technology was originally developed on sapphire substrates. Yet, even though scientists worldwide managed to turn GaN epitaxial layers grown on sapphire and SiC substrates into devices with exceptionally high performance, these substrates are lacking in two important aspects: first, the threading dislocation density of epitaxial layers grown by hetero-epitaxy are very high (in the order of $10^8/cm^2$ and above); secondly, these substrates are expensive by themselves and are incompatible with the wide range of low-cost enabling processing techniques of the Si semiconductor industry. This chapter discusses the answer to these shortcomings, respectively the two ends of the string: the growth of bulk GaN crystals by the ammonothermal method and by hydride vapour phase epitaxy (HVPE) on the one hand, and hetero-epitaxial growth of GaN on silicon substrates (GaN-on-Si) by the more industrial metal-organic vapour phase deposition (MOCVD) on the other hand. We will focus here on the challenges of each technique and their possible solutions, and outline the implications of the technological choices that can be made. As substrate material for GaN, silicon is not perfect. It has a large mismatch in the lattice parameter and the thermal expansion coefficient that result in large strain build-up in III-nitride epilayers that are grown on top. The development of suitable strain management techniques has allowed overcoming these technical hurdles to some degree, and GaN-on-Si epitaxial wafers with 200 mm diameter are commercially available today. On the other hand, having a native GaN substrate would pave the way to ultra-low defect densities in the material and enable new (e.g. vertical) device structures.

Chapter 2 (Chang Soo Suh – *Texas Instruments*) describes the basic GaN-based HEMT device, including the polarization effects and surface states responsible for the formation of the 2DEG in AlGaN/GaN heterostructures. With the push for performance progression greater than the incremental steps, coupled with increasing demand for power conversion systems operating at higher frequencies, group III-nitride family of semiconductors, capable of delivering superior performances beyond the limits of Si, appears poised to become the next power-switching device material of choice with the gallium nitride (GaN) high-electron mobility transistor (HEMT) device at its forefront. GaN HEMT device structure innovations for increasing the channel mobility, reducing the current collapse phenomena, achieving high-voltage breakdown and enabling normally-off operation are presented.

Chapter 3 (Srabanti Chowdhury and Dong Ji – ECE Department, University of California, Davis) titled "Vertical GaN Transistors for Power electronics" takes the reader through the research and development cycle of GaN vertical technology, detailing out the three-terminal devices developed over the last decade. Power converters rely on solid-state devices featuring diodes and transistors as their basic building blocks. GaN technology is an ever-expanding topic for R&D, proving its potential to solve several challenges in power conversion that cannot be addressed by Si. Medium voltage (650–900V) devices using the HEMT configuration have been able to reduce form factor at the system level by driving circuits at higher frequencies (100KHz–1MHz) and eliminating heat sinks or reducing cooling requirements. This alone sparked the interest in GaN device research to address power conversion needs. However, in power conversion the demand of high current (50A and higher) from a single chip for a rated voltage (1KV and higher) is a standard requirement. Particularly when the market is favorable toward electrification of cars and other means of transportations, GaN must expand its scope to provide high power solutions with higher power density compared to Si, and even SiC. Vertical devices have been the choice of power device engineers for economic use of the material and maximum use of its physical properties (which allow highest possible blocking field, field mobility, etc.). In this chapter, we discuss vertical transistors first in its normally on form (CAVETs) and then in its normally off design (MOSFET). The advantages and disadvantages are discussed for each type besides describing their operation principles. We have tried to make this chapter scholastic and informative by use of modeling and experimental data for each device we describe. The chapter will help the reader to realize the most recent status of GaN vertical transistors and appreciate its potential in power conversion.

Chapter 4 (Isabella Rossetto et al. - *University of Padova*) describes the main limitation, in terms of parasitic and reliability, of GaN-based transistors. No new product is possible without reliability: this is especially true for new and emerging technology, such as gallium nitride-based devices. For GaN power transistors, breakdown mechanisms play a significant role. The reduction of the robustness and the long-term reliability still represent a serious issue that must be taken into consideration. The first part of the chapter deals with the abovementioned aspects and mainly focuses on the permanent degradation induced in GaN-based devices by off-state time-dependent mechanisms. The second part of the chapter analyses the degradation mechanisms, which affect HEMTs with a p-type gate submitted to a high positive bias. An overview of the main results reported in the literature concerning the origin of the permanent degradation is discussed. Results on the recoverable trapping mechanisms are furthermore provided. The third part analyses the instabilities in MISHEMT structures. A detailed analysis of negative bias threshold voltage instabilities (NBTI) is discussed in terms of dependence on the applied temperature and performance-worsening induced by a cascode configuration.

Chapter 5 (Kenichiro Tanaka, Ayanori Ikoshi, and Tetsuzo Ueda - *Panasonic*) discusses how to evaluate the robustness of GaN power transistors. In recent years, as GaN power transistors come into widespread use as switches for power converter applications, it is crucial and inevitable to guarantee their reliability. In the switching of GaN power transistors, they can be subject to the so-called current collapse that is a specific phenomenon for GaN in which the ON state resistance is increased once the device is exposed to a high voltage. Since the current collapse induces instability of the device in the form of the increase in the temperature, non-uniform internal electric field distribution and so on, it may lead to the reliability issue. Therefore, the robustness of GaN transistors should be examined under switching operation besides the conventional reliability tests standardized for Si power transistors. Since current collapse is crucially dependent on the drain current-voltage locus curve during the switching event, the switching reliability of GaN transistor depends on the switching locus. Accordingly, a concept of Switching Safe Operating Area (SSOA) is proposed to define the switching conditions wherein the device can be switched safely. As an example, we define the SSOA for our hybrid-drain-embedded gate injection transistor (HD-GIT) that is now commercially available. Furthermore, we propose the long-time SSOA (LSSOA) in which we guarantee the robustness of HD-GIT under long-time switching operations (e.g. 10 years). The proposed method for confirming the robustness of GaN power transistors can be utilized to estimate the devices' lifetime when they are employed in a given switching application.

Chapter 6 (Johan T. Strydom - *Texas Instruments*) describes how parasitics impacts the performances of the power conversion based on GaN devices. The concept of circuit "parasitics" has, for the most part, been an attempt to vilify the unwanted or unexpected device and system-level characteristics whenever they were found to be limiting the system operation or performance in some manner.

The current approach has always been to mitigate the effects of these unwanted parasitics through design improvements, be it on a device, package or system level. As these parasitics are, for the most part, determined by their geometry, the methods for their improvements are also been spatial in nature. To start, the relevant GaN power conversion parasitics can, roughly, be broken up into three distinct categories, namely: (i) GaN device level parasitics, internal to the die itself; (ii) package and board level (interconnect) parasitics that directly impact device performance; and (iii) external board and system-level components and their parasitics that impact the overall system performance and operation. In this chapter, we will discuss the impact of a wide range of the so-called "parasitics" with respect to GaN-based power conversion. We will conclude with a brief speculative discussion on the future impact of these parasitics, as GaN technology continues to improve over the coming years, including some thoughts on relevant areas of research for future system improvements.

Chapter 7 (Fred Wang, Bo Liu - *University of Tennessee, Knoxville*) describes the advantages and challenges of applying GaN devices in AC/DC converters. With high-switching speed and low-conduction loss, GaN device technology is expected to benefit future power electronics in several aspects: improved efficiency and power density, simplified converter topology, new system-level functions and new applications. In this chapter, these benefits for GaN-based AC/DC converters in both single-phase and three-phase systems are described. Main challenges stemming from high-switching frequency, high di/dt and dv/dt, and low device size are analysed, and special design considerations are addressed. For single-phase ac/dc converters, the advantage of topology simplification by adopting GaN devices is highlighted in the hard-switching PV and power factor correction (PFC) applications. New topologies and modulation schemes are reviewed in emerging applications of GaN such as wireless power transfer and medical power supply. New challenges and solutions when applying GaN in high-frequency converter design are analysed in soft-switching totem-pole PFC converters. For three-phase ac/dc, attempted application of GaN in PV, motor drive and battery charger systems is examined. Control and cooling design challenges associated with GaN parasitic capacitance, switching transition and low-profile package are then analysed in detail, especially for high-frequency compact converters. Approaches from modulation compensation and sampling scheme, as well as different thermal solutions, are explored to address these new issues.

Chapter 8 (Dave Perreault - MIT, Juan Rivas-Davila - Stanford, Charles Sullivan - Dartmouth) describes the advantages and opportunities in applying GaN into switched-mode power amplifiers.

Radio-frequency (rf) power is important to a wide range of applications, including radio transmitters, plasma generation, medical imagers (e.g. MRI), power converters and wireless power transfer (WPT) among myriad other applications. Advances in power semiconductor devices, magnetics and circuit design are opening the door to much more efficient generation and delivery of power at radio frequencies. This chapter presents an overview of switched-mode power amplifiers – or radio-frequency inverters – encompassing their design, control and construction.

We focus on the high-frequency (HF, 3–30 MHz) and very-high-frequency (VHF, 30–300 MHz) ranges. We explore key aspects of rf power conversion, including power circuit architecture and design, selection and efficient drive of power devices at rf and control methods for modulating power and managing load variations. We also address circuit construction, including the design and application of passive components at radio frequencies. Magnetics for power applications at HF and VHF pose a special challenge when compactness and high efficiency are desired. We explore the design of cored and coreless inductors and transformers for this frequency range, including winding design, core material evaluation and selection, and application of magnetic cores.

Department of Information Engineering
University of Padova - DEI, Padova, Padova, Italy

Gaudenzio Meneghesso

Contents

Chapter 1
Taking the Next Step in GaN: Bulk GaN Substrates and GaN-on-Si Epitaxy for Electronics

Joff Derluyn, Marianne Germain, and Elke Meissner

1.1 Introduction

As the size of the global market for LED lighting is expected to surpass 50 billion dollars by 2022 [1], it is not unfair to say that gallium nitride (GaN) is the second most important semiconductor material, after silicon.

Unfortunately, nature does not provide GaN bulk crystals. Therefore, in the late 1980s and early 1990s, GaN technology for optoelectronics lacked a native substrate and needed to turn to heteroepitaxy on sapphire. Many of the critical breakthroughs in technological skills and understanding of GaN material, concerning, for instance, the development of hetero-epitaxial growth techniques themselves and the behavior of Mg as p-type dopant, were obtained in home-built equipment on small-diameter sapphire substrates. Still today, a large part of the production of LEDs is done on 100 mm sapphire substrates in dedicated GaN foundries.

Piggybacking on the rapid developments of the optoelectronics, soon also a superior transistor concept was demonstrated in GaN, in the form of the high-electron-mobility transistor (HEMT) based on the concept of the formation of a two-dimensional electron gas (2DEG) at the interface between GaN and AlGaN layers [2]. Due to (Al)GaN's high critical breakdown field combined with the excellent transport properties of the 2DEG, it was soon understood that such transistors could operate at much higher bias voltages than competing GaAs or Si transistors, leading to a tenfold increase in RF output power density. The

J. Derluyn · M. Germain
EpiGaN nv, Hasselt, Belgium
e-mail: joff.derluyn@epigan.com; marianne.germain@epigan.com

E. Meissner (✉)
Fraunhofer Institute for Integrated Systems and Device Technology IISB, Erlangen, Germany
e-mail: elke.meissner@iisb.fraunhofer.de

G. Meneghesso et al. (eds.), *Gallium Nitride-enabled High Frequency and High Efficiency Power Conversion*, Integrated Circuits and Systems,
https://doi.org/10.1007/978-3-319-77994-2_1

substrate of choice for the GaN RF technology, which was quickly adopted for high-end applications in the space and defense fields, is silicon carbide (SiC). Although the SiC is definitely not a cost-effective solution, its crystal structure and parameters are relatively closely matched to those of GaN which enables high-quality heteroepitaxy. Important for high power electronics, SiC has an excellent thermal conductivity so that devices made on top of GaN on SiC can more easily handle high power densities.

Yet, even though scientists worldwide managed to turn GaN epitaxial layers grown on sapphire and SiC into devices with exceptionally high performance, e.g., as expressed in luminous efficacy for LEDs or power density and drain efficiency for RF amplifiers, these substrates are lacking in two important aspects: first, the threading dislocation density of epitaxial layers grown by heteroepitaxy is very high (in the order of $10^8/cm^2$ and above); second, these substrates are expensive by themselves and are incompatible with the wide range of low-cost-enabling processing techniques of the Si semiconductor industry. On the other hand, having a native GaN substrate, this would pave the way to ultralow defect densities in the device structures. At the time being, this is however the most expensive route. In this chapter, we will discuss the answer to these shortcomings, respectively, the two ends of the string: the growth of bulk GaN crystals on the one hand and hetero-epitaxial growth of GaN-on-silicon substrates on the other hand.

1.2 Brief Overview of the State of the Art of GaN Bulk Crystal Growth and Native Substrates

For the production of a modern semiconductor device, the control and mastering of thin film techniques are essential. This includes metallization, dielectric, or passivation layers but also crystalline functional layers of high crystal quality. The crystalline layers are deposited by epitaxy which means the layer-by-layer expansion of the crystal on the surface of a substrate wafer. In this sense, the growth of a bulk crystal and the epitaxy of a crystalline layer are no different. However, mostly in epitaxy, the substrate and the crystalline layer are of different materials and as such have different lattice geometries and thermal expansion coefficients. This situation is usually called heteroepitaxy and has naturally, due to the differences of the materials, severe physical consequences like the formation of dislocations at the interfaces, buildup of strain, and influences on the reliability of the devices that are fabricated in this material. If the materials are of the same kind, these issues are considered not to apply, and the case is usually referred to as homoepitaxy.

For GaN, the heteroepitaxy case is the most abundant due to the scarcity of native substrate wafers. There are today nevertheless some suppliers for freestanding GaN wafers, but the costs are unacceptably high (approximately 2500$ per 2″ piece), and the available diameters are only 2″ at the maximum. Moreover, the quality differs strongly, and homoepitaxy is not easy to perform due to too many growth defects, still small differences in lattice parameters, imperfect surface finish of the wafers, and problems with lattice and wafer bow. Anyway, it is very desirable to have a substrate wafer made out of a pure and perfect GaN crystal, as it is the best choice for making a GaN-based device with the highest possible performance and best reliability. The details of epitaxial production of a GaN electronic device will be discussed in the second half of this chapter.

So great the desire, so difficult is the crystal growth of a bulk GaN crystal. A bulk crystal basically grows following the same physical principles as in epitaxy, but as a bulk crystal, we consider a larger crystal body which can be sliced into wafers. In order to attain a several millimeters or in the ideal case centimeters long crystal boule, there are other conditions needed. Most prominently the growth rate which should be achieved with the chosen method is decisive and should be high enough for volume crystal growth but is less critical for a thin film growth.

In the last 15 years, there have been various attempts to find the right technique for the growth of GaN bulk crystals. A good compilation can be found here [3]. In this chapter we will give a brief overview and point out the major difficulties and challenges which need to be overcome in order to achieve true bulk crystals of GaN.

GaN is quite difficult to grow as a crystal. The compound is strongly covalent bonded and contains a volatile element, nitrogen, so the material cannot be melted due to its extreme high melting point above 2000 °C [4, 5] and an extremely high equilibrium pressure of approximately 60 kbar [4–8] of nitrogen as binary constituent. Below the equilibrium curve, the GaN would decompose incongruently. The closest to equilibrium growth is the high nitrogen pressure growth technique which is performed in a high pressure – high temperature system under 10–20 kbar or more of nitrogen pressure and elevated temperatures about 1400–1600 °C [9–11, see, e.g.]. Due to the extreme conditions and the relatively small growth volumes, this method cannot be envisioned for industrial usage. The incongruent decomposition of GaN without high nitrogen overpressure, as a fact, makes it impossible to grow GaN crystals from any of the classical methods like Czochralski pulling, vertical gradient freeze technique, or others, which are based on the formation of a melt [12]. Moreover, the solubility of nitrogen in a pure gallium melt is very low [13], so the classical detour using a solution growth method instead of a melt-based process is not straightforward. The threefold bond of the nitrogen atom in N_2 gas makes it additionally low-reactive, and a more activated nitrogen species like ammonia or nitrogen plasma can be applied, but control of reaction rates and growth kinetics is difficult in that case. Solution growth basically means that the compound that is desired to be crystallized is dissolved in a solvent and after the full saturation of the solution is attained, a supersaturation is created at a place of crystallization and the compound may start to crystallize [14]. Since the solubility is a function of temperature, the supersaturation can, e.g., be established by changing the temperature in the system at the place of crystallization. If the solubility of the component is low, the addition of an additive enhancing the solubility is needed. Good examples for solution growth of GaN with solubility-enhancing additives are the low-pressure solution growth (LPSG) [15] and the sodium flux method [16, 17]. However, problems are still manifold: the growth rates are relatively low (approximately 2 μm/h for LPSG and 20 μm/h for the Na-flux method) as a result of a method which operates close to thermodynamic equilibrium. Nevertheless, one of the solution growth methods, namely, the ammonothermal crystal growth, has become very promising for the growth of GaN crystals in the last years [18–20, e.g.] and is able to grow crystals of the best structural quality with low dislocation content down to $10^4/cm^2$ at growth rates of about 100–200 μm per day. In some case much higher growth rates can be expected, but demonstrations are yet to be published.

For solution growth the solubility of the compound has to be a function of the temperature, and in the ideal case, the solubility of the compound is on the order of some percent. If the solubility is smaller than that, then one has to add solvents which help to attain a higher solubility of the desired compound. The crystallization itself can be seen as a phase change taking place as a result of the supersaturation.

In the ammonothermal method, the GaN is the compound which is dissolved in supercritical ammonia. The ammonothermal growth is a pendant to the well-developed hydrothermal technique which is used to produce tons of quartz crystals per year. In both methods an autoclave is brought under pressure and temperature in order to create a supercritical fluid. While the temperatures are very moderate (300–600 °C), the corresponding pressures are high, in the rage of 100 MPa to 300 MPa. Additionally, so-called mineralizers are added which help to enhance the solubility of the GaN feedstock in the ammonothermal fluid by creating intermediate complexes. The feedstock is dissolved by natural convection. In fact the action of the intermediates is not fully understood, but is more than just helping to solve the GaN. The complex mediates the mass transport, and the reactions of the complex at the surface of the seed help govern the growth kinetics. The analysis of the ammonothermal process through in situ measurements in the autoclave is very challenging. Although the temperatures are quite moderate between 300 and 600 °C, the pressure inside the autoclave is high (up to 300 MPa), and only a small number of publications exist addressing that [21, 22]. When successful, such observations are extremely valuable. Figure 1.1a shows an example of a simple laboratory autoclave for ammonothermal crystal growth, placed in a furnace with two separate heater zones. The insulation material was removed for better visualization. A schematic drawing of the setup is illustrated in Fig. 1.1b for an ammonoacidic constellation.

The crystallization takes place at nearly constant temperature, and supersaturation is created by a temperature gradient. Transport happens by convective flow and can be steered by the geometry of the inner parts like the baffle plates. Usually the growth rates are small under solution growth conditions due to the fact that the system is held close to equilibrium, temperature is relatively low, and gradients are comparatively shallow; however as mentioned before, the crystalline quality which can be achieved is high. There are in principle two routes for ammonothermal growth of GaN. In one case the chemistry is chosen such that the fluid can be seen as an acidic system ("ammonoacidic"), and in the second case, the process is called "ammonobasic" where the fluid condition is considered basic. Whether it is an ammonobasic or an ammonoacidic condition is determined by the mineralizers which are added to the supercritical NH_3. There are a number of mineralizers

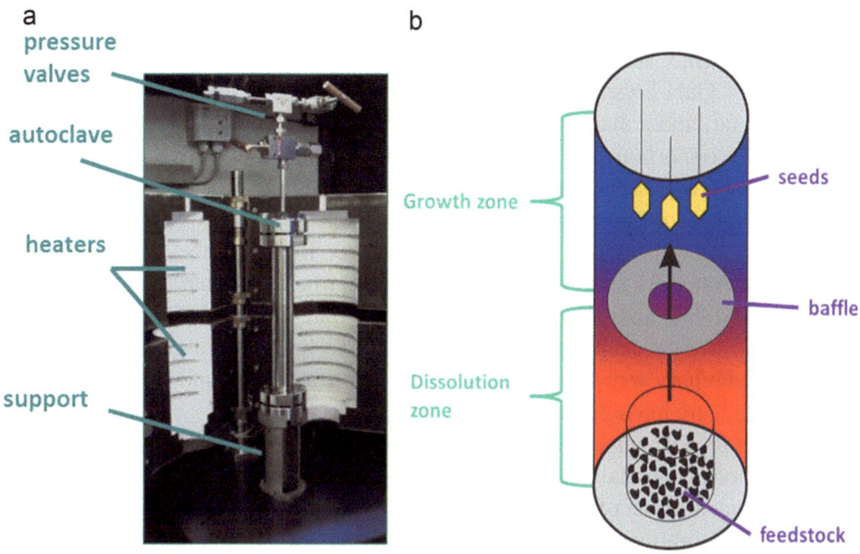

Fig. 1.1 (**a**) (Left side) photograph of a simple laboratory ammonothermal autoclave with two-zone heater system. (**b**) illustrates the principle of an ammonoacidic setup. For an ammonobasic setup, the temperature zone should be switched

Table 1.1 Typical mineralizers in ammonothermal growth of GaN

Mineralizer	System reaction	Reference
$LiNH_2$	Basic	1
$NaNH_2$	Basic	1, 2, 9
KNH_2	Basic	1, 3, 4, 5, 9
Sr (NH2)2	Basic	6
Ba (NH2)2	Basic	6, 7
NH4F	Acidic	8
NH4Cl	Acidic	8, 10, 11
NH4Br	Acidic	10, 11
NH4I	Acidic	10, 11, 12

investigated, but the identification of the intermediate phase is difficult and still in infantile state. The fundamental difference between the mineralizers is not only their basic or acid reaction. The resulting solubility can be quantitatively different as well as its behavior as a function of temperature. Basic mineralizers tend to exhibit a retrograde solubility, whereas acid mineralizers are thought to behave in a regular way. As a result, the required temperature gradients are opposite in the two cases. The knowledge of the intermediates is very important; however only little is known about. Very recently there were a small number of publications related to the identification of intermediates in such systems [23–25]. Table 1.1 summarizes a number of typically applied mineralizers for the ammonothermal growth of GaN; however the list is not complete.

The results of the last years show that there is a big potential for the ammonothermal growth of GaN and that this could be the method of choice for the large-scale production of GaN crystals. The quality of the crystals is the best which can be achieved so far, and a diameter up to 2″ was demonstrated. A large size autoclave will offer the opportunity to not only grow larger but also many crystals in one batch. It can be also seen as one of the advantages of the ammonothermal process that even complex or novel nitrides can be synthesized, which were not accessible before with other methods [26–28]. For instance, Be_3N_2, LaN, and Cu_3N [29] were demonstrated by ammonothermal synthesis, whereas ternary nitrides such as $LiSi_2N_3$, $NaSi_2N_3$, and even $K_3P_6N_{11}$ [30–32] were shown using ammonobasic mineralizers. Very recently the ammonothermal technique is used to extensively explore novel binary, ternary, or even quaternary nitride compounds like $CaGaSiN_3$, some of which are having interesting new properties [33, 34].

The great disadvantage of the ammonothermal as well as of any other solution growth method is the relatively complex growth setup, low growth rates, and the risk of a high concentration of impurities due to corrosion from the autoclave walls and potential incorporation of species from the solute. An ammonothermal autoclave is fabricated from a special alloy which is resistant against the extreme corrosive media (supercritical ammonia plus mineralizers) and at the same time applicable to high pressures and temperatures. Only a small number of technical alloy, namely, Ni-based ones, are usable for that purpose. The production of big alloy bodies as half-finished product in larger dimensions is quite difficult. Alternatively, one can use other alloys for the autoclave and apply a liner made from precious metals at the inner walls against corrosion. But, such a precious metal liner in larger dimensions is also not readily available, and both of the alternatives are quite expensive as well. So, envisioning the ammonothermal technique for large diameter crystals in big autoclaves is not likely to be cheap.

From an industrial point of view, the hydride vapor phase epitaxy (HVPE) is considered as a potential candidate for the large-scale production of GaN crystals to gather native substrates. In the HVPE process, the GaN is crystalized from the vapor phase. The basis of the process is the reaction:

$$NH_3 + GaCl \rightarrow GaN + HCl + H_2$$

The GaCl is formed by flowing HCl over the Ga source; subsequently the Ga is transported in the form of GaCl to the seed position and there brought into contact with ammonia in order to form GaN from the reaction between GaCl and the NH_3. The unwanted by-product is ammonium chloride, which is transported downstream into the reactor exhaust lines. The HVPE reactor is a relatively complex system where reaction rates, species flows, and crystallization kinetics are delicate to balance. The growth rates observed in HVPE are much higher, some hundred μm/h compared to the solution growth variants and ammonothermal growth of GaN. Yoshida et al. [35] demonstrated even 1870 μm/h growth rate at 1060 °C growth temperature. Figure 1.2 shows a schematic illustration of a horizontal HVPE system.

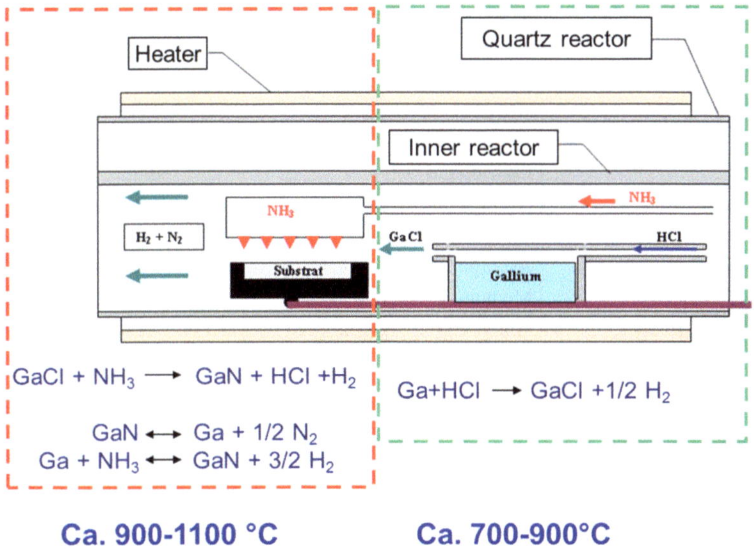

Fig. 1.2 Schematic illustration of a horizontal HVPE reactor. The temperature typical ranges for source and growth zone temperatures

Normally the growth rates which can be attained in a gas phase growth are relatively high and can be, as said before, several hundreds of microns per hour. However, the transport rates have to be adjusted to the kinetics of the growing interface. If the growth rates are too high, the crystal quality usually degrades. So, normally the bulk growth step is performed at medium growth rates clearly below 200 μm/h. The HVPE process was studied intensively. As early as 1969, first results were published [36]. During the last years, huge progress was made with regard to the overall quality of the crystals (by, e.g., [36–44] and others).

It can be seen in literature that the HVPE crystals mainly suffer from background impurities which are silicon and oxygen coming from the reactor part materials [38, 45] making the material unintentionally doped and n-type conducting. So controlled doping remains an issue although huge efforts were made to get the growth reactors cleaner. Only very recently Fujikura et al. [46] showed impressive GaN crystals from HVPE with extremely low residual impurity concentration, which supports hopes with regard to the quality of the material. The structural perfection of the HVPE GaN is quite good although the seed that is used can still be considered a hetero seed. Dislocation densities down to $10^6/cm^2$ can be achieved depending on the thickness of the crystal and the quality of the seed. Table 1.2 gives a short comparison of HVPE GaN versus GaN stemming from the ammonothermal method.

A fundamental problem for all of the crystal growth processes arises, if no native seed is available. The quality of the seed is utmost important for a good crystal growth result. In case the seed diameter is smaller than the later crystal, the dislocation microstructure of the crystal can be modified upon growth from the seed.

Table 1.2 Short comparison of the most relevant criteria for GaN crystals from HVPE and ammonothermal crystal growth

Criteria	HVPE GaN	Ammonothermal GaN
Growth method	Gas phase growth	Solution growth
Temperature	Around 1050 °C	300–700 °C
Pressure	Flowing gas	1000–5000 bar
Growth rates	Average 200 up to 1800 μm/h	Up to 200 μm/day
Major background impurities	Oxygen, silicon	Oxygen, metal elements
Point defect concentration	Average 10E17 and down to 10E15/cm^2	Average 10E18/cm^2
Structural quality	Medium, good	High
Dislocation density	10E6/cm^2	10E4/cm^2

With diameter expansion, mechanisms of dislocation reduction can be provoked. If the lateral growth rates are small compared to the vertical growth velocity, as in the case of GaN, the crystal has to be seeded with the full diameter. In such a case, the dislocation microstructure of the seed will be transferred into the crystal, and only a very few actions are possible to reduce the dislocation content, mostly by increasing the thickness.

For GaN, usually there are no native seeds other than those which were self-generated. They are the treasures of a crystal grower, and the knowledge about native seed generation and preparation for a particular material is not disclosed in detail and nowhere published. However, the only way to grow a crystal where no native seed exists is (i) the natural self-nucleation of small crystals from a supersaturation and later expansion of the crystal size or (ii) the seeding with the full diameter applying a foreign seed, which is closely matched in terms of lattice and thermal expansion.

The seeding with a foreign seed has physically many consequences. The crystal growth processes run at elevated temperatures irrespective of the type of the growth method itself. Usually a number of different temperature steps, respectively, heat-up and cooling cycles, are involved. Thus, a thermal mismatch of the two materials tightly grown together will cause a lot of strain in the growing crystal, breakage, or additional dislocation generation. Even if the crystal does not break upon cool down, the strain will cause a bowing of the crystal lattice planes, which is still apparent after the separation of the foreign seed. Moreover, even upon seeding with a native seed, the materials are slightly different because of, e.g., different background impurity concentrations or issues with the surface preparation and thus the bowing, however drastically reduced, still exists when using native seeds [47]. Another fundamental problem with the growth of GaN on a full diameter seed is the fact that the growth rates of the crystal are different in the different crystallographic directions. The c-direction grows fastest and lateral expansion is not possible in the HVPE process. In the best case, the diameter could eventually be maintained, and even that is not reported but protected by intellectual property. Usually the pyramidal facets start to develop over the process time, and the crystals decrease their diameter with the thickness of the crystal body. In turn, the wafers sliced out of a crystal are

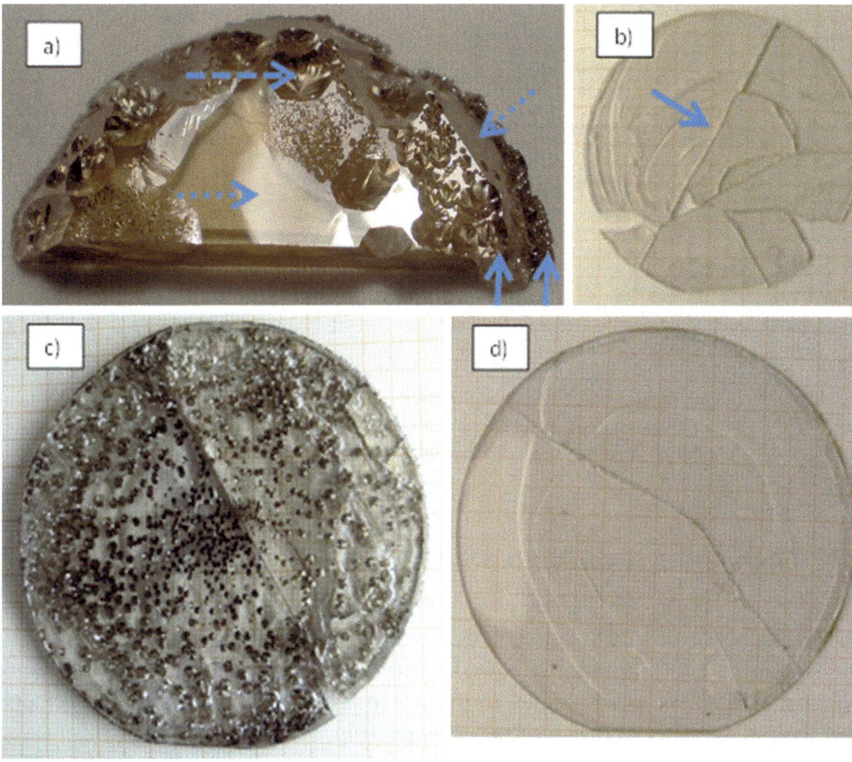

Fig. 1.3 Different crystals grown by HVPE, exhibiting the majority of the growth defects and problems associated with the HVPE growth and foreign seeding. All the crystals are freestanding GaN. Except (**a**) all the crystals are 3″ in diameter. (**a**) shows a 2″, 8 mm thick GaN crystal with growth V-pit on the surface (striped arrow), a clear facet (dotted arrows), and the decrease in diameter (two neighbored arrows). (**b**) shows the typical cracking, in this case of a 3″ 1.5 mm thick crystal. (**c**) shows excess Ga in surface pits, clearly visible from the black coloration of the V-pits. (**d**) shows a nice 3″ crystal, 1.5 mm thick and with good quality with only one crack

not all the same in diameter, but the latter rather decreases with height in the crystal. Also a lot of growth defects are usually reported, e.g., V-pits, which may make the crystal useless for further processing. Figure 1.3 shows GaN crystals grown by the HVPE method which expresses the majority of the common problems associated with HVPE-grown GaN crystals, namely, V-pits on the surface, development of rhombohedral facets, high impurity content, or dark coloration, respectively, Ga droplets and cracking. On the other hand, the last picture shows a 3″ GaN crystal made by HVPE which is quite nice in quality and only exhibiting one single crack.

The quality of the crystal shown in Fig. 1.3d would be the desired one, and such a crystal can be grown with some reasonable thickness up to some mm in length. However, it is possible to slice wafers from such a crystal. But, as it was explained before, the lattice of the crystal is usually strongly bowed. Figure 1.4 illustrates this situation.

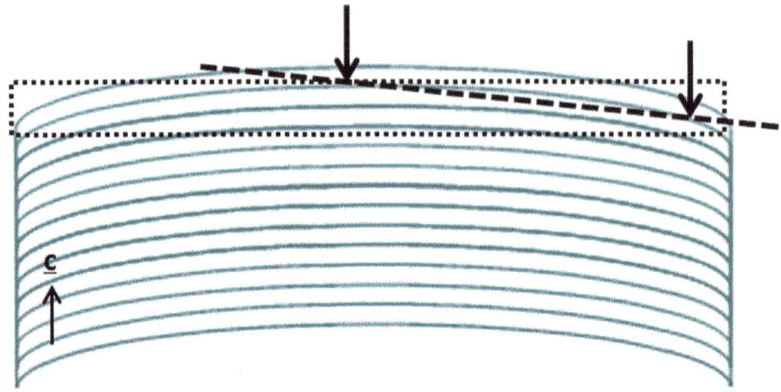

Fig. 1.4 Schematic illustration of the crystal lattice as appearing due to the strong bowing as a result of the growth situation. A cutting of such a crystal into a wafer (dotted box) will result in inclination (indicated by the arrows and the dashed line) of the desired lattice plane with regard to the wafer surface

1.3 Homoepitaxy on Native Substrates

The consequence of such a lattice bow is dramatic. The wafering and polishing of the crystal may cut through the bowed lattice. As a result a desired lattice plane which should be prepared parallel to the wafer surface will have different offcuts over the full diameter of the substrate obtained (Fig. 1.4).

Epitaxy means a strongly oriented and correlated growth of two structures. The epitaxial material takes over the orientation of the substrate wafer and grows further with the same lattice and orientation to the desired thickness. Theoretically, if the lattice on both sides of the interface is the same, epitaxial growth will start with a classical layer-by-layer regime without misfit dislocations. However, in a real homoepitaxial situation, the lattices may vary slightly; the magnitude of the variation between the lattices is the major difference in the case of a native substrate compared to GaN on a foreign substrate. The small divergence of the two GaN materials, the substrate and the epitaxial GaN, is a result of the different growth processes that the material experiences, for instance, in terms of amount and type of background impurities, point defects, and residual strain. Epitaxy thus is very critically mirroring strain states in a substrate, as will be discussed later for the case of GaN on silicon further in this chapter. In essence, any epitaxy is mainly sensitive to the lattice constants and crystal orientation, no matter if the reason for those is the use of a different substrate lattice, the presence of large amounts of point defects and impurities, or the presence of strain. Directly at the growing interface of the epitaxial lattice, the differences in lattice constants and crystal orientation have to be mediated. So, it's not easy to perform epitaxy of GaN on GaN even if native substrates would be readily available. For some applications the usage of a native substrate is mandatory in order to achieve a reliably functioning device, like in the case of blue and ultraviolet semiconductor lasers.

1.4 Heteroepitaxy of GaN

When bulk crystals to make native substrates are not easily available, the only alternative is to look for substrates in other materials that have lattice parameters and crystal structures and orientations that are reasonably close to those of the material that is to be synthesized. For the epitaxial growth of GaN, in practice four types of substrate are used.

For optoelectronics, sapphire is the substrate of choice. It has an acceptable match to GaN in terms of lattice parameter and thermal expansion coefficient, is relatively cheap, and is transparent. As drawbacks, it has a poor thermal and no electrical conductivity, making it less suited for electronic applications with high power densities. For laser diodes, GaN bulk substrates have to be used because the epilayers grown on top need to have the best crystalline quality ($10^6/cm^2$ and below). However, the GaN bulk substrates are only available in small sizes (2″ diameter and less) and are prohibitively expensive for most other applications. For a high demanding application like lasers, this can be accepted, but costs are an important issue for widespread applications like LEDs or transistors.

Therefore, SiC is commonly used as a substrate in electronic and in high-end LED applications. It is available both as n-type and semi-insulating material and has mechanical properties that are suitably matched to the ones of GaN. Moreover it has a very good thermal conductivity, making it a perfect heat sink for devices that are operated at high power density. As drawback, SiC is very energy-intensive to fabricate, making it an expensive material (more than 2000$ for a 6″ semi-insulating wafer). At the same time, the supply chain is dominated by a limited number of monopolists, and the material is subject to ITAR restrictions because of its use in nuclear technologies.

This leaves the fourth candidate: silicon. It is the material of choice for the bulk of the semiconductor industry. Silicon substrates are, as a derivate of SiO_2 (sand), abundantly available, in large sizes (up to 300 mm diameter) and at low costs (around 50$ for 200 mm substrate). Processing of silicon has been intensively studied, yielding an extremely mature platform for a wide range of front-end technologies (e.g., extreme scaling now at the 14 nm node), back-end technologies (such as multiple metal levels or through-silicon vias), micromachining (for MEMS or microfluidics), and most importantly the co-integration of these various technologies on die or wafer level. Moreover these technologies are available in mass volume production at an extremely low cost. In terms of manufacturing capabilities, the silicon industry remains far ahead of all other semiconductors, not at least the compound semiconductors, both in terms of sheer production capacity and resulting cost structure and in terms of technological capabilities such as the lithographically defined minimum feature size and the level of (co-)integration density.

To bridge the gap between the two fundamentally different worlds of Si and GaN, researchers [48–50] in the late 1990s and early 2000s started looking into the epitaxial growth of GaN directly on Si substrates. Even though this approach is significantly more challenging than heteroepitaxy on sapphire due to large lattice and thermal mismatches between the two materials and the existence of Ga melt-back, it has eventually allowed the entry of GaN epiwafers into the large capacity Si semiconductor fabs. Moreover, the mechanical properties of the silicon substrate in combination with well-established techniques such as aggressive wafer thinning, through-silicon vias, and in general 2.5D and 3D integration open up the possibility to combine the best of the silicon and GaN worlds. In the next paragraphs, we will describe how the epitaxy of GaN on Si is performed and what are its constraints and boundary conditions. We will focus on the layer structure of a HEMT device, which is the workhorse in any electronic application.

As substrate material for GaN, silicon is not perfect. It has a large mismatch in the lattice parameter and the thermal expansion coefficient that result in large strain buildup in III-nitride epilayers that are grown on top, leading to wafer deformation (bow and warp) and layer cracking and in the worst case wafer breakage. So from the first glance, this is very similar to what was explained before in the case of a foreign material seed for a crystal growth process. However, modern advanced epitaxial techniques offer different options compared to the bulk crystal growth. The development of suitable strain management techniques has allowed overcoming these technical hurdles to some degree, and GaN-on-Si epiwafers with 200 mm diameter are commercially available today. Once these are mastered, GaN on silicon offers a low-cost substrate, compatible with a wide variety of well-established processing techniques in combination with a unique family of semiconductor materials with superior properties and the novel device concepts that result from them.

Even though all substrates have their pros and cons, the authors strongly believe that only the use of Si substrates (which guarantees a low cost and the possibility of co-integration with Si-based electronics) can lead to a broad adoption of GaN in electronics, whereas the native substrates will have a more limited space for high-performance applications, due to the inherent higher costs and lower-volume processing technologies.

1.5 Heterostructures Made from III-Nitrides: A Family of Semiconductors as a Powerful Toolbox for Device Design by Material Engineering

While the superior material properties of GaN as a wide bandgap semiconductor (such as high electrical strength, high temperature capability, etc.) can easily be recognized, in fact a much larger, unique, and complete family of semiconductors is available under the denominator of the so-called III-nitrides. These cover a very large range of bandgap energies [82], starting from the infrared (InN having a bandgap of 0.7 eV) and reaching into the extreme ultraviolet (EUV, with AlN having a bandgap of 6.2 eV). For optoelectronic applications, because of their direct bandgap transition, this means that the III-nitride family covers the full

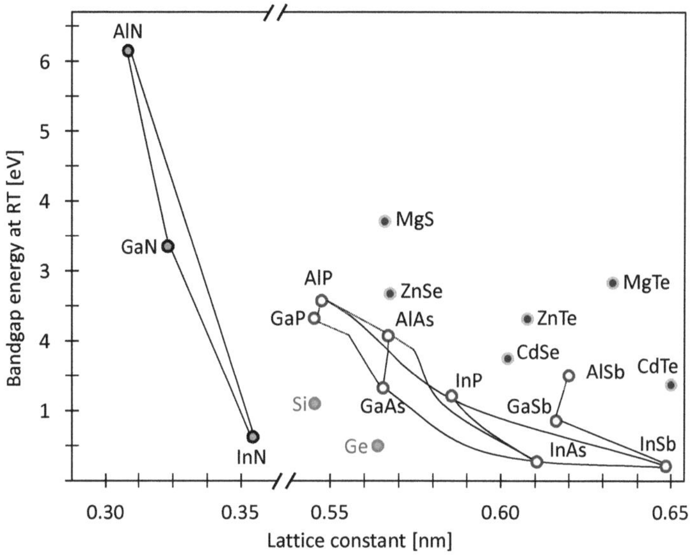

Fig. 1.5 Bandgap energy versus lattice constant for various IV, III–V, and II–VI semiconductors

visible spectrum. Figure 1.5 shows the variation of the bandgap of the III-nitride semiconductors versus their crystal lattice. As a reference, some other III-V and II-VI compounds as well as silicon and germanium are also shown, which all show lower bandgap energies than GaN and AlN.

In theory, the relative concentrations of the metal group-III elements (Al, Ga, In) can be varied to grow any ternary or quaternary crystal alloy with a composition $Al_xIn_yGa_{1-x-y}N$ with x, y and x + y included between 0 and 1. In practice, some limitations exist, e.g., due to issues with phase separation occurring in the indium-rich compounds.

This offers the opportunity to epitaxial growers to not only vary the composition and bandgaps of the III-nitride semiconductors but also to stack these different layers of semiconductors with their various compositions and as such engineering complex heterostructures, thereby creating new device concepts. Examples of such heterostructures include InGaN/GaN quantum wells for the active areas of light-emitting devices, separate confinement heterostructures (SCH) in laser diodes to guide both the light and the carriers, or high-mobility transistor structures (both in GaAs and GaN technology).

1.6 Piezoelectric Field in III-Nitrides

An important property of III-nitrides is the presence of strong piezoelectric fields that exist intrinsically in these semiconductors. The large difference in size between the group III and the nitrogen atoms leads to distortion of the crystal lattice, which combined with a large difference in Pauling electronegativity induces spontaneous polarization inside the material. Even though this effect is detrimental for optoelectronics (causing quantum-confined Stark effect in the quantum wells, spatially separating the hole and electron wave functions) [83, 84], it can be taken advantage of in electronic applications. By adopting a heterostructure consisting of a thin (In)AlGaN "barrier" layer grown pseudomorphically on top of a thicker GaN "channel" layer, the strain that is induced in the barrier layer adds an additional piezoelectric component to the total polarization charge in the barrier layer. This heterostructure is often referred to as the high-electron-mobility transistor (HEMT) structure. This structure was first demonstrated by Khan [2] and is well described elsewhere, e.g., by Ambacher [51]. It is the epitaxy of such heterostructure that we will explore in the next paragraphs.

1.7 Metal-Organic Vapor Phase Deposition: The Most Suited Technique for III-Nitride Epilayer Deposition

Epitaxial crystal growth of compound semiconductor layers can be performed by different techniques, such as molecular beam epitaxy (MBE). In practice, the technique of choice for depositing epitaxial layers on a production scale is metal-organic chemical vapor deposition (MOCVD), also called metal-organic vapor phase epitaxy (MOVPE). In this technique [52], the constituent atoms of the growing crystal are introduced in a reactor chamber under carefully controlled thermodynamic conditions in the form of precursor molecules in gaseous form, transported by a carrier gas. By providing (thermal) energy, these molecules are forced to dissociate and react in the gas phase and on the substrate's surface, thereby enlarging ("growing") the substrate's crystal with additional semiconductor material. The carrier gas can be either nitrogen (N_2) or hydrogen (H_2) or a mixture thereof.

The precursor molecules for the group III elements are a combination of the group III metals with organic groups such as methyl or ethyl groups, hence giving the name to the MOVPE technique. The source for gallium is typically trimethylgallium $(CH_3)_3Ga$ (TMGa) or triethylgallium $(C_2H_5)_3Ga$ (TEGa). The sources for aluminum and indium are typically, respectively, trimethylaluminum $(CH_3)_3Al$ (TMAl) and trimethylindium $((CH_3)_3Al$ (TMIn). The precursor for the group V element, nitrogen, is a so-called hydride, ammonia (NH_3), where the constituent nitrogen atom is directly bonded to hydrogen atoms.

The epitaxial growth process is guided by steering the thermodynamical conditions (temperature, pressure, partial pressures of species), of both the ambient gas phase in the reactor and of the substrate. These conditions determine the free Gibbs energy of the substrate and the gas phase, with the difference between the two energies determining if the system will provide etching of the solid, (dynamic) equilibrium, or growth of the solid semiconductor (consisting of substrate and epilayers). A second important consideration is that of the surface kinetics, which explains how the growth process is affected by the adsorption onto and desorption from the surface. The rate of desorption is influenced by the lateral migration of the gas species on the semiconductor surface and their local microscopic arrangements (near atomic steps or nucleation islands).

It is important to note that there is always equilibrium between the gas phase composition and the vapor pressure of the constituents of the solid phase, implying that any contamination of the gas phase induces contamination of the solid semiconductor material. The metal-organic precursors themselves are a source for residual contamination impurities that are incorporated into the materials. Intrinsically, carbon and hydrogen atoms are present in the gas phase, and it requires careful tuning of the growth conditions (controlling the respective vapor pressures) to minimize (but never eliminate) inclusion of said atoms into the solid phase. At the same time, the purity of the source materials needs to be guaranteed for exactly the same reason, something which is especially important for sources of oxygen (e.g., contamination by O_2, H_2O, CO, and CO_2).

In the III-nitrides, similarly to the other III-V materials (such as phosphides or arsenides), the equilibrium pressure of the group V element is higher than that of the group III element. Therefore one always needs an excess of group V material in the gas phase, which means that the V/III ratio is always larger than 1.

The MOCVD process for GaN typically occurs at temperatures above 1000 °C (but as low as 700 °C for InGaN and as high as 1200 °C for AlN) and at reactor pressures between 10 mbar and 200 mbar (but as high as 500 mbar for certain layers). Because the process is operated at finite pressures, it means that beside the thermodynamic considerations, the MOCVD process also needs to be optimized in terms of fluid dynamics and thermophoresis effects [85]. Various tool vendors each have their own philosophies to address this, e.g., by using a distributed "showerhead" injection [53], a lateral laminar injection with a fivefold "penta" injector head [54], or a high-speed rotation-induced pumping effect of the susceptor [55].

Although the most successful electronic device concept (HEMT) doesn't require any doping of the active structure, doping is mandatory for creating bipolar devices. Additionally, doping of III-nitrides can be used either for reducing ohmic contact resistance on the topmost layer of the structure (n-type doping), for compensating the intrinsic n-type character of the III-nitride buffer layers (p-type compensation), for increasing the resistivity of a given material layer, or for inserting a p-type layer under the gate to fabricate normally off JFET devices [56].

The doping method that is typically used for introducing n-type conductivity in GaN is introducing a group IV atom on a group III site, typically Si provided in the form of silane gas (SiH_4).

For p-type conductivity, the most well-known doping element is magnesium, which is widely used in GaN LED or laser technology and lately also in GaN JFETs. Besides creating an energy level that is relatively deep in the bandgap (125 to 215 meV), it has the important drawback that the magnesium forms a complex with hydrogen (used as carrier gas in MOCVD) that renders the magnesium electrically inactive. The latter can be counteracted by an activation step (annealing in nitrogen) for layers at the top of the layer stack but is ineffective for buried layers. Magnesium is also relatively fast diffusing in GaN, so it may also cause a problem with regard to temperature sequences in epitaxy. As such, magnesium is not easy to handle in the preparation of GaN heterostructures. Some groups are using Fe doping of the GaN buffer (group II on group III, Fe_{Ga}), whereas others control the incorporation of the carbon impurities (group IV on group V site C_N) using either residual carbon from the methyl groups of the metal-organic precursor or by using an independent doping precursor such as methane (CH_4). This leads to p-type conductivity that compensates the n-type background carriers of GaN and makes for perfectly semi-insulating buffer layers with a resistivity higher than 10^{12} Ohm/sq.

MOCVD allows a certain level of in situ monitoring to control the growth process. It is more limited than, for instance, MBE, where electron diffraction techniques such as "reflection high-energy electron diffraction" (RHEED) are possible, thanks to the vacuum operating condition. MOCVD is limited to optical techniques, one of which is laser interferometry. The growth rate of the semiconductor, as well as an indication of the surface roughness, is provided by the interferometry pattern of the semiconductor epitaxial growth. With a little modification, the functionality of such tools can be extended to also measure the mechanical wafer deformation, which is a measure of the strain buildup during the epitaxy process. Further extensions yield data on material composition by measuring also the spectral reflection of broadband white light [57].

1.8 Buildup of Typical AlGaN/GaN Layer Structure on a Si Substrate and Pitfalls to Avoid

In an epitaxy process, one aims to deposit the layers with a perfect continuity of the crystal lattice [58]. As shown in Fig. 1.6, starting from a silicon substrate, the typical layer stack consists of the following parts: (1) a nucleation layer to initiate the growth of the III-nitride on the foreign Si substrate; (2) a strain management buffer stack to accommodate for the differences in mechanical properties between Si and GaN; and finally (3) the active part of the HEMT device, consisting of a GaN channel layer, an (In)AlGaN barrier layer. Last but not least, a final capping or passivation layer is usually deposited to protect the surface or passivate surface states. Each subpart has its own functionality in the final wafer and electronic device as will be described below.

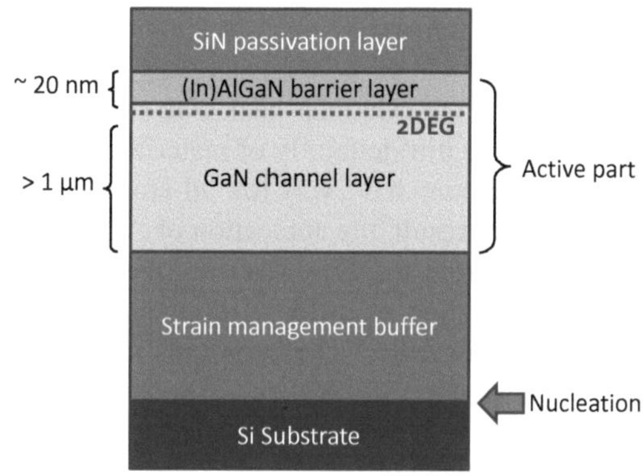

Fig. 1.6 Composition of an epitaxial layer stack for an AlGaN/GaN HEMT device grown on a Si substrate

1.8.1 The Nucleation Layer

Crystal Orientation of the Silicon Substrate

Silicon crystals have a cubic crystal structure, the same as diamond, made up by two interpenetrating face-centered cubic primitive lattices. GaN on the other hand is most commonly found as a hexagonal wurtzite crystal. A cubic variety can also be formed but with less promising semiconductor properties. To match the different crystal structures, GaN is most of the times grown on <111> oriented Si substrates, where the atoms on the surface are arranged in a triangular pattern along the threefold axis of the cubic lattice (although some work has been performed on <110> and even <100> oriented Si substrates). This imposes a limitation on the side-by-side integration of GaN devices and Si CMOS devices, as the formation of a gate oxide on Si <111> yields a poorer quality compared to the <100> counterpart.

Gallium Melt-Back

Unfortunately, one cannot start the epitaxial growth of GaN directly on the silicon substrate [59]. At elevated temperatures, Ga species easily diffuse into the silicon wafer as well as attack the surface of the silicon substrate. The result would be the creation of three-dimensionally extended defects such as inverted pyramids and causing a roughening of the substrate's surface. This effect is called Ga melt-back and disrupts the epitaxial relationship between the substrate and growing layers, thus ruining the latter's quality. It has been shown that it occurs already at very low molar concentrations of Ga in the gas phase, well below the molar concentrations that are required to obtain crystal growth.

AlN Microstructure

The solution to the Ga melt-back is to start the epitaxial growth on Si using another member of the nitride family of materials, aluminum nitride [60]. During MOCVD epitaxy, aluminum has a very low ad-atom surface mobility as it is chemically highly reactive. As a result, the nucleation of AlN on the Si surface occurs simultaneously at many different locations, causing the formation of separated nucleation islands with a distribution of different orientations. Eventually, these islands coalesce as the growth progresses. The layer thus includes a large amount of grain boundaries and other extended 1D, 2D, and 3D defects. A typical threading dislocation density for AlN grown on Si is in the order $10^{10}/cm^2$. Moreover, the low surface mobility combined with the difference in lattice constants of Si and AlN, respectively, causes AlN epitaxy to easily shift into a three-dimensional growth mode which leads to roughening of the surface. To minimize the roughness, typically the thickness of the AlN nucleation layer is limited to 100–200 nm, at which point the top part of the AlN layer is already fully relaxed.

The interface between the Si and the III-nitrides has been shown to be the weakest part in the heterostructure regarding breakdown voltage [61, 62] as well as regarding RF losses [63]. This effect is attributed by different groups to either diffusion of group III elements (mainly Ga) into the Si or the creation of an inversion layer due to the combination of band offset and polarization charges. Even though the HEMT is a lateral device, it was discovered early that the breakdown voltage of devices made in GaN on Si does not continuously scale with the lateral separation of the device contacts but saturates at a given value depending on the thickness of the epilayer stack. The explanation for this is that beyond the saturation point, the electrical breakdown does not occur laterally in the III-nitride layers anymore, but along a path consisting of a vertical breakdown from the surface to Si/AlN interface, a lateral conduction along the Si/AlN interface followed by another vertical breakdown from this interface to the surface [61]. This implies that the voltage rating of the buffer is directly linked to the thickness of the epitaxial layers. Today, buffers of about 4–5 μm thickness are used to allow device operation at 650 V. In RF applications, there is a capacitive coupling between the RF waveguides and the potentially conductive AlN/Si interface that leads to RF signal dissipation and reduced transistor efficiencies.

1.8.2 The Buffer for Strain Management

As the MOCVD process occurs at relatively high temperatures, above 1000 °C for Al-containing compounds, one of the main issues of growing GaN-based layers on Si is the mismatch of the thermal expansion coefficients (>54%): during cooldown after the process, the III-nitride materials shrink much faster than the Si substrate generating a large tensile strain in the grown layers as well as significant wafer bow, layer cracking, and in the worst case wafer breakage. The amount of tensile strain is proportional to the thickness of the epitaxial layers. As the thickness of the III-nitride layer stack determines its voltage handling capability, this thermal mismatch becomes more and more severe as the required voltage rating increases.

To mitigate the bowing of the wafer, there is a need to include compressively strained layers in the buffer stack between the AlN nucleation layer and the active GaN/AlGaN HEMT part of the stack. In its simplest form and as first proposed as early as 1999 [64], this is achieved by introducing one or more step-graded AlGaN layers between the AlN nucleation layer and the GaN-based active part. Starting from an AlN or AlGaN layer with a given Al content which is fully relaxed, a second AlGaN layer is grown with lower Al content. Such second layer has a larger lattice constant than the layer below, and as long as the layer grows pseudomorphically on the underlying layer, compressive strain is built up. When the growth front of this layer is fully relaxed, a next layer can be grown with an even lower Al content, repeating the above, until finally a GaN layer can be grown. A typical AlGaN step graded strain management buffer transition from AlN to GaN uses two to five steps in the Al concentration with individual layers. The thickness of the singular layers ranges from 250 nm to 1000 nm, depending on the targeted thickness of the full III-nitride layer stack in general and of the GaN channel layer in particular.

A beneficial side effect of using discrete, strained layers in the stack is that perpendicularly propagating threading dislocations tend to bend at the interfaces between the different layers under the influence of the strain that is present there. This increases the probability that those threading dislocations eventually meet and annihilate. As a result, the crystal quality improves as the number of interfaces and thus the total thickness increases.

Besides the step-graded buffer, different approaches exist to the strain management. One variant is to grade the Al content continuously instead of in steps. The most commonly used approach, especially for thick layers, is that of the strained AlGaN/GaN superlattice structures, in which thin GaN and AlGaN layers are periodically repeating several tens of times, where the tensile cooldown stress is decoupled from the substrate [65, 66]. Alternatively, the strain is managed by growing thick GaN layers that are interrupted by low-temperature AlN interlayers [67].

Poor control of the strain management in the structure may lead to wafer deformation, causing warp or bow that makes the substrate unacceptable for further processing because of lithography or chucking problems. In the worst case, the strain causes cracks in the III-nitride epilayer that may cause the wafer to break.

A special aspect of the strain management is the situation of the wafer edge which forms naturally a discontinuity in even a perfectly balanced strain profile. The bevel of the wafer is typically also rounded, so that different crystal orientations compared to that of the top surface are exposed on which the behavior of the epitaxial process will differ. (Remember that this may also happen due to the lattice bow in case of native (or any) substrate as mentioned before.) The edge discontinuity may also disrupt the gas flow pattern during the MOCVD process, and it complicates the uniform heating of the substrate. These features make it hard to avoid small cracks at the wafer edge that can trigger the nucleation and propagation of cracks extending through the whole wafer.

1.8.3 The Active Part of the AlGaN/GaN HEMT

After nucleation layer and strain management buffer are in place, the active part of the device can be grown. GaN electronics are predominantly based on the HEMT structure, in which a thin (In)Al(Ga)N barrier layer is grown pseudomorphically on a thick(er) GaN channel layer. The barrier layers have a higher bandgap in combination with a larger polarization charge than the GaN layer, which leads to the formation of a two-dimensional electron gas at the interface of the two layers. The exact density of electrons in the 2DEG depends on the thickness, composition, and relative strain state of the barrier as well as the surface potential of the barrier. Typical values are 20 nm for the thickness and an aluminum content of 25%, which combined with a 3 nm thick GaN cap layer yields a carrier concentration of around 0.8 $10^{13}/cm^2$. Depending on the smoothness of the interfaces and the crystal quality of the heterostructure, the associated electron mobility can reach values of over 2000 $cm^2/V.s$, yielding a sheet resistance of the 2DEG around 400 Ohm/sq.

To further increase the carrier density, one can either increase the barrier thickness or increase the aluminum content of the AlGaN barrier. At a certain point however, the critical thickness of the strained AlGaN layer will be reached [68], at which point the layer will start to relax. The relaxation will give rise to a reduced carrier density, trapping issues, reduced reliability, or gate leakage current through extended defects. This imposes an upper limit to the 2DEG formation. For instance, for pure AlN barriers [69], the critical thickness at which relaxation starts occurring is only 5–8 nm.

One alternative configuration is to replace part of the gallium atoms by indium. In its extreme case, this leads to InAlN material [70], which at an indium concentration of 17% can be grown lattice matched to GaN but which still leads to a high carrier density in the 2DEG (even up to 2.5 $10^{13}/cm^2$), due to its large bandgap and large spontaneous polarization. As it is lattice matched, the absence of strain relative to the GaN channel layer may have a beneficial effect on the reliability of the heterostructure. Unfortunately, III-nitride layers containing indium typically suffer from indium segregation effects that cause excessive leakage currents through these layers. Figure 1.7 shows the electron density in the 2DEG for the three types of barrier layers described, as function of the barrier thickness.

In the AlGaN/GaN heterostructure, an important role is also taken by the GaN channel layer. It is instrumental in obtaining high electron mobility. Its surface roughness will determine the roughness of the interface with the barrier layer. Its background contamination levels and dislocation structure will determine the amount of electron scattering centers.

At the same time, the GaN channel layer needs to be sufficiently resistive to prevent leakage current outside of the 2DEG, for instance, under high electrical field conditions, like at the edge of a short gate. This effect is called punch-through [71]. There are several different counteractions possible to better confine electrons in the 2DEG. The first is to replace the bottom part of the GaN channel layer by a low Al content AlGaN "back-barrier" layer [72]. Because of the larger bandgap of

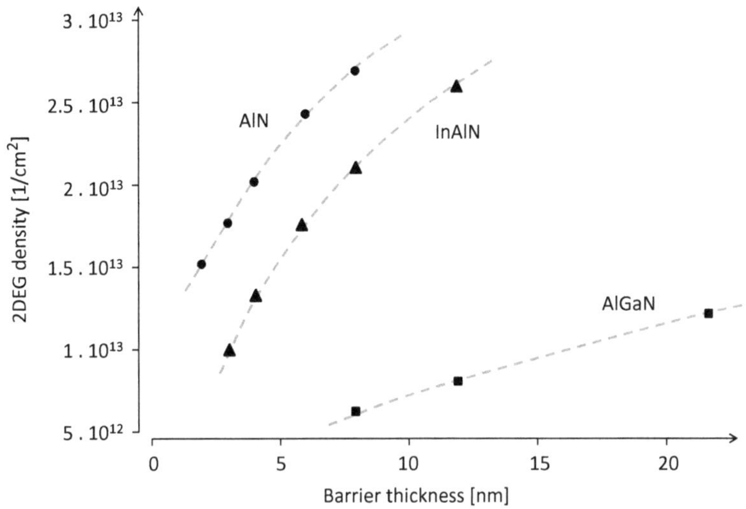

Fig. 1.7 Density of electrons in the 2DEG as function of the thickness and composition of the barrier layer capped with an in situ SiN layer

this layer but mainly due to the polarization effect, this will cause bending of the band diagram and create a barrier for electrons to penetrate into the lower layers. The downside of this approach is that the AlGaN material has a significantly lower thermal conductivity than GaN which can lead to elevated channel temperatures and that AlGaN material is more prone to point defects that could cause trapping effects. The back-barrier can also consist of InGaN material, in which case the confinement solely depends on the effect of the polarization charge offset and not the bandgap offset [72]. Alternative approaches are doping the lower part of the GaN channel layer with iron or carbon impurities. This makes the layer p-type which again creates a barrier for 2DEG electrons to penetrate into the lower layers. These approaches need to be well controlled because the charging and de-charging of these impurities will impact on the 2DEG density and may cause current collapse effects [73]. Moreover, especially Fe impurity doping has a significant memory effect leading to a long tail in the doping profile that may affect electron mobility when the iron ions come too close to the active area.

1.8.4 Capping and Surface Passivation Layers: From GaN Cap to In Situ SiN

GaN-based electronics owe their success to the superb properties of the AlGaN/GaN heterostructure with the formation of a 2DEG at its interface and the resulting HEMT devices [51]. The combination of an offset in polarization charge and an offset in the bandgap of the two layers leads to the formation of a quantum well in the conduction band, which dips below the Fermi level. In this two-dimensional plane just below the interface, electrons can freely move. Because these electrons are collected in this electron gas without any need for impurity doping that may cause scattering of the free carriers, the electron mobility is high, even beyond 2000 cm^2/Vs.

However, because the carrier density in the 2DEG is not determined by thermalization of impurity doping as in other semiconductors but by polarization and conduction band engineering, it is very sensitive to the surface potential of the structure. In the early days of the technology, several papers discussed the origin of the so-called virtual gate and DC-to-RF dispersion, where charged surface states in the (ungated) access regions of a transistor caused a slowly decaying depletion of the 2DEG [74]. It was shown (but at the time poorly understood) that a passivation layer consisting of SiN could mitigate the dispersion effect [75], because the ionized Si atoms can compensate the surface charge and thus stabilize the surface potential. Typically such SiN layers are deposited by plasma-enhanced chemical vapor deposition (PECVD) or low-pressure chemical vapor deposition (LPCVD) as part of the process. Alternatively, the SiN can also be deposited by MOCVD [76], in which case the sensitive AlGaN barrier is never exposed to atmosphere (not even when unloading the epiwafers from the MOCVD reactor) except during a small number of processing steps. In that case, the top surface of the barrier layer cannot be as easily contaminated or oxidized, preventing modification of its surface charge state.

Further investigations of the SiN grown in situ in the MOCVD reactor revealed some interesting properties. First, it was established by elastic recoil detection analysis (ERDA) that the layer has a very high density with less than 5% of hydrogen in the layer, which is much lower than typical values, for instance, in PECVD SiN (up to 30% hydrogen). Second, TEM analysis showed how the first monolayers of the SiN grow epitaxially on the AlGaN barrier, thereby drastically reducing the number of dangling bonds with associated trapping states at this interface. A similar configuration was reported for SiN layers deposited by catalytic CVD [77].

Thirdly, it was shown that in situ SiN can reduce the strain-induced relaxation of the (In)Al(Ga)N barrier material. From Hall and XRD measurements, it was shown that HEMT heterostructures with relatively high Al concentrations in the barrier and capped with in situ SiN have a lower relaxation degree of the AlGaN barrier material in conjunction with a higher electron density in the 2DEG. In a follow-up paper [78], detailed growth studies show how the AlGaN barrier relaxes through the formation of grooves caused by out-diffusion of Ga from the barrier during a growth interruption or the cooldown process after epitaxial growth. It is shown that a single monolayer of in situ SiN can prevent this mechanism. In this way, the in situ SiN enables a route toward AlGaN/GaN heterostructures with higher Al concentration and resulting higher 2DEG density. One notable possibility is the use of pure AlN as barrier material [69]. Of all possible $Al_xIn_yGa_{1\text{-}x\text{-}y}N$ alloys, AlN has both the largest bandgap energy and the largest polarization offset to GaN. This layer can be kept very thin (e.g., 5 nm) while maintaining a high electron density significantly above $10^{13}/cm^2$. Because of the much enhanced capacitive coupling between gate and 2DEG, a transistor made on such a structure will thus have a higher transconductance which is one of the main parameters determining the frequency response of RF devices. Moreover, the so-called short channel parasitic effects, i.e., reduction of the transconductance due to a poor gate length versus gate-to-channel distance aspect ratio (ideally higher than 15) when scaling transistor gates below 0.15 μm, are drastically suppressed.

The in situ SiN also has a beneficial impact on the thermal stability of the HEMT as indicated by the thermal stability. Medjdoub *et al.* investigated the degradation of HEMT structures in thermal storage tests [79]. It was clearly shown that under the test conditions (up to 950 °C), the in situ SiN-capped samples did not show any degradation, whereas uncapped or GaN-capped samples degraded significantly as of 700 °C.

1.9 Conclusions

In epitaxial processes, the structure and quality of the newly grown crystal is a reflection of the underlying substrate lattice. Ideally, crystal growers want to start from substrates made out of the same material, in order to minimize any crystal imperfections in the growing layers. It is expected that the physical performance as well as the reliability of such a device is outstanding and better than in the hetero-epitaxial case. The synthesis of GaN crystals which would lead to the fabrication of the necessary native GaN substrates is, unfortunately, not readily available. So, the crystal growers worldwide are forced to intensify their activities in growing high-quality GaN crystals by (hopefully) cost-effective methods. The two most promising methods are the ammonothermal growth and the HVPE technique. Ammonothermal GaN crystals have an outstanding quality with a dislocation density as low as $10^4/cm^2$, whereas the HVPE crystals are reaching $10^6/cm^2$. Some of the physical difficulties appearing in bulk crystal growth and seeding as well as in the homoepitaxy on native substrates will come up again in the heteroepitaxy of GaN on foreign substrates. In the second part of the chapter, it can be seen that the fundamental physical issues are the same, but for thin layer stacks, they can be managed differently.

Limitations in bulk GaN crystal quality and in the wafer diameters that can be obtained from those crystals still push crystal growers today toward heteroepitaxy, as it is the most suitable crystal growth method for obtaining the desired products for optoelectronic or electronic applications. This has so far been quite successful: besides widespread use of GaN on sapphire in modern lighting solutions, GaN technology is also setting a foot in the arena of electronic applications. High-end RF applications for space and defense are already dominated by GaN on SiC, a technology which has an important cost disadvantage that is holding back the true potential of GaN. This hurdle can only be overcome by adopting the much more cost-effective GaN-on-Si technology. This is happening at the time of this writing: several providers are introducing GaN-on-Si power switches [80, 81] on the market with superior performance compared to their Si and SiC counterparts. The adoption of GaN-on-Si technology is also expected to happen in RF applications with the deployment of the 5G standards in mobile communication. To keep up with the promises of the 5G mobile network ("anywhere, anytime, anyone (any object)"), using GaN is a must for high-efficiency, high-bandwidth, high power amplifiers. This will create additional challenges for crystal growers to master the hetero-epitaxial processes, heading for the next-generation GaN-on-Si technology on 300 mm diameter substrates.

GaN itself as a material is a strong candidate for many new applications which were not even mentioned here. Discussing all the potential of GaN for electronic and other applications would definitely fall aside the scope of this chapter. But, finally, both of the routes presented here, the native, homoepitaxial approach and the heteroepitaxy of electronic structures, represent physically strong approaches and an intense scientific activity worldwide. It is also not necessary to give a final weighting which approach is considered stronger. GaN on Si is closer to market, but GaN on GaN, even though expensive, offers additional options like vertical structures, lasers, or other opportunities. Both ways are attractive and will coexist in the future with different application focus.

References

1. https://www.zionmarketresearch.com/sample/led-lighting-market
2. M. Asif Khan, J.N. Kuznia, D.T. Olson, Appl. Phys. Lett. **65**(9), 1121–1123 (1994)
3. D. Ehrentraut, E. Meissner, M. Bockowski (eds.), *Technology of Gallium Nitride Crystal Growth* (Springer, Heidelberg, 2010)
4. S. Porowski, B. Sadovyi, S. Gierlotka, S.J. Rzoska, I. Grzegory, I. Petrusha, V. Turkevich, D. Stratiichuk, The challenge of decomposition and melting of gallium nitride under high pressure and high temperature. J. Phys. Chem. Solid **85**, 138–143 (2015)
5. W. Utsumi, H. Saitoh, H. Kaneko, T. Watanuki, K. Aoki, O. Shimomura, Congruent melting of gallium nitride at 6GPa and its application to single-crystal growth. Nat. Mater. **2**, 735–738 (2003)
6. J. Karpiński, J. Jun, S. Porowski, Equilibrium pressure of N2 over GaN and high pressure solution growth of GaN. J. Cryst. Growth **66**, 1–7 (1984)
7. A.G. Sokol, Y.N. Palyanov, N.V. Surovtsev, Incongruent melting of gallium nitride at7.5GPa. Diam. Relat. Mater. **16**, 431–434 (2007)
8. K. Harafuji, T. Tsuchiya, K. Kawamura, Molecular dynamics simulation for evaluating melting point of wurtzite-type GaN crystal. J. Appl. Phys. **96**, 2501–2512 (2004)
9. I. Grzegory, High pressure growth of bulk GaN from solutions in gallium. J. Phys. Condens. Matter **13**, 6875 (2001)
10. J. Karpinski, S. Porowski, J. Cryst. Growth **66**, 11 (1984)
11. M. Bockowski, High nitrogen pressure solution growth of GaN. Jpn. J. Appl. Phys. **53**, 100203 (2014)
12. P. Rudolph (ed.), Bulk crystal growth, in *Handbook of Crystal Growth*, 2nd edn., (Elsevier Verlag., ISBN: 978-0-444-63303-3, 2015)

13. G. Sun, E. Meissner, P. Berwian, G. Müller, Application of a thermogravimetric technique for the determination of low nitrogen solubilities in metals: Using iron as an example. Thermochim. Acta **474**(1–2), 36–40 (2008)
14. D. Elwell, H.J. Scheel, *The Growth of Crystals in Solution, Advances in Colloid and Interface Science*, vol 10 (Academic Press, London/New York, 1979), pp. 215–252
15. S. Hussy, E. Meissner, J. Friedrich, Low-pressure solution growth (LPSG) of GaN templates with diameters up to 3 in. J. Cryst. Growth **310**, 738–747 (2008)
16. Y. Mori, M. Imade, M. Maruyama, M. Yoshimura, Growth of GaN crystals by Na flux method. ECS J. Solid State Sci. Technol **2**(8), N3068–N3071 (2013)
17. D. Ehrentraut, E. Meissner, A review on the Na-flux method toward growth of large-size GaN crystal, in *Technology of GaN Crystal Growth*, ed. by D. Ehrentraut, E. Meissner, M. Bockowski (Eds), (Springer Verlag, Heidelberg, 2010)
18. R. Doradziński, R. Dwiliński, J. Garczyński, L.P. Sierzputowski, Y. Kanbara, Ammonothermal growth of GaN under Ammono-basic conditions, in *Technology of Gallium Nitride Crystal Growth*, pp. 137–160
19. D. Ehrentraut, Y. Kagamitani, T. Fukuda, F. Orito, S. Kawabata, K. Katano, S. Terada, Reviewing recent developments in the acid ammonothermal crystal growth of gallium nitride. J. Cryst. Growth **310**, 3902–3906 (2008)
20. M.P. D'Evelyn, D. Ehrentraut, W. Jiang, D.S. Kamber, B.C. Downey, R.T. Pakalapati, H.-D. Yoo Ammonothermal, Bulk GaN substrates for power electronics. ECS Trans. **58**(4), 287–294 (2013)
21. N.S. Alt, E. Meissner, et al., Development of a novel in situ monitoring technology for ammonothermal reactors. J. Cryst. Growth **350**, 2–4 (2012)
22. T.G. Steigerwald, N.S.A. Alt, B. Hertweck, E. Schluecker, Feasibility of density and viscosity measurements under ammonothermal conditions. J. Cryst. Growth **403**, 59 (2014)
23. S. Zhang, F. Hintze, W. Schnick, R. Niewa, Intermediates in Ammonothermal GaN crystal growth under Ammono-acidic conditions. Eur. J. Inorg. Chem. **32**, 5387 (2013)
24. S. Zhang, N.S.A. Alt, E. Schlücker, R. Niewa, Novel alkali metal Amidogallates as intermediates in Ammonothermal GaN crystal growth. J. Cryst. Growth **403**, 22 (2014). https://doi.org/10.1016/j.jcrysgro.2014.06.015
25. T.M.M. Richter, R. Niewa, Chemistry of Ammonothermal synthesis. Inorganics **2**, 29 (2014). https://doi.org/10.3390/inorganics2010029
26. R. Juza, H. Jacobs, H. Gerke, Ammonothermalsynthese von Metallamiden und Metallnitriden. Ber. Bunsenges. Phys. Chem. **70**, 1103–1105 (1966)
27. H. Jacobs, H. Scholze, Untersuchung des Systems Na/LaNH3. Z. Anorg. Allg. Chem. **427**, 8–16 (1976)
28. U. Zachwieja, H. Jacobs, Synthesis of Bariumimide from the elements and orientational disorder of anions in BaND, studied by neutron diffraction from 8 K to 294 K. J. Less-Common Met **161**, 175–184 (1990)
29. U. Zachwieja u, H. Jacobs, Ammonothermalsynthese von Kupfernitrid, CU3N. J. Less-Cemmen Met **161**, 175–184 (1990)
30. S. Kaskel, M. Khanna, B. Zibrowius, H.-W. Schmidt, D. Ullner, Crystal growth in supercritical ammonia using high surface area silicon nitride feedstock. J. Cryst. Growth **261**, 99–104 (2004)
31. H. Jacobs, H. Mengis, Preparation and crystal structure of a sodium silicon nitride, NaSbN3, Eur. J. Solid state. Inorg. Chem. **30**, 45–53 (1993)
32. H. Jacobs, R. Nymwegen, Darstellung und Kristallstruktur eines Kaliumnitridophosphats, K3P6N11. Z. Anorg. Allg. Chem. **623**, 429–433 (1997)
33. J. Hausler, L. Neudert, M. Mallmann, R. Niklaus, A.-C.L. Kimmel, S.A. Nicolas, E.S. Alt, O. Oeckler, W. Schnick, Ammonothermal synthesis of novel nitrides: Case study on CaGaSiN3. Chem. Eur. J. **23**, 2583–2590 (2017)
34. J. Hertrampf, N.S.A. Alt, E. Schlücker, R. Niewa, SrBa2(NH2)6: A new ternary amide from ammonothermal synthesis. Z. Kristallogr. Suppl **35**, 80 (2015)

35. T. Yoshida, Y. Oshima, K. Watanabe, T. Tsuchiya, T. Mishima, Ultrahigh-speed growth of GaN by hydride vapor phase epitaxy. Phys. Status Solidi C **8**(7–8), 2110 (2011)
36. H.P. Maruska, J.J. Tietjen, The preparation and properties of vapor-deposited single-crystal GaN. Appl. Phys. Lett. **15**(10), 327 (1969)
37. T. Bohnen, H. Ashraf, G.W.G. van Dreumel, S. Verhagen, J.L. Weyher, P.R. Hageman, E. Vlieg, Enhanced growth rates and reduced parasitic deposition by the substitution of Cl2 for HCl in GaN HVPE. J. Cryst. Growth **312**, 2542 (2010)
38. K. Fujito, S. Kubo, H. Nagaoka, T. Mochizuki, H. Namita, S. Nagao, Bulk GaN crystals grown by HVPE. J. Cryst. Growth **311**, 3011 (2009)
39. E. Richter, M. Gründer, C. Netzel, M. Weyers, G. Tränkle, Growth of GaN boules via vertical HVPE. J. Cryst. Growth **350**, 89 (2012)
40. F. Lipski, M. Klein, X. Yao, F. Scholz, Studies about wafer bow of freestanding GaN substrates grown by hydride vapor phase epitaxy. J. Cryst. Growth **352**, 235 (2012)
41. H. Aida, K. Koyama, D. Martin, K. Ikejiri, T. Aoyagi, M. Takeuchi, S.W. Kim, H. Takeda, N. Aota, N. Grandjean, Optical characteristics of InGaN/GaN light-emitting diodes depending on wafer bowing controlled by laser-treated grid patterns. Appl. Phys. Express **6**, 035502 (2013)
42. I. Grzegory, B. Łucznik, M. Boćkowski, B. Pastuszka, M. Kryśko, G. Kamler, G. Nowak, S. Porowski, Growth of bulk GaN by HVPE on pressure grown seeds. Proceedings of SPIE 6121, Gallium Nitride Materials and Devices,612107, 3 March 2006. https://doi.org/10.1117/12.645976
43. B. Łucznik, B. Pastuszka, I. Grzegory, M. Boćkowski, G. Kamler, E. Litwin-Staszewska, S. Porowski, Deposition of thick GaN layers by HVPE on the pressure grown GaN substrates. J. Cryst. Growth **281**, 38–46 (2005)
44. H. Teisseyre, C. Skierbiszewski, B. Łucznik, G. Kamler, A. Feduniewicz, M. Siekacz, T. Suski, P. Perlin, I. Grzegory, S. Porowski, Free and bound excitons in GaN/AlGaN homoepitaxial quantum wells grown on bulk GaN substrate along the nonpolar (11-20) direction Appl. Phys. Lett. **86**, 162112 (2005)
45. Y. Oshima, T. Yoshida, T. Eri, M. Shibata, T. Mishima, Thermal and electrical properties of high-quality freestanding GaN wafers with high carrier concentration. Jpn. J. Appl. Phys. **45**, 7685 (2006)
46. H. Fujikura et al., Hydride-vapor-phase epitaxial growth of highly pure GaN layers with smooth as-grown surfaces on freestanding GaN substrates. Jpn. J. Appl. Phys. **56**, 085503 (2017)
47. K. Yamane, T. Matsubara, T. Yamamoto, N. Okada, A. Wakahara, K. Tadatomo, Origin of lattice bowing of freestanding GaN substrates grown by hydride vapor phase epitaxy. J. Appl. Phys. **119**, 045707 (2016)
48. A. Krost, A. Dadgar, GaN-based devices on Si. Phys. Status Solidi A **194**, 361–375., r (2002)
49. J. Brown et al., Performance of AlGaN/GaN HFETs fabricated on 100 mm silicon substrates for wireless base station applications. Microwave Symposium Digest, 2004 IEEE MTT-S International, Fort Worth, 2004
50. K. Cheng et al., AlGaN/GaN high electron mobility transistors grown on 150 mm Si(111) substrates with high uniformity. Jpn. J. Appl. Phys. **47**(3R), 1553 (2008)
51. O. Ambacher et al., Two-dimensional electron gases induced by spontaneous and piezoelectric polarization charges in N-and Ga-face AlGaN/GaN heterostructures. J. Appl. Phys. **85**, 3222–3233 (1999)
52. J. Stringfellow, *Organometallic Vapor-Phase Epitaxy: Theory and Practice*, 2nd edn. (Academic Press, 1998.) ISBN-13), pp. 978–0126738421
53. https://www.aixtron.com/fileadmin/documents/Technologien/AIX_Broschuere_AIX_R6_low_DS.pdf
54. https://www.aixtron.com/fileadmin/documents/Technologien/AIX_G5Plus_200mm_Low.pdf
55. http://www.veeco.com/products/turbodisc-maxbright-m-gan-mocvd-system-for-led-production
56. Y. Uemoto et al., Gate injection transistor (GIT)—A normally-off AlGaN/GaN power transistor using conductivity modulation. Trans. Electr. Dev **54-12**, 3393–3399 (2007)

57. http://www.laytec.de/epiras/
58. M.R. Leys et al., Growth of multiple thin layer structures in the GaAs-AlAs system using a novel VPE reactor. J. Cryst. Growth **68**, 431 (1983)
59. H. Ishikawa et al., Thermal stability of GaN on (111) Si substrate. J. Cryst. Growth **189–190**, 178–182 (1998)
60. A. Watanabe et al., The growth of single crystalline GaN on a Si substrate using AlN as an intermediate layer. J. Cryst. Growth **128**, 391–396 (1993)
61. D. Visalli et al., Experimental and simulation study of breakdown voltage enhancement of AlGaN/GaN heterostructures by Si substrate removal. Appl. Phys. Lett. **97**, 113501 (2010)
62. H. Yacoub et al., The effect of the inversion channel at the AlN/Si interface on the vertical breakdown characteristics of GaN-based devices. Semicond. Sci. Technol. **29**, 115012 (2014)
63. T.T. Luong et al., RF loss mechanisms in GaN-based high-electron-mobility-transistor on silicon: Role of an inversion channel at the AlN/Si interface. Phys. Status Solidi A **214**, 1600944 (2017)
64. H. Ishikawa et al., GaN on Si with AlGaN/GaN intermediate layer. Jpn. J. Appl. Phys **38**, L492–L494 (1999)
65. E. Feltin et al., Stress control in GaN grown on silicon (111) by metalorganic vapor phase epitaxy. Appl. Phys. Lett. **79**, 3230 (2001)
66. A. Ubukata et al., GaN growth on 150-mm-diameter (111) Si substrates. J. Cryst. Growth **298**, 198–201 (2007)
67. A. Reiher, Efficient stress relief in GaN heteroepitaxy on Si(111) using low-temperature AlN interlayers. J. Cryst. Growth **248**, 563–567 (2003)
68. J.W. Matthews, A.E. Blakeslee, Defects in epitaxial multilayers: I. Misfit dislocations. J. Cryst. Growth **27**, 118–125 (1974)
69. K. Cheng et al., AlN/GaN heterostructures grown by metal organic vapor phase epitaxy with in situ Si3N4 passivation. J. Cryst. Growth **315**(1), 204–207 (2011)
70. F. Medjdoub, Barrier-layer scaling of InAlN/GaN HEMTs. IEEE Electron Device Lett, 29–25 (2008)
71. J. Wurfl et al., Device breakdown and dynamic effects in GaN power switching devices: Dependencies on material properties and device design. ECS Trans. **50**(3), 211–222 (2013)
72. T. Palacios et al., AlGaN/GaN high electron mobility transistors with InGaN back-barriers. IEEE Electron Device Lett, 27–21 (2005)
73. M. Uren et al., Intentionally carbon-doped AlGaN/GaN HEMTs: Necessity for vertical leakage paths. IEEE Electron Device Lett, 35–33 (2014)
74. J. Ibbetson et al., Polarization effects, surface states, and the source of electrons in AlGaN/GaN heterostructure field effect transistors. Appl. Phys. Lett. **77**, 250–252 (2000)
75. T.R. Prunty, Passivation of AlGaN/GaN heterostructures with silicon nitride for insulated gate transistors. Proceedings of the 2000 IEEE/Cornell conference on high performance devices, Ithaca, 2000
76. J. Derluyn et al., Improvement of AlGaN/GaN high electron mobility transistor structures by in situ deposition of a Si3N4 surface layer. J. Appl. Phys. **98**, 054501 (2005)
77. M. Higashiwaki, Effects of SiN passivation by catalytic chemical vapor deposition on electrical properties of AlGaN/GaN heterostructure field-effect transistors. J. Appl. Phys. **100**, 033714 (2006)
78. K. Cheng et al., Formation of V-grooves on the (al,Ga)N surface as means of tensile stress relaxation. J. Cryst. Growth **353**, 88–94 (2012)
79. F. Medjdoub, GaN-on-Si HEMTs Above 10 W/mm at 2 GHz together with high thermal stability at 325°C. Proceedings of the 5th European Microwave Integrated Circuits Conference, Paris, 27–28 Sept 2010
80. https://www.infineon.com/cms/en/product/promopages/gallium-nitride/
81. http://gansystems.com/transistors.php
82. S.L. Rumyantsev, M.S. Shur, M.E. Levinshtein, Materials properties of nitrides. Int. J. High Speed Electron Syst **14**(1), 1 (2004)

83. J.-H. Ryou, W. Lee, J. Limb, D. Yoo, J.P. Liu, R.D. Dupuis, Z.H. Wu, A.M. Fischer, F.A. Ponce, Control of quantum-confined stark effect in multiple quantum well active region by -type layer for III-nitride-based visible light emitting diodes. Appl. Phys. Lett. **92**, 101113 (2008)
84. M. Leroux, N. Grandjean, M. Laugt, J. Massies, B. Gil, P. Lefebvre, P. BigenwaldM, Quantum confined stark effect due to built-in internal polarization fields in Al,Ga . . . N/GaN quantum wells. Phys. Rev. B **58**(20) (1998)
85. W. Udo, *Pohl, Epitaxy of Semiconductors: Introduction to Physical Principles* (Springer Science & Business Media, 2013)

Chapter 2
Lateral GaN HEMT Structures

Chang Soo Suh

2.1 Introduction

Transistors used for power switching typically operate as an "on-off" type device where the voltages applied to its electrodes alternate between the extreme ends of its functional spectrum. As illustrated in Fig. 2.1, power switching device in the off-state should exhibit negligible current flow through the device to minimize off-state conduction loss. The device should be able to withstand large off-state voltage (V_{OFF}) with sufficient margin to breakdown voltage (V_{BD}) to allow for non-ideal effects such as signal ringing and surge to ensure reliable operation over the lifetime of the switching circuit or system. In the on-state, the device should be capable of flowing large maximum current (I_{MAX}) with low on-resistance (R_{ON}) to minimize on-state conduction loss. Furthermore, the device should be able to rapidly transition back and forth between the two states to minimize switching loss. While each individual trait described above can be achieved for any semiconductor via device design, simultaneously optimizing multiple traits requires trade-offs, and the theoretical boundaries are determined by fundamental properties of the material.

Silicon (Si) has been the material of choice over the past several decades for semiconductor power switching devices, but the progression of performance improvement has slowed in recent years as each generation of devices is getting nearer to the theoretical limits determined by fundamental material properties. With the push for performance progression greater than the incremental steps, coupled with increasing demand for power conversion systems operating at higher frequencies, group III-nitride family of semiconductors, capable of delivering

C. S. Suh (✉)
Texas Instruments, Dallas, 13121 TI Blvd, MS 364 Dallas, TX 75243, USA
e-mail: c-suh@ti.com

G. Meneghesso et al. (eds.), *Gallium Nitride-enabled High Frequency and High Efficiency Power Conversion*, Integrated Circuits and Systems,
https://doi.org/10.1007/978-3-319-77994-2_2

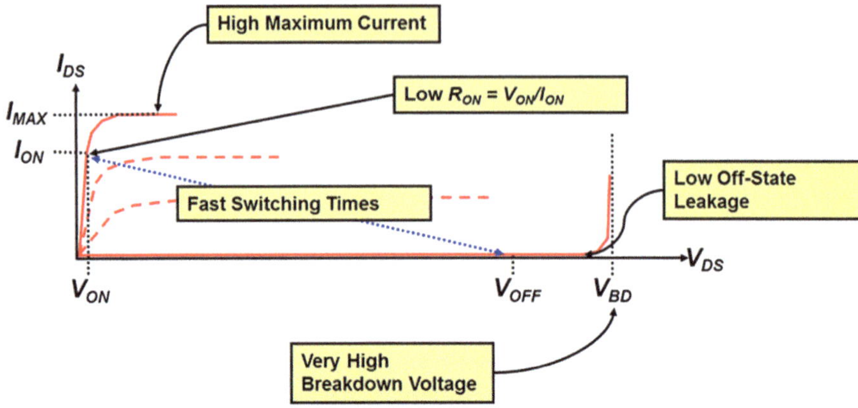

Fig. 2.1 Characteristics of an ideal switching device

superior performances beyond the limits of Si, appears poised to become the next power switching device material of choice with the gallium nitride (GaN) high-electron-mobility transistor (HEMT) device at its forefront.

Table 2.1 lists various material properties of relevant semiconductors and figures of merit (FOM) pertinent to power switching applications. Note that the FOM values are normalized to the values for GaN. Baliga's figure of merit (BFOM) defines material parameters to minimize conduction losses, thus primarily relevant for low-frequency applications. Baliga's high frequency figure of merit (BHFFOM) defines material parameters to minimize total power loss in high-frequency switching applications [1]. While both FOMs reflect the advantages of group III-nitrides (GaN, AlN) over other semiconductors, superior performances are expected particularly in high-frequency applications. Since the first demonstration of GaN-based HEMT devices in 1993 by Khan et al. [2], tremendous advances have been made in multiple fronts of GaN development, enabling various device structures leading to performances beyond the material limits of Si. In this chapter, the basic GaN-based HEMT device is introduced, followed by sections on structural innovations for increasing channel mobility, reducing the current collapse phenomena, increasing breakdown voltages, and achieving normally off operation.

2.2 The Basic GaN HEMT Device: Polarization, Surface States, and the 2DEG

One of the most unique and important properties which sets GaN apart from other semiconductor materials is its ability to form an ultrathin, highly conductive channel, without the use of any extrinsic dopants or applied electric field. As shown in Fig. 2.2, when a relatively thin layer of aluminum gallium nitride (AlGaN) is

Table 2.1 Material properties of semiconductors at room temperature and key figures of merit (relative to GaN) for power switching

Material	E_g(eV)	ϵ_r	μ_n(cm^2/Vs)	E_c(MV/cm)	υ_s(10^7cm/s)	κ_{th}($E_c \upsilon_s / 2\pi$)	BFOM($\epsilon \mu_n E_g^3$) [1]	BHFFOM($\mu_n E_c^2$) [1]
Si	1.12	11.7	1350	0.3	1	1.3	0.051	0.007
GaAs	1.42	12.9	8500	0.4	2	0.55	0.716	0.076
4H-SiC	3.23	9.66	900	2.5	1.9	3.7	0.668	0.316
GaN	3.39	8.9	1265	3.75	2.5	2.5	1	1
AlN	6.2	8.5	300	11.7	1.4	2.85	1.386	2.309

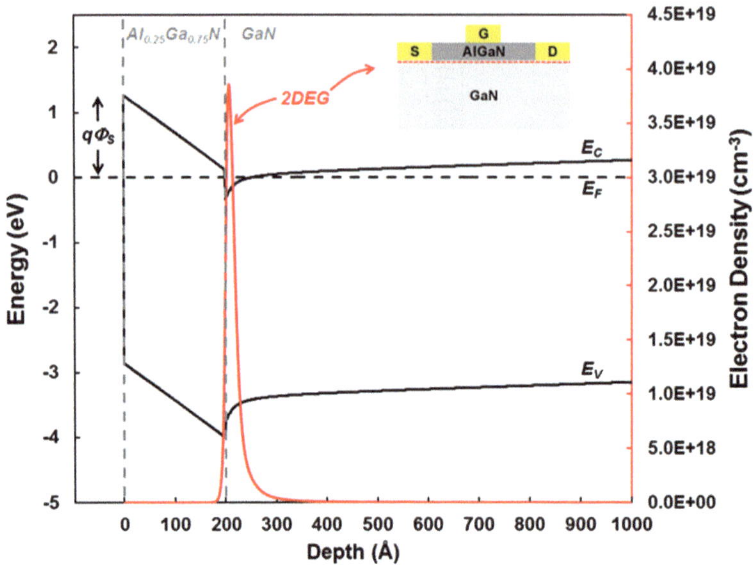

Fig. 2.2 Energy-band diagram of an AlGaN/GaN heterostructure along the vertical direction and the 2DEG electron profile. (Simulation performed using BandEng, self-consistent 1D Poisson-Schrodinger solver developed by Dr. Michael Grundmann)

grown on top of a much thicker GaN layer (typically 20–25 nm thick), a dense "cloud" of electrons forms immediately beneath the AlGaN/GaN hetero-interface due to polarization fields and donor-like surface states [3]. This electron cloud is vertically confined within a thin triangular potential well, similar to that of an inversion channel in a silicon (Si) metal-oxide-semiconductor (MOS) structure. Due to likeness to a sheet of charge, the electron cloud is referred to as two-dimensional electron gas (2DEG) and is utilized as the channel of the GaN HEMT device. Thus, the AlGaN/GaN heterostructure is commonly referred to as the GaN HEMT structure. While there are numerous combinations of group III-nitride heterostructures that will form a 2DEG at its hetero-interface, the AlGaN/GaN heterostructure is most commonly used at present.

While GaN exists in both "wurtzite" (WZ) and "zinc-blende" (ZB) phases, the WZ form, which exhibits strong polarization properties is more stable and predominantly used. As illustrated by the stick-and-ball model of the WZ GaN structure in Fig. 2.3, each gallium atom is bonded to four nitrogen atoms and vice versa in a tetrahedral bond configuration. Due to all gallium-to-nitrogen bonds parallel to the [0001] or [000$\bar{1}$] axis pointing in the same direction, uniaxial anisotropy about this axis is present, resulting in net spontaneous polarization along the same axis. Spontaneous polarization is presence of built-in polarization field in an unstrained crystal at equilibrium. Crystal structures with asymmetry such as WZ structures exhibit spontaneous polarization along the crystal direction lacking inversion symmetry, and its magnitude is dependent on the ionicity of

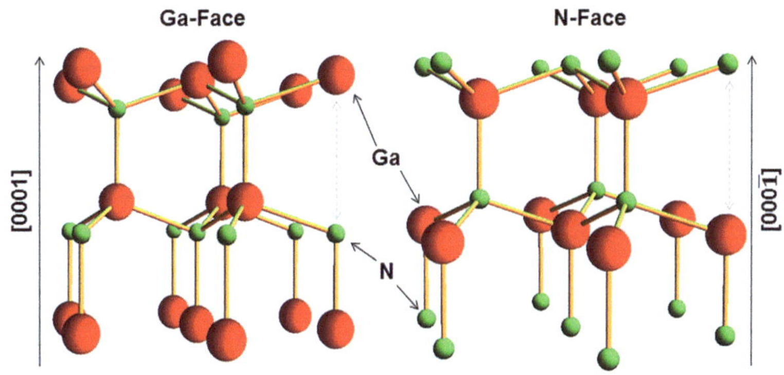

Fig. 2.3 Three-dimensional stick-and-ball illustration of "wurtzite" GaN grown in the Ga-face (left) and N-face (right) direction

the constituent atoms. Due to the strong ionicity of GaN, there is an additional electrostatic attraction of significant magnitude between gallium and nitrogen atoms that are positioned above/below each other without forming bonds as marked by dashed lines in Fig. 2.3. The added forces from this attraction result in structural deformations that further displace gallium and nitrogen atoms from ideal structure locations and contribute to the large spontaneous polarization within GaN [4].

In the GaN HEMT structure, the much thinner AlGaN layer is pseudomorphically grown above the GaN layer. Due to the lattice constant differences between the two materials, the AlGaN layer is compressively strained, and the added structural deformation caused by the strain gives rise to piezoelectric polarization. Piezoelectric polarization is presence of polarization field resulting from distortion of the crystal lattice in response to mechanical stress. Unlike spontaneous polarization, piezoelectric polarization occurs in both WZ and ZB structures. However, piezoelectric polarization in WZ structures is nearly an order of magnitude larger than in ZB structures.

Due to canceling of dipoles within the bulk of the material, polarization is modeled as sheets of fixed surface charge (σ_{POL}) with opposite signs at the top and bottom surfaces as shown in Fig. 2.4 [5]. When comparing material grown in the Ga-face and N-face direction, only the sign of the polarization charge is opposite. Currently, majority of the GaN materials used by the power switching industry are grown in the Ga-face direction. As shown in Fig. 2.5, within the thicker, fully relaxed GaN layer, only spontaneous polarization is present, whereas within the pseudomorphically grown AlGaN layer, both spontaneous and piezoelectric polarization are present. The polarization charge densities exceed well beyond 10^{13} cm^{-2}, leading to presence of large built-in electric fields that play a critical role in the formation of the 2DEG.

Along with polarization, surface properties of nitride semiconductors drive the formation of 2DEG at the AlGaN/GaN interface. At the surface of $Al_XGa_{1-X}N$ in the as-grown state, partially filled positively charged donor states exist at a level

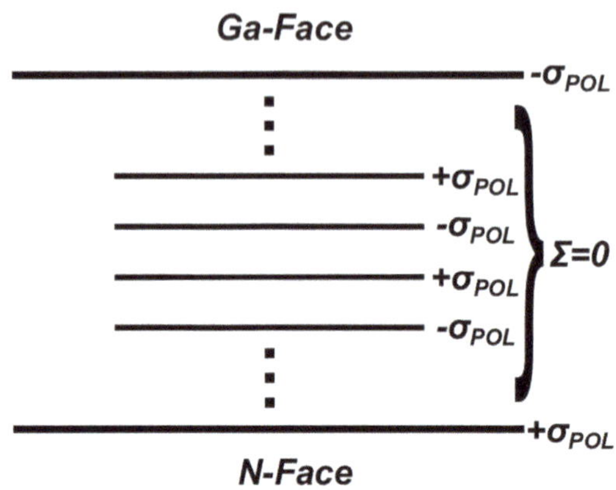

Fig. 2.4 Modeling of the polarization. Internal dipoles cancel each other, leaving only sheets of fixed charge at the top and bottom surfaces with opposite polarity

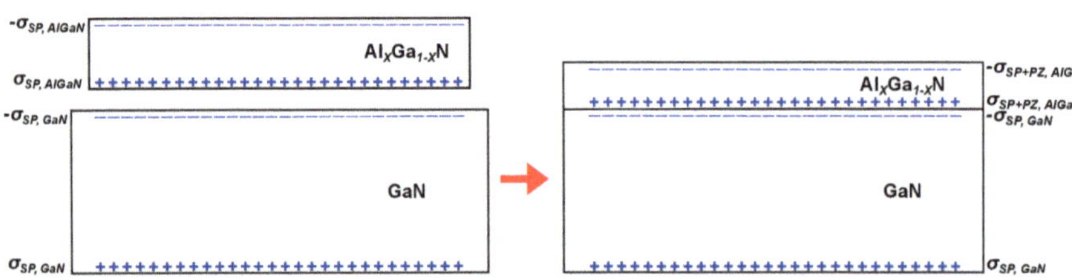

Fig. 2.5 Illustration of polarization induced charges. As separate layers, only spontaneous polarization (SP) charges are present. When the thinner AlGaN layer is pseudomorphically grown on the much thicker GaN layer, charge induced from both spontaneous and piezoelectric polarization (PZ) is present in the AlGaN layer

$\Phi_S(x) \approx (1 + x)$ eV below its conduction band edge [6]. The electrons in the 2DEG are supplied from these surface states, and as illustrated in Fig. 2.6, its presence is necessary to offset the negative charge of the 2DEG for charge neutrality conditions to be met. Collectively, this surface state and the polarization fields give rise to the 2DEG, and its density (n_S) is determined as a function of aluminum composition and AlGaN thickness (t_{AlGaN}).

2.3 Structures for Higher Mobility

Switching times become increasingly important in high power switching applications as operating frequencies are increased. As shown in Fig. 2.7, with increasing frequency, the ratio of switching time to conduction time increases (within a given time span, number switching events increase). Increasing the device area reduces conduction losses, but switching losses increase due to slower switching times resulting from increased capacitance and vice versa. For a given frequency, the total

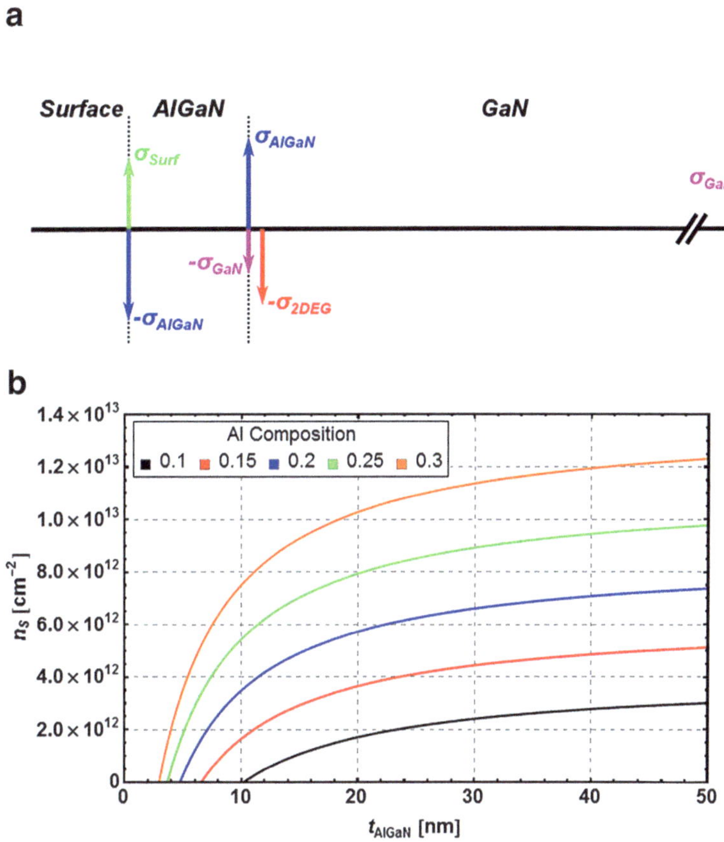

Fig. 2.6 (**a**) Illustration of various charges within the AlGaN/GaN HEMT structure. The electrons in the 2DEG are supplied by the partially filled donor states at the surface. (**b**) 2DEG density (n_s) for various aluminum composition and AlGaN thickness (t_{AlGaN})

power loss is minimized when the device area is designed for equal conduction and switching losses. Under equal conduction and switching loss conditions, minimum power loss follows:

$$P_{\text{Loss,min}} \propto \frac{\sqrt{f}}{\sqrt{\mu} E_C} \tag{2.1}$$

where μ is the channel mobility and E_C is the critical electric field [7]. Since E_C is determined by the material and the frequency is determined by the application, higher channel mobility is desired for minimization of power loss.

Lack of ionized donors in the AlGaN/GaN HEMT structure contributes to the high mobility of the 2DEG channel. However, alloy scattering due to randomly distributed gallium and aluminum within the AlGaN crystal is the dominant factor

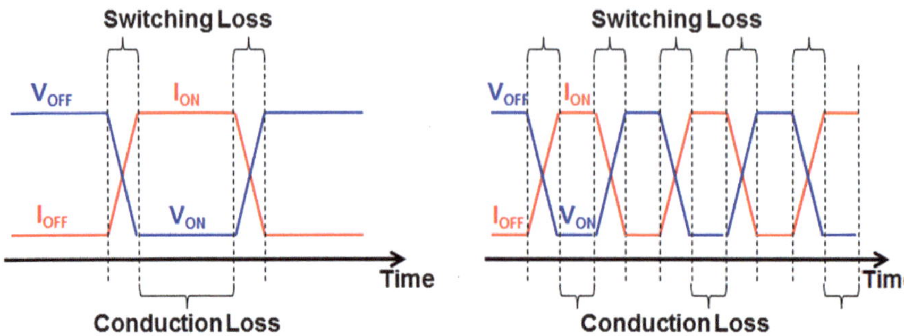

Fig. 2.7 Current and voltage vs. time illustration of low- and high-frequency switching. At low frequencies (left), conduction losses dominate, whereas switching losses become significant as operating frequency is increased (right)

that limits mobility in the AlGaN/GaN HEMT structures [6]. Hsu and Walukiewicz first proposed the introduction of an AlN interlayer at the AlGaN/GaN hetero-interface to mitigate the alloy scattering limitation [8]. When a thin AlN layer is inserted between the AlGaN and GaN layer, the 2DEG penetration into the barrier layer is reduced due to larger conduction band offset (ΔE_C) between AlN and GaN than between AlGaN and GaN. The conduction band diagrams (E_C) and 2DEG electron distributions for an AlGaN/GaN HEMT structure with and without 1 nm-thick AlN interlayer are shown in Fig. 2.8. With the insertion of AlN interlayer, the 2DEG density increases slightly due to larger polarization coefficients of AlN than AlGaN, but the 2DEG penetration into the barrier layer is limited to within just the AlN layer, thus eliminating alloy scattering that normally occurs in the ternary AlGaN layer.

Although improved 2DEG mobility and excellent device performances are widely reported with the use of an AlN interlayer [9–12], it may not be suitable for applications where large positive voltages may be applied to the gate. Thickness fluctuations of the AlN layer, as small as single monolayer, can significantly change the effective barrier height of the AlGaN/AlN layer. As shown in Fig. 2.9, the effective barrier height (ϕ_b) between the 2DEG and the $Al_{0.25}Ga_{0.75}N$ layer can range from less than 0.1 eV without AlN to approximately 0.75 eV with 1-nanometer (nm)-thick AlN when the gate is forward biased by 1 V. Since typical gate widths of power switching devices can range between tens and several thousands of millimeters, presence of localized AlN layer thickness fluctuations below the gate regions is inevitable even with the most advanced material growth techniques such as molecular beam epitaxy (MBE) and metal-organic chemical vapor deposition (MOCVD). With the gate under forward bias, locations beneath the gate where the AlN is thinner will conduct more current than the rest of the gate due to localized barrier height reduction, thus leading to premature degradation and breakdown of the gate [13].

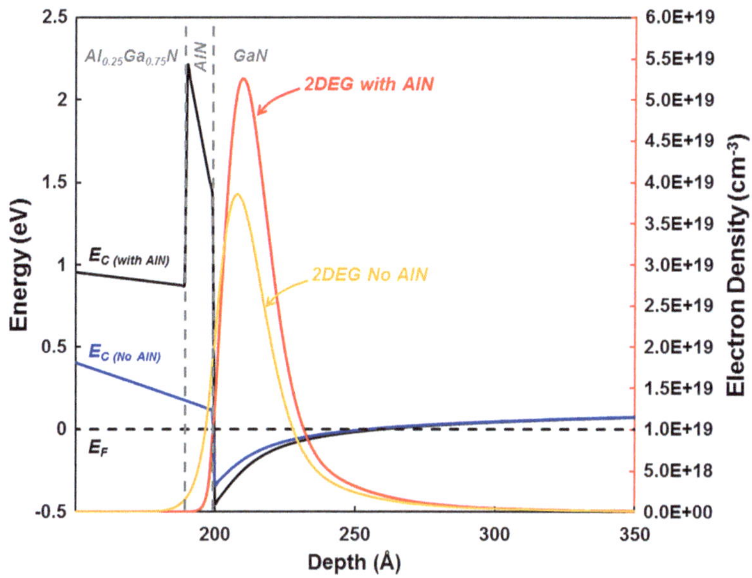

Fig. 2.8 Energy-band diagram (only conduction band shown) and 2DEG distribution near the AlGaN/GaN hetero-interface with and without a 1 nm-thick AlN interlayer. Despite the increased 2DEG density, electron penetration into the barrier layer is limited to within the AlN interlayer thus eliminating alloy scattering caused by the AlGaN layer. (Simulation performed using BandEng, self-consistent 1D Poisson-Schrodinger solver developed by Dr. Michael Grundmann)

2.4 Structures for Current Collapse Mitigation

When GaN HEMT devices are stressed under high drain voltage in the off-state and switched on, the resulting output current is less than that of the value prior to the high voltage stress. This behavior is typically referred to as current collapse, and the severity is increased with the magnitude of the off-state stress voltage and switching speeds as shown in Fig. 2.10. Relative to the fresh "DC" output currents in black, pulsed I-V output currents are decreased with shorter pulse widths, analogous to higher frequency switching. Although the collapse is not permanent, time to full recovery is on the order of seconds, thus problematic for power switching applications. Current collapse is also referred to as dispersion, DC-to-RF dispersion, knee-walkout, or dynamic R_{ON}.

One of the primary causes of current collapse in GaN HEMTs is trapping of electrons at the AlGaN surface near the drain side of the gate [3]. A widely accepted model of the surface trapping effects is illustrated in Fig. 2.11. During off-state stress under high V_D stress (state I), positive surface states capture electrons originating from the gate metal, forming a reverse-biased "virtual gate." Immediately upon removal of the high V_D stress and switching to on-state (state II), a significant portion of the charges forming the "virtual gate" remain due to long detrapping time

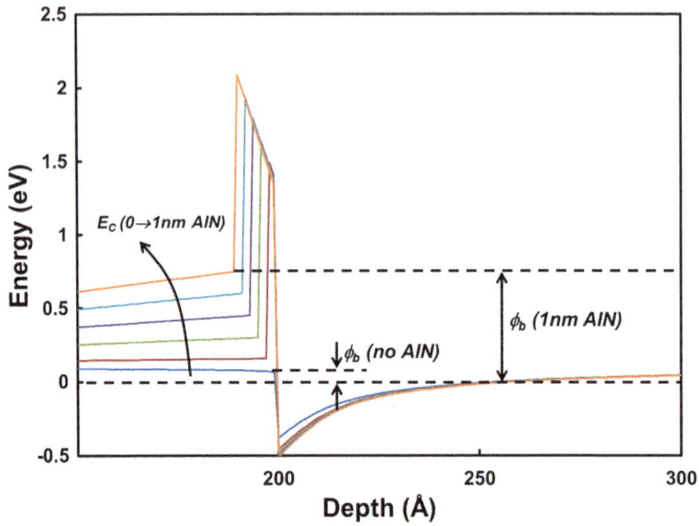

Fig. 2.9 Energy-band diagram (showing conduction band of AlGaN/GaN hetero-interface only) of AlGaN/AlN/GaN HEMT structure under forward bias. The effective barrier height (ϕ_b) between the 2DEG and the AlGaN barrier toward the gate increases with thickness of the AlN. (Simulation performed using BandEng, self-consistent 1D Poisson-Schrodinger solver developed by Dr. Michael Grundmann)

Fig. 2.10 Output current of an unstressed GaN HEMT device (DC) vs. pulsed output current after off-stress at high drain voltage showing effects of current collapse

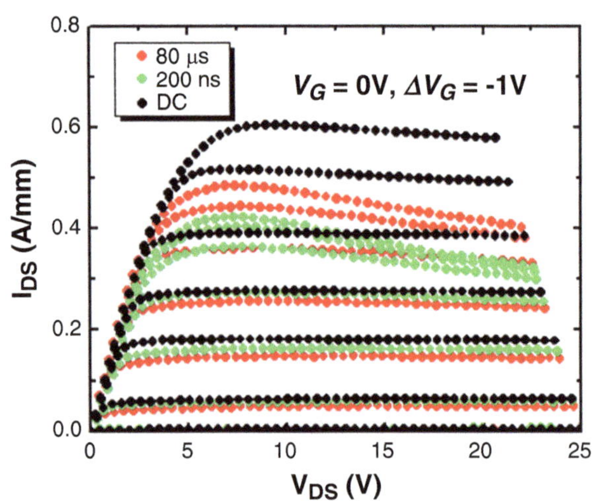

constants, reducing the 2DEG concentration beneath. Thus, current flow is reduced and R_{ON} is increased relative to prestress values. Only upon complete detrapping of the charged surface states (state III), current flow and R_{ON} fully recover.

Current collapse due to trapping at the AlGaN surface can be reduced significantly by passivating the AlGaN surface with a layer of silicon nitride (SiN_X)

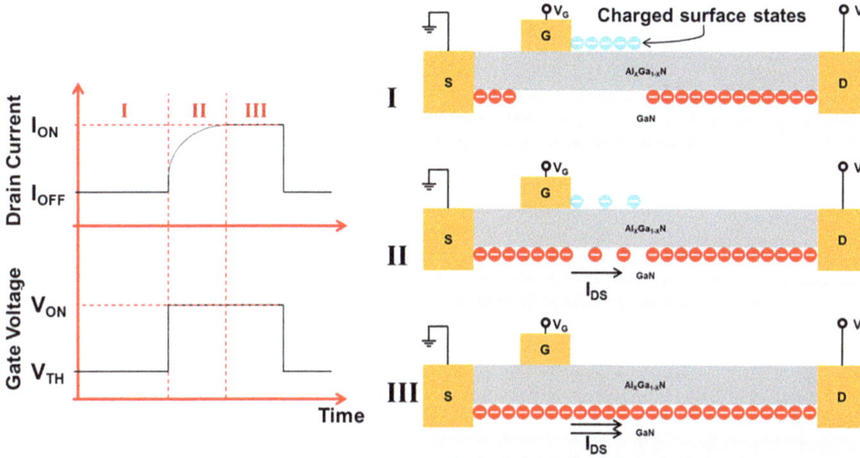

Fig. 2.11 Illustration of the surface trapping mechanism for current collapse in AlGaN/GaN HEMT devices

film [14]. While exact mechanism by which SiN_X prevents current collapse is not fully understood, passivating the surface early as possible in the fabrication process flow helps prevent degradation of the AlGaN surface and helps maintain a high-quality interface between the AlGaN and the SiN_X film. The use of GaN HEMT structure with an in situ SiN_X capping layer prevents the AlGaN surface from ever being exposed to the air during the fabrication steps and has demonstrated reduced current collapse and improved uniformity compared to conventional SiN_X deposited external to the epi deposition tool [15–18]. Because the in situ SiN_X is deposited in the same tool as the epitaxial deposition, the SiN_X can be deposited at higher temperatures than the typical deposition temperatures of other commonly used deposition tools to achieve superior quality. As shown in Fig. 2.12, the in situ SiN_X film can be used as the lone passivation film of a Schottky gate device or as part of a multi-stack passivation film in a metal-insulator-semiconductor (MIS) gate device.

An alternative approach to SiN_X passivation is using a HEMT structure with large distances between the surface and the 2DEG. Due to the close proximity of the 2DEG and the surface of AlGaN HEMTs (typically 15–25 nm), any changes to the surface potential impart strong response to the 2DEG. Since the ability of surface potential to modulate channel charge is inversely proportional to the distance between the surface and the channel, increasing the distance between the 2DEG and the AlGaN surface can help reduce current collapse. However, due to the lattice mismatch between the AlGaN layer and GaN, increasing the AlGaN thickness too much will lead to relaxation of the strain. Although growing a thick GaN cap layer is possible without changing the mechanical stress in the AlGaN layer, a Si-doped graded-AlGaN layer between the AlGaN barrier and the GaN cap is needed

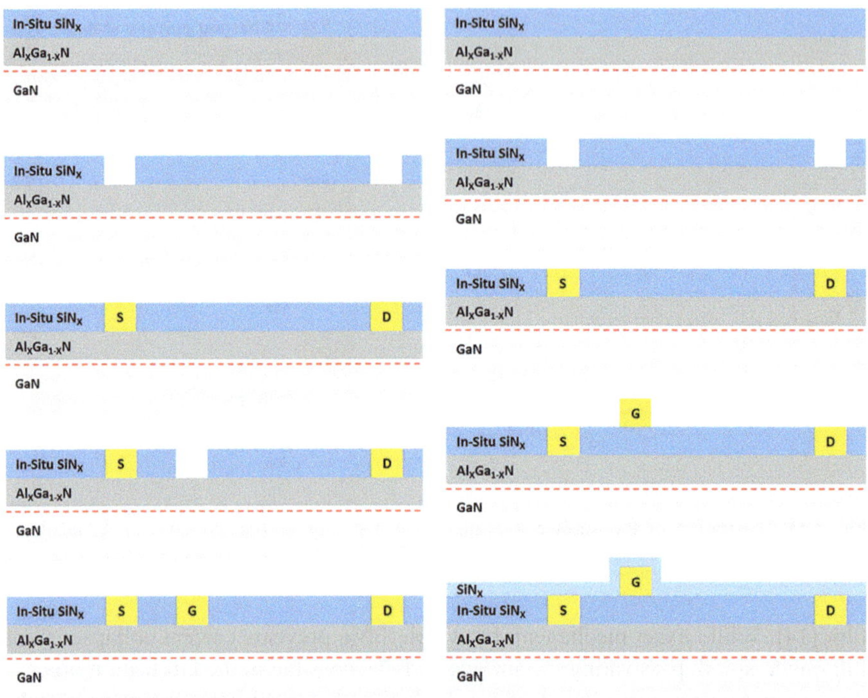

Fig. 2.12 Cross-sectional schematic showing GaN HEMT process flow (top to bottom) utilizing in situ SiN_X as passivation film of a Schottky gate device (left) and as part of a multi-stack passivation film of a MIS gate device (right)

to compensate for the polarization fields within the GaN cap layer and prevent accumulation of holes at the top GaN/AlGaN interface. As shown in Fig. 2.13, this approach adds process complexities as source, drain, and gate contacts must be formed after performing a deep recess etch to reach the AlGaN barrier layer, but the resulting device as demonstrated by Shen et al. [12] exhibits excellent pulsed-IV characteristics without the use of any SiN_X passivation films.

Electron trapping at the AlGaN surface responsible for current collapse can also occur due to injection of hot electrons from the channel to the surface along the entire gate-to-drain access regions [19]. As presented in the previous section, the use of AlN interlayer increases the barrier height between the 2DEG and the surface significantly. As demonstrated by Lee et al., in addition to providing the benefit of increased channel mobility, the use of AlN interlayer also improves current collapse as evidenced by the pulsed-IV output curves of devices with and without AlN interlayer in Fig. 2.14 [20].

Off-state stress under very high drain voltages, especially at high temperatures, can cause a sharp potential drop or high concentration of electric field at the drain-side edge of the gate-drain access region. This leads to hole emission from

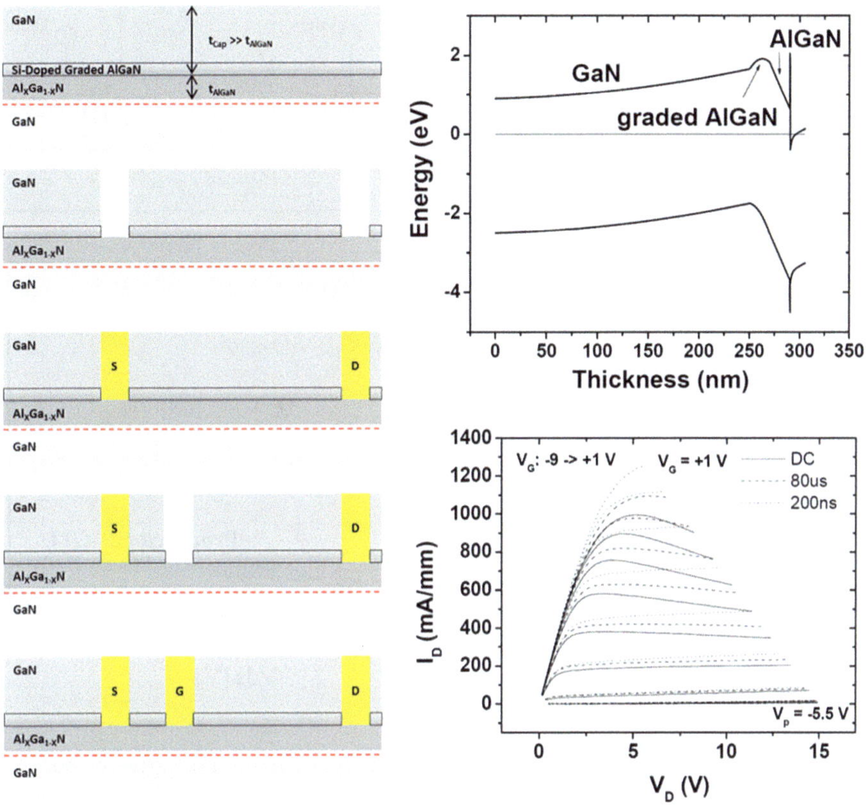

Fig. 2.13 Cross-sectional schematic showing process flow for passivation-free thick GaN-cap HEMT structure (left) and the energy-band diagram of the epi structure and the resulting pulsed-IV output curves [12]. (© TMS-The Minerals, Metals and Materials Society 2004)

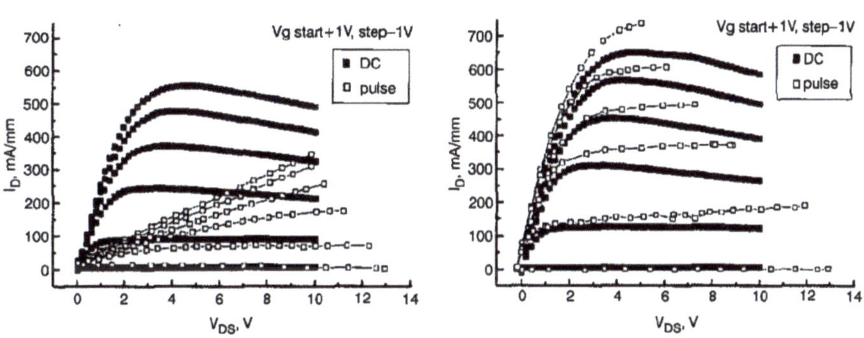

Fig. 2.14 DC and pulsed-IV output characteristics of a device without AlN interlayer (left) and with 1 nm-thick AlN interlayer (right) [20]. (Reproduced by permission of the Institution of Engineering & Technology)

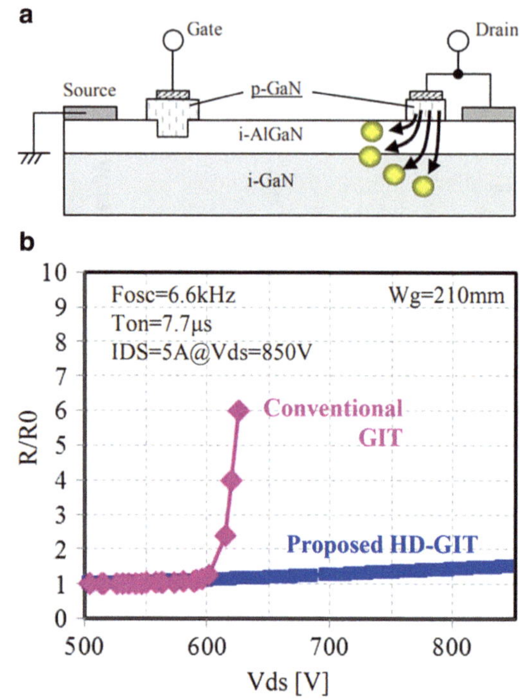

Fig. 2.15 (**a**) Cross-sectional schematic of a GaN HEMT device with a hybrid drain contact consisting of a normal n-type ohmic contact and a p-GaN hole injection contact. With the hybrid drain contact, (**b**) current collapse effects are drastically reduced even at 850 V [21]. (© IEEE 2015)

the GaN and other buffer layers beneath the 2DEG, leaving a negatively charged state. After the device turns on, the neutralization of these negatively charge states occurs at a much slower rate than typical power switching time scales, resulting in current collapse. As shown in Fig. 2.15, through the introduction of a hybrid drain consisting of a normal n-type ohmic contact and a p-GaN hole injection contact, Kaneko et al. demonstrated drastic reduction of dynamic R_{ON} effects up to 850 V. Injection of holes from the drain side neutralizes the effects of the hole emission, thus eliminating the current collapse [21].

2.5 Structures for High Voltage Operation

In the off-state, the maximum electric field within AlGaN/GaN HEMT devices occur at the drain side edge of the gate due to fixed positive charges in the gate-to-drain depletion region imaging to the gate metal corner. Because the electric field profile is not very uniform across the depletion region, the V_{BD} does not scale well with gate-to-drain spacing (L_{GD}). Devices without passivation typically exhibit higher V_{BD}, which scale linearly with L_{GD} because the "virtual gate" created by the surfaces states increase in length with increasing drain voltage and distribute the

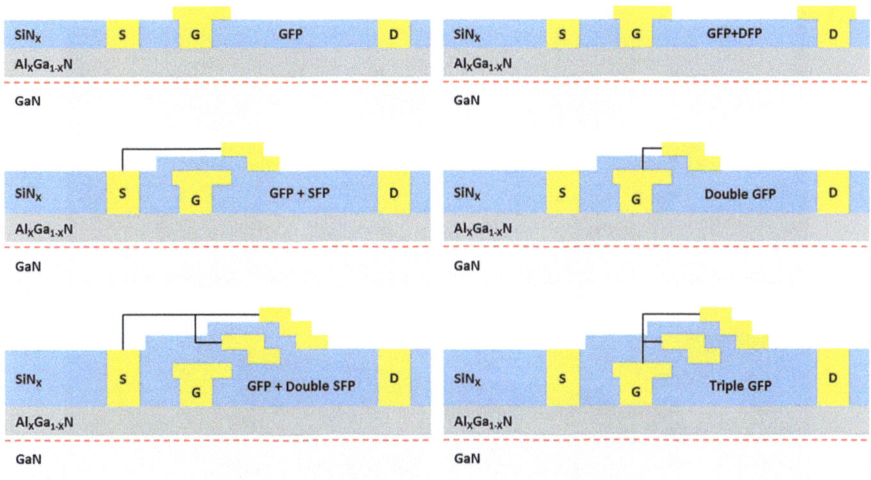

Fig. 2.16 Cross-sectional schematics of various field-plate configurations. Optimal configurations are determined by the required operating voltages and circuit requirements

electric field within the depletion region uniformly [22]. As previously mentioned, the slow recovery of the charges trapped in the "virtual gate" renders the device useless in power switching applications.

High V_{BD} in AlGaN/GaN HEMTs have been largely achieved by utilization of field plates in various configurations [22–27]. As illustrated in Fig. 2.16, a field plate is an extension of metal electrodes above the gate-drain access region. Field plates offer additional edges for the electric field lines to terminate as drain voltages are increased to distribute the electric fields at the cost of increased capacitances. Electric field management can be further improved by controlling the sidewall slopes of the field plates. Analytical models predict significant improvements with sidewalls sloped to less than 30 degrees (measured from the surface) [28]. With the use of approximately 15 degree slant field plate, depletion-mode (D-mode) devices with greater than 2 kV V_{BD} (Dora et al. [22]), enhancement-mode (E-mode) devices with greater than 1.4 kV V_{BD} (Suh et al. [29]), and the technology for producing an asymmetric slant field plate with as low as 6 degree angle for both D-mode and E-mode devices (Wong et al. [31]) have been demonstrated. Fig. 2.17 shows the cross-sectional images of the slanted field plates.

An alternative approach to electric field management was demonstrated by Nakajima, et al. through the use of a polarization super-junction (PSJ) in the gate-drain access region as shown in Fig. 2.18 [32]. The PSJ utilizes compensation effects of polarization charges at the top and bottom hetero-interfaces of a GaN/AlGaN/GaN double heterostructure to achieve charge balance, enabling uniform distribution of electric field. The 2DEG density in the PSJ region however is reduced due to the top GaN/p-GaN layers lifting the surface potential which results in depletion of the 2DEG. Conceptually, this approach is very similar to silicon-based super-junction MOS devices.

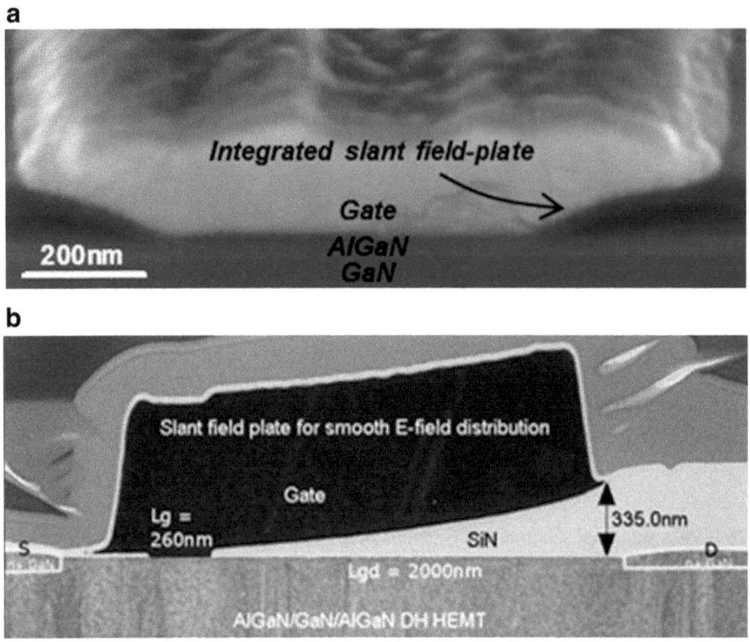

Fig. 2.17 Cross-sectional image of (**a**) symmetric [30] and (**b**) asymmetric [31] slanted field plates on AlGaN /GaN HEMT structures. (© IEEE 2017)

2.6 Structures for Normally Off Operation

High power switching applications typically require normally-off or enhancement-mode (E-mode) devices for simplicity of biasing requirements and added safety when compared to a normally-on or depletion-mode (D-mode). D-mode devices require a negative gate bias source, and the bias range required by GaN-based D-mode high voltage device is typically larger than the range required by E-mode devices. Loss of gate control in D-mode devices could be detrimental to the entire circuit or module. Also, a threshold voltage (V_{TH}) greater than 1 V is desired for immunity from unexpected conduction due to gate signal noise and shoot-through. Since GaN HEMTs in its simplest form is a D-mode device, the 2DEG below the gate region must be removed to achieve E-mode operation.

Figure 2.19 illustrates three main approaches used for achieving E-mode operation in GaN devices. The first approach is to use a metal-insulator-semiconductor (MIS) gate structure where the thickness of the AlGaN layer below the gate is thinned to a thickness below which the 2DEG is no longer present, or completely remove the AlGaN layer [30, 33, 34]. Various challenges such as V_{TH} stability/uniformity and reduced channel mobility remain for this approach to be a viable option for the power switching industry. The second approach utilizes implantation

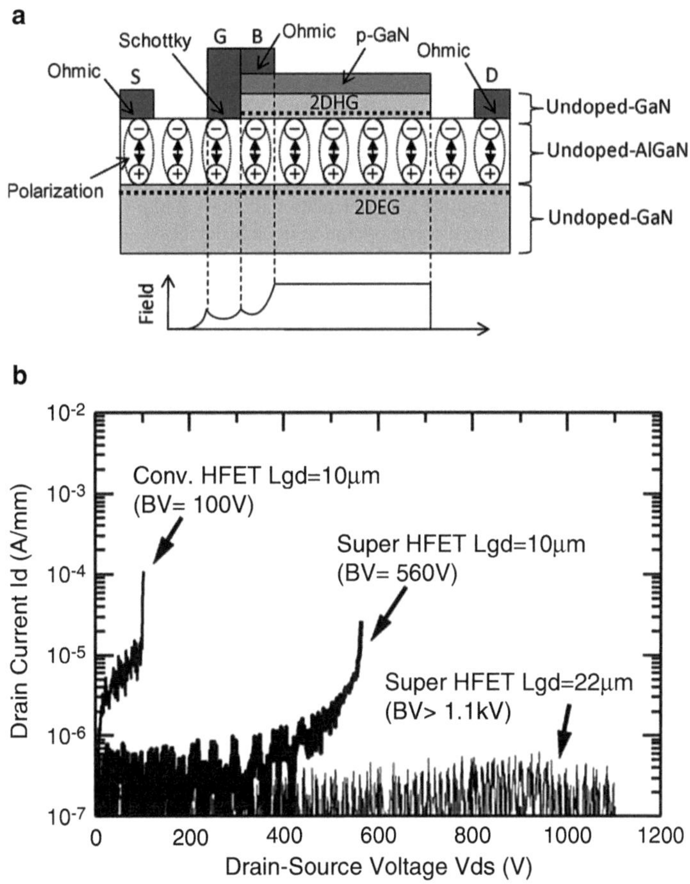

Fig. 2.18 (**a**) Cross-sectional schematic of a GaN HEMT device with a PSJ field management structure. (**b**) Use of PSJ structure increased the V_{BD} from 100 V to 560 V [32]. (© IEEE 2011)

of negatively charged ions such as fluorine or hydrogen into the AlGaN layer below the gate [35, 36]. While E-mode operation has been successfully demonstrated, reports of V_{TH} instability under high temperature and electric field stresses have hindered industry adoption [30, 36]. The last approach is to use a p-type GaN layer between the gate metal and the AlGaN barrier to deplete the channel [30, 37–40]. Currently, this approach is the most widely used method in the industry.

Conventional process flow for p-GaN-capped HEMT devices starts with p-GaN/AlGaN/GaN stack and the p-GaN in the access, and the source/drain ohmic contact regions are etched off to form the 2DEG in those regions. One drawback of this device structure is the trade-off between the 2DEG density and the V_{TH}. As shown in Fig. 2.20, to increase the V_{TH}, the AlGaN barrier thickness must be reduced, resulting in lower 2DEG density. Furthermore, the thickness of AlGaN

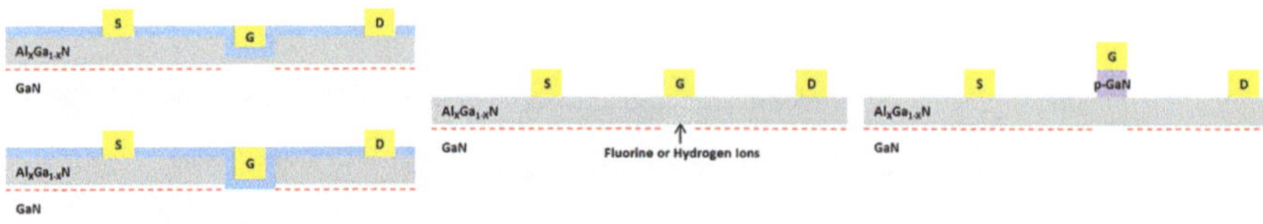

Fig. 2.19 Gate recessing (left), negative ion implantation (center), and p-GaN gate (right) are the main approaches used for achieving E-mode operation in GaN devices

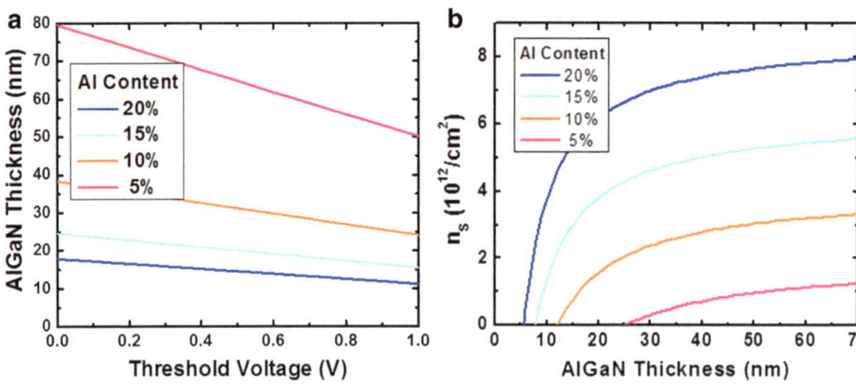

Fig. 2.20 (**a**) Target V_{TH} vs. required AlGaN thickness in a p-GaN/AlGaN/GaN structure with [Mg] of 10^{18} cm^{-3} and (**b**) AlGaN thickness vs. maximum possible 2DEG density in the access region after removing the p-GaN layer

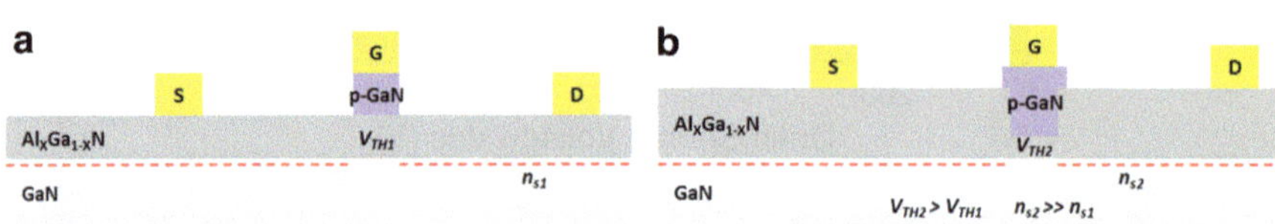

Fig. 2.21 (**a**) Conventional p-GaN/AlGaN/GaN E-mode device structure with a V_{TH} and n_s trade-off (left) and (**b**) improved device structure without the trade-off

barrier required for E-mode operation fall on the region of the 2DEG density vs. AlGaN thickness curve with larger slope. Thus, if any AlGaN layers are etched off during the p-GaN etch steps, the sheet resistance of the 2DEG in the access regions is rapidly increased.

A device structure with a thinner AlGaN barrier below the p-GaN gate region and thicker AlGaN barrier in the access regions as shown in Fig. 2.21 is required to break the trade-off between the V_{TH} and 2DEG density in the access regions. Production of such device requires a regrowth step, but due to high levels of donor impurities typically associated with the regrowth interfaces off III-nitrides, use of regrowth has been primarily limited in n-type ohmic contacts [41]. In 2016, Okita et al. [42] developed a breakthrough technology enabling regrowth of p-GaN/AlGaN layer and demonstrated high voltage devices with very uniform V_{TH} and low R_{ON}. Furthermore, the regrowth technology enabled simultaneous demonstration of hybrid p-GaN drain contact discussed in the previous section.

References

1. B.J. Baliga, Power semiconductor device figure of merit for high frequency applications. IEEE Electron Device Lett **10**(10), 455–457 (1989)
2. M.A. Khan et al., High electron mobility transistor based on a GaN-AlXGa1-XN heterojunction. Appl. Phys. Lett **63**, 1214 (1993)
3. R. Vetury, Polarization induced 2DEG in AlGaN/GaN HEMTs: On the origin, DC, and transient characterization, D.Phil. dissertation, Department of Electrical and Computer Engineering, University of California at Santa Barbara, Santa Barbara, CA, 2000
4. T. Hanada, Basic properties of ZnO, GaN, and related materials, in *Oxide and Nitride Semiconductors. Advances in Materials Research*, ed. by T. Yao, S. K. Hong (Eds), vol. 12, (Springer, Berlin/Heidelberg, 2009)
5. U. Mishra, J. Singh, 8.6 polar materials and structures, in *Semiconductor Device Physics and Design*, Dordrecht, The Netherlands (Springer, 2007)
6. D. Jena, Polarization induced electron populations in III-V nitride semiconductors transport, growth, and device applications, D.Phil. dissertation, Department of Electrical and Computer Engineering, University of California at Santa Barbara, Santa Barbara, CA, 2003
7. A.Q. Huang, New unipolar switching power device figures of merit. IEEE Electron Device Lett **43**(10), 1717–1731 (1997)
8. L. Hsu, W. Walukiewicz, Effect of polarization fields on transport properties in AlGaN/GaN heterostructures. J. Appl. Phys. **89**, 1783–1789 (2001)
9. L. Shen et al., AlGaN/AlN/GaN high-power microwave HEMT. IEEE Electron Device Lett **22**, 457–459 (2001)
10. C.S. Suh et al., High-breakdown enhancement-mode AlGaN/GaN HEMTs with integrated slant field-plate, IEDM Technical Digest, pp. 911–913, Dec 2006
11. Y. Hao et al., High-performance microwave gate-recessed AlGaN/AlN/GaN MOS-HEMT with 73% power-added Efficienty. IEEE Electron Device Lett **32**, 626–628 (2011)
12. L. Shen et al., Unpassivated GaN/AlGaN/GaN power high electron mobility transistors with dispersion controlled by epitaxial layer design. J. Electron. Mater. **33**, 422–425 (2004)
13. R. Coffie et al., Impact of AlN Interlayer on Reliability of AlGaN/GaN HEMTs. IEEE 44th Annual international reliability physics symposium technical digest, pp. 99–102, 2006
14. B.M. Green et al., The effect of surface passivation on the microwave characteristics of undoped AlGaN/GaN HEMTs. IEEE Electron Device Lett **21**, 268–270 (2000)
15. J. Derluyn et al., Improvement of AlGaN/GaN high electron mobility transistor structures by in situ deposition of Si3N4 surface layer. J. Appl. Phys. **98**, 054501 (2005)
16. B. Heying et al., In situ SiN passivation of AlGaN/GaN HEMTs by molecular beam epitaxy. IEEE Electron Lett **43**(14), 779–780 (2007)
17. H. Jiang et al., Investigation of in situ SiN as gate dielectric and surface passivation for GaN MISHEMTs. IEEE Trans. Electron Devices **64**(3), 832–839 (2017)
18. P. Moens et al., An insdustrial process for 650V rated GaN-on-Si power devices using in-situ SiN as a gate dielectric. Proceedings of the 26th international symposium on power semiconductor devices & IC's, pp. 374–377, June 2014
19. A. Brannick et al., Modeling of hot electron effects in GaN/AlGaN HEMT with AlN interlayer, in *Simulation of Semiconductor Processes and Devices*, vol. 12, Springer, Vienna, pp. 281–284 (2007)

20. J.S. Lee et al., Reduction of current collapse in AlGaN/GaN HFETs using AlN interfacial layer. IEEE Electron Lett **39**(9), 750–752 (2003)
21. S. Kaneko et al., Current-collapse-free Operations up to 850V by GaN-GIT utilizing Hole Injection from Drain. Proceedings of the 27th international symposium on power semiconductor devices & IC's, pp. 41–44, May 2015
22. Y. Dora, Understanding material and process limits for high breakdown voltage AlGaN/GaN HEMTs, D.Phil. dissertation, Department of Electrical and Computer Engineering, University of California at Santa Barbara, Santa Barbara, CA, 2006
23. N.-Q. Zhang et al., High breakdown GaN HEMT with overlapping gate structure. IEEE Electron Device Lett **21**, 421–423 (2000)
24. W. Saito et al., High breakdown voltage AlGaN-GaN power-HEMT design and high current density switching behavior. IEEE Trans. Electron Devices **50**, 2528–2531 (2003)
25. N. Ikeda et al., High power AlGaN/GaN HFET with a high breakdown voltage of over 1.8 kV on a 4 inch Si substrates and the suppression of current collapse. Proceedings of the 20th international symposium on power semiconductor devices & IC's, pp. 287–290, May 2008
26. H. Xing et al., High breakdown voltage AlGaN-GaN HEMTs achieved by multiple field plates. IEEE Electron Device Lett **25**, 161–163 (2004)
27. R. Chu et al., 1200-V normally off GaN-on-Si field-effect transistors with low dynamic on-resistance. IEEE Electron Device Lett **32**, 632–634 (2011)
28. R. Coffie, Slant field plate model for field-effect transistors. IEEE Trans. Electron Devices **61**, 2867–2872 (2014)
29. C.S. Suh et al., High-breakdown enhancement-Mode AlGaN/GaN HEMTs with integrated slant field-plate. Proceedings of the 2006 international electron device meeting, Dec 2006
30. C.S. Suh, Enhancement-mode GaN-based HEMTs for high-voltage switching applications, D.Phil. dissertation, Department of Electrical and Computer Engineering, University of California at Santa Barbara, Santa Barbara, CA, 2008
31. J. Wong et al., Novel asymmetric slant field plate Technology for High-Speed low-Dynamic R_{ON} E/D-mode GaN HEMTs. IEEE Electron Device Lett **38**, 95–98 (2017)
32. A. Nakajima, GaN-based super Heterojunction field effect transistor using the polarization junction concept. IEEE Electron Device Lett **32**, 542–544 (2011)
33. W. Saito et al., Recessed-gate structure approach toward normally off high-voltage AlGaN/GaN HEMT for power electronics applications. IEEE Trans. Electron Devices **53**, 356–362 (2006)
34. T. Oka, T. Nozawa, AlGaN/GaN recessed MIS-gate HEFT with high-threshold-voltage normally-off operation for power electronics applications. IEEE Electron Device Lett **29**, 668–670 (2008)
35. Y. Cai et al., High-performance enhancement-mode AlGaN/GaN HEMTs using fluoride-based plasma treatment. IEEE Electron Device Lett **26**, 435–437 (2005)
36. B. Lu et al., Enhancement-mode AlGaN/GaN HEMTs with high linearity fabricated by hydrogen plasma treatment. Device Research Conference, University Park, PA, pp. 59–60 (2009)
37. X. Hu et al., Enhancement mode AlGaN/GaN HFET with selectively grown pn junction gate. IEEE Electron Lett **36**, 753–754 (2000)
38. Y. Uemoto et al., Gate injection transistor (GIT) – A normally-off AlGaN/GaN power transistor using conductivity modulation. IEEE Trans. Electron Devices **54**, 3393–3395 (2007)
39. O. Hilt et al., Normally-off AlGaN/GaN HFET with p-type GaN Gate and AlGaN Buffer. Proceedings of The 22nd international symposium on power semiconductor devices & ICs, pp. 347–350, June 2010
40. K.-Y. Rong Wong et al., A next generation CMOS-compatible GaN-on-Si transistors for high efficiency energy systems. Proceedings of the 2006 international electron device meeting, pp. 229–232, Dec 2015

41. S. Heikman, MOCVD growth technologies for applications in AlGaN/GaN high electron mobility transistors, D.Phil. dissertation, Department of Electrical and Computer Engineering, University of California at Santa Barbara, Santa Barbara, CA, 2002
42. H. Okita et al., Through recessed and regrowth gate technology for realizing process stability of GaN-GITs. Proceedings of The 28th international symposium on power semiconductor devices & ICs, pp. 23–26, June 2016

Chapter 3
Vertical GaN Transistors for Power Electronics

Srabanti Chowdhury and Dong Ji

3.1 Introduction

Wide bandgap (WBG) semiconductors present a pathway to push the limits of power conversion efficiency beyond that available from silicon (Si)-based devices, enabling significant energy savings. Recent progress made in gallium nitride (GaN)-based power electronic devices is compelling. Reducing conversion losses is not only critical for minimizing consumption of limited resources, it simultaneously enables new compact architectures, the basis for a new industry offering increased power conversion performance at reduced system cost. This is because GaN devices enable power electronics with (1) higher efficiency at a higher frequency of operation and (2) higher efficiency over a wider range of operating temperature, compared with what is possible with Si, which is approaching its physical material limit in power conversion. High-efficiency operation at higher operating frequency reduces the size, weight, and cost of the overall system by reducing the size of the passive components and the heat sink. Lateral GaN-on-Si HEMTs has enabled over 99% efficient converters at 800 V at 100KHz [1]. Recent results have shown that an average of 5% improvement in the overall system efficiency was possible by GaN (lateral) switched inverter used for driving an induction motor [2]. While lateral GaN devices are more matured in technology and have entered the medium power conversion market (up to 10 kW), vertical GaN devices are evolving to address high power conversion (10 kW–10 MW). Device researchers are exploring vertical GaN devices and a viable manufacturing technology to be able to define and sustain a roadmap with these devices.

S. Chowdhury (✉) · D. Ji
ECE Department, UC Davis, Davis, CA, USA
e-mail: chowdhury@UCDAVIS.EDU

G. Meneghesso et al. (eds.), *Gallium Nitride-enabled High Frequency and High Efficiency Power Conversion*, Integrated Circuits and Systems,
https://doi.org/10.1007/978-3-319-77994-2_3

Similar to Si-based vertical devices, GaN vertical devices are of different flavors. In this chapter we will focus on two very complimentary architectures of GaN vertical devices – (A) normally on CAVET and (B) normally off MOSFETs. Both device topologies are being widely studied and present us the opportunity to discuss them with our readers using our recently achieved experimental results, as well as other group's reported work.

3.2 Current Aperture Vertical Electron Transistors (CAVETs)

CAVETs (shown in Fig. 3.1) are GaN vertical devices that take full advantage of the polarization-based two-dimensional electron gas (2DEG) in its design. Comprised of a 2DEG induced channel to carry the current, a CAVET needs a thick homoepitaxially grown drift region to hold the blocking voltage. There have been two types of approach in fabricating CAVETs: (1) planar CAVET and (2) trench CAVET. We will discuss both of these designs with relevant results. But before we go into the approaches, let us take a look at how CAVETs operate.

3.2.1 Operation Principle of the CAVET

Due to the positive polarization charge locates at the AlGaN/GaN interface, a triangular quantum well is formed, and electrons are attracted (shown in Fig. 3.1). The high-density electrons are confined in the triangular quantum well, forming the 2DEG. Under a zero-gate bias, the channel is conductive due to the 2DEG, and the device is in the on-state. During on-state, the electrons from the source flow through the 2DEG channel to the drift region and are finally collected by the drain electrode. The overlap between the gate electrode and the current blocking layer (CBL) is the effective channel length.

In order to pinch off the 2DEG channel, a negative gate bias is required. At the off-state, the electrons in the channel are depleted by the gate potential (shown in

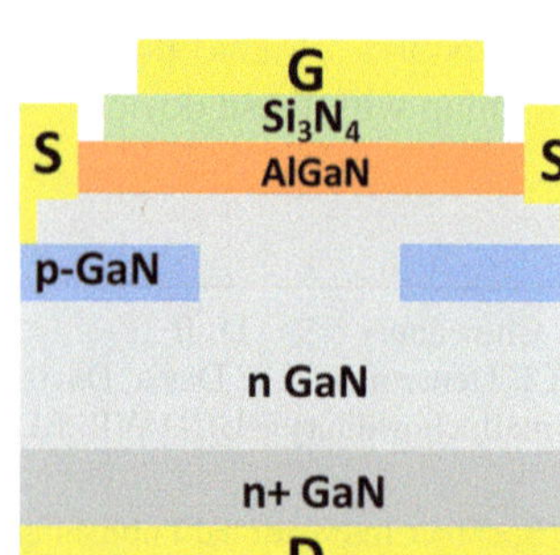

Fig. 3.1 Cross section of an on-state GaN CAVET and the conduction band diagram

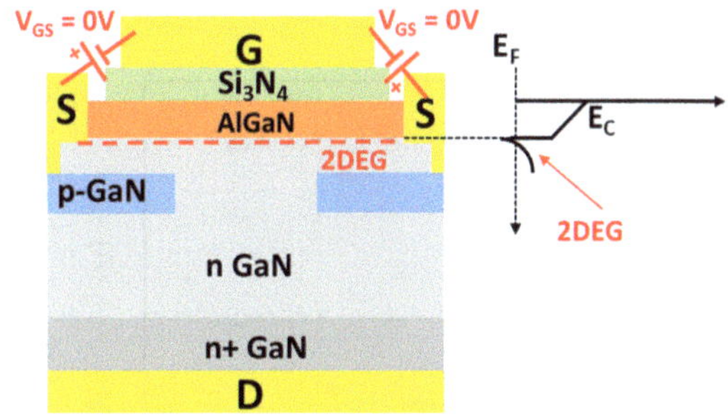

Fig. 3.2 Cross section of an off-state GaN CAVET and the conduction band diagram

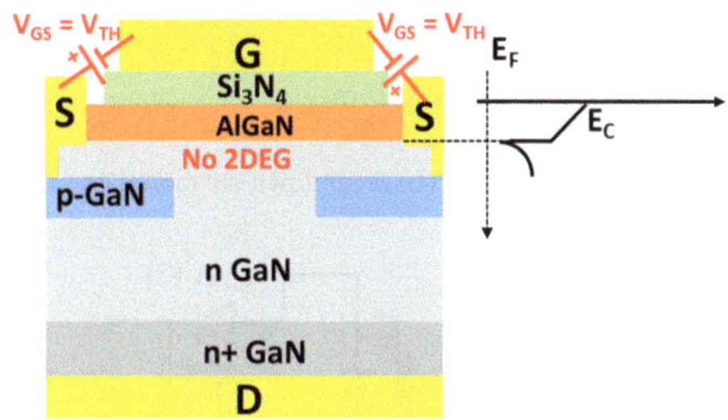

Fig. 3.3 Cross section of a GaN CAVET

Fig. 3.2). The *p-n* junction formed by *p*-type GaN (or the CBL) (or Mg-implanted) CBL and drift region blocks the off-state voltages.

Planar CAVETs

The first GaN CAVET grown on a sapphire substrate was developed in 2004 [3] (shown in Fig. 3.3). The device structure began with a *p*-GaN/*n*-GaN/*n* + GaN epilayers. An aperture region was etched, and an AlGaN/GaN HEMT structure was regrown by MOCVD to refill the aperture region and served as the channel as well. The design of the first GaN CAVET was for radio-frequency (RF) applications. Several years later in 2008, the first power GaN CAVET based on bulk GaN substrate was reported with the CBL formed by Mg ion implantation [4]. The device process began with the MOCVD-grown *n*-GaN on conductive bulk GaN substrate; then, the selective area Mg ion implantation was implemented to form the CBL to block current flowing from any other paths besides the aperture region. The top AlGaN/GaN structure was regrown by MBE. The first high-voltage CAVET with switching characteristics was demonstrated by Chowdhury et al. with a 300 V breakdown voltage and a $R_{on,sp}$ of 2.2 mΩ·cm^2 (Fig. 3.4) [5]. A significant progress in the device development was realized with the demonstration of a 1.5 kV JFET (a variant of CAVET) on bulk substrate by H. Nie et al. in 2014 from Avogy Inc.

CAVETs are normally on devices similar to GaN HEMTs; however power switching demands normally off operation. The techniques that are typically applied to HEMTs can be applied to CAVETs to render it normally off. To date, three

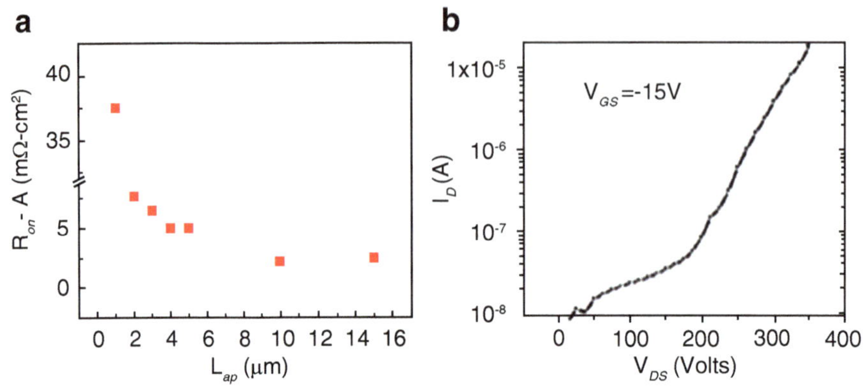

Fig. 3.4 (**a**) $R_{on,sp}$ and (**b**) breakdown characteristics of the CAVET in Ref. [5]

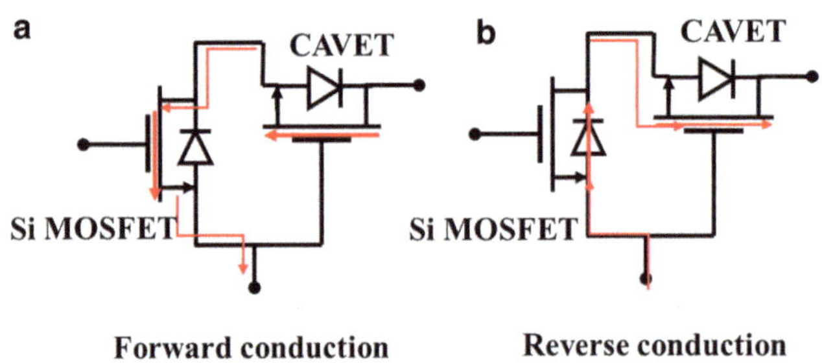

Fig. 3.5 (**a**) Forward conduction of the cascoded CAVET; (**b**) reverse conduction of the cascoded CAVET

methods have been proposed to achieve normally off CAVETs: (1) the cascode configuration [6], (2) the p-GaN gate layer [7, 8], and (3) the gate recess structure [9].

Figure 3.5 shows the cascode configuration designed with a low-voltage normally off Si MOSFET as the input and a high-voltage normally on CAVET as the output. The current paths of the forward conduction are labeled with the arrows in Fig. 3.5a. In the forward conduction, the total on-state resistance (R_{on}) is the sum of the MOSFET R_{on} and the CAVET R_{on}. In the reverse conduction cycle (shown in Fig. 3.5b), the current flows through the body diode of the Si MOSFET and the CAVET channel. It is important to note that the body diode of the CAVET does not conduct current at any time.

In the second method of achieving normally off CAVET or CAVET-like devices, a p-GaN gate layer is applied. Under zero gate bias, electrons in the 2DEG channel are depleted by the p-GaN gate layer. Using a p-GaN gating layer, positive threshold voltages of 0.5 V [7] and 2.5 V [8] were reported (Fig. 3.6).

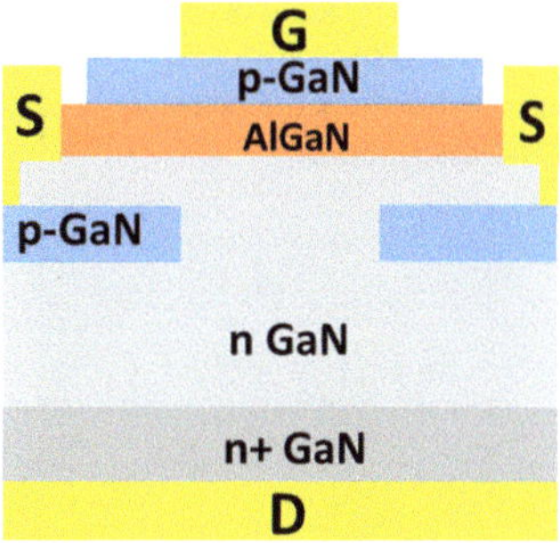

Fig. 3.6 Structure of vertical CaN transistor with p-GaN gate layer [7]

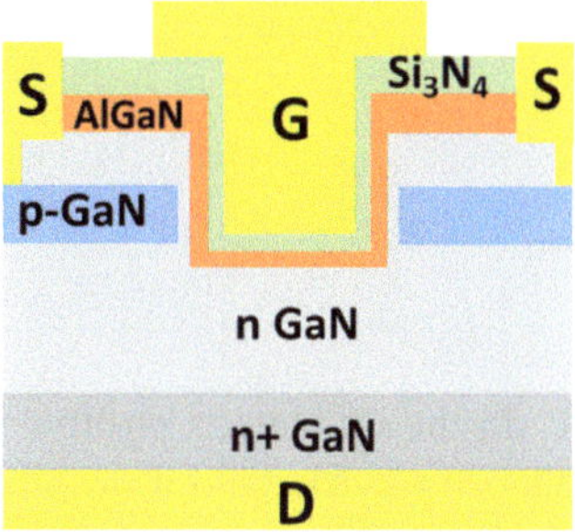

Fig. 3.7 Structure of a recessed-gate CAVET

Trench CAVET

The recessed-gate CAVET structure is shown in Fig. 3.7. The electrons in the sidewall channel are depleted by the p-GaN base region, providing a normally off operation. The threshold voltage is dependent on the slant of the sidewall. From a simulation study, >1 V threshold voltage can be obtained for a straight sidewall (90°).

3.2.2 *CAVET as a Power Switch*

As a power switch CAVETs are designed to offer high blocking voltage and low R_{on}. In the follwing sections we will examine what makes a CAVET a good power swicth.

High Breakdown Voltage

Ideally, the CAVET uses p-n junction formed by the p-GaN base region and the n-GaN drift region to block the high voltage. For simplicity, a p-n diode is used in Fig. 3.8 to represent the off-state CAVET.

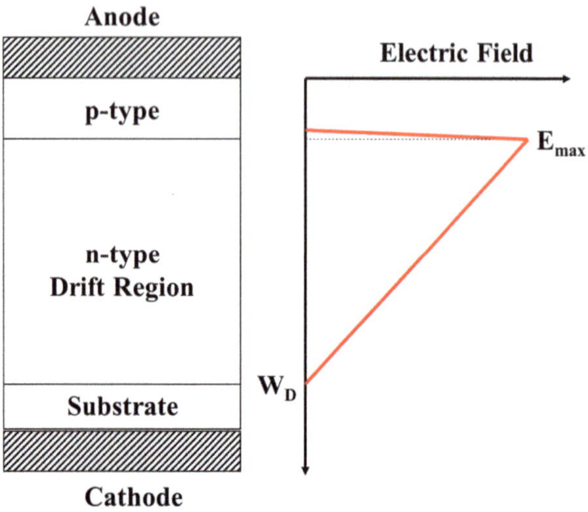

Fig. 3.8 The ideal drift region and the triangular-shaped electric field distribution

In the off-state, a positive voltage is applied onto the cathode, and a triangular-shaped electric field distribution along the depletion region is obtained. According to Poisson's equation, the maximum electric field E_{max} can be written as

$$E_{\text{max}} = \frac{qN_{\text{D}}}{\varepsilon_{\text{r}}\varepsilon_0} W_{\text{D}}$$

The breakdown voltage, V_{BR}, can be written as

$$V_{\text{BR}} = \frac{1}{2} W_{\text{D}} E_{\text{C}}$$

Figure 3.9 shows the ideal breakdown voltage as a function of drift region doping concentration; comparison of several reported >1 kV vertical GaN devices is also presented [7, 8, 10–12]. Device engineers design structures for electric field management to make sure E_C is not reached during device operation.

Low On-State Resistance

The ideal $R_{\text{ON,TOT}}$ expressed as specific resistance of the GaN CAVET can be written as

$$R_{\text{ON,TOT}} = \rho_{\text{2DEG}} L_{\text{G}} p + \frac{W_{\text{D}}}{q\mu_n N_{\text{D}}}$$

From Fig. 3.10, for low breakdown voltage device design (BV < 2000 V) for a given charge concentration, the on-state resistance is limited by the channel electron mobility; while for BV > 2000 V, the on-state resistance is limited by the bulk GaN electron mobility in the drift region.

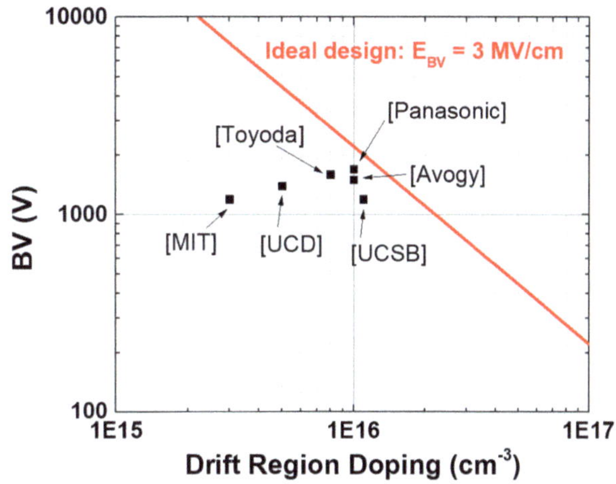

Fig. 3.9 Breakdown voltage of vertical GaN transistor as a function of drift region doping concentration [7, 8, 10–12]

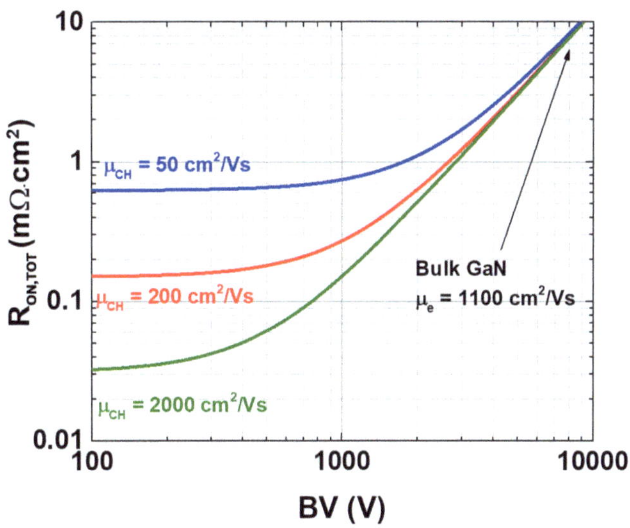

Fig. 3.10 Device figure of merit of vertical GaN transistors

Considering the current spreading from the aperture into the drift region (shown in Fig. 3.11), the on-state resistance gets limited by the aperture region. There are two major components that contribute to the device $R_{\mathrm{ON,TOT}}$. The resistance offered by the AlGaN/GaN channel, R_{CH}, and drift region resistance, R_{DR}, can be written as

$$R_{\mathrm{CH}} = \frac{L_{\mathrm{gs}} + L_{\mathrm{go}}}{q\mu_{\mathrm{2DEG}} n_{\mathrm{2DEG}}} \left(2\left(L_{\mathrm{gs}} + L_{\mathrm{go}}\right) + L_{\mathrm{ap}}\right)$$

and

$$R_{\mathrm{DR}} = \frac{T_{\mathrm{drift}}\left(2\left(L_{\mathrm{gs}} + L_{\mathrm{go}}\right) + L_{\mathrm{ap}}\right)}{q\mu_n N_{\mathrm{D,drift}} 2\left(L_{\mathrm{gs}} + L_{\mathrm{go}}\right)} \ln\left(\frac{2\left(L_{\mathrm{gs}} + L_{\mathrm{go}}\right) + L_{\mathrm{ap}}}{L_{\mathrm{ap}}}\right)$$

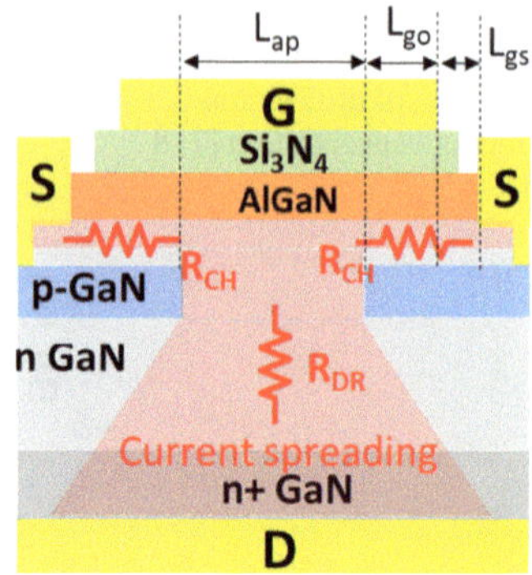

Fig. 3.11 On-state resistance distribution in the CAVET

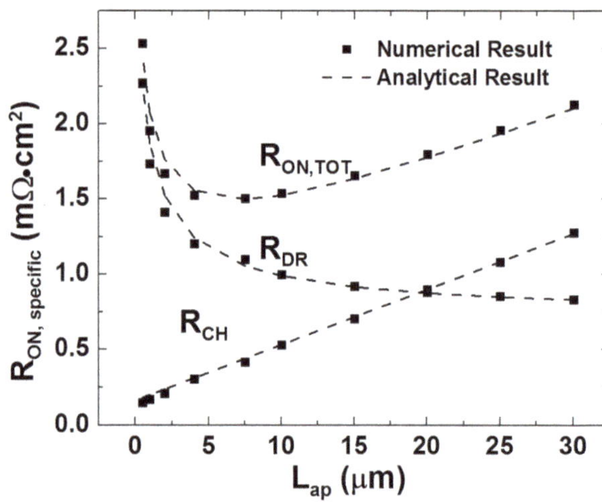

Fig. 3.12 $R_{ON,TOT}$, $R_{DR,}$ and R_{CH} as a function of L_{ap}. The solid squares indicate the results obtained numerically, while the dash curves show the analytical result (channel electron mobility, 1500 cm^2/Vs; bulk GaN electron mobility, 900 cm^2/Vs) [6]

where μ_{2DEG} and μ_n are electron mobilities in 2DEG channel and bulk GaN, respectively, the values of which used in the simulation are 1500cm^2/Vs and 900cm^2/Vs, n_{2DEG} is the 2D electron concentration the channel, and $N_{D,drift}$ is the doping density in the drift region.

Figure 3.12 shows the $R_{ON,TOT}$, R_{DR}, and (R_{CH}) versus aperture length L_{ap}. The lowest $R_{ON,TOT}$ is obtained between a L_{ap} of 4 μm and 10 μm, at a minimum value of 1.5 mΩ·cm^2 [6].

3.3 Switching Performance of the CAVET

Using the device-circuit-integrated model built in Silvaco's Mixed Mode platform [6, 15], the switching performance of the CAVET was analyzed. The simulation methodology is shown in Fig. 3.13. Using this device-circuit-integrated model, one can start with 2D drift-diffusion modeling of the device and build all the way up to

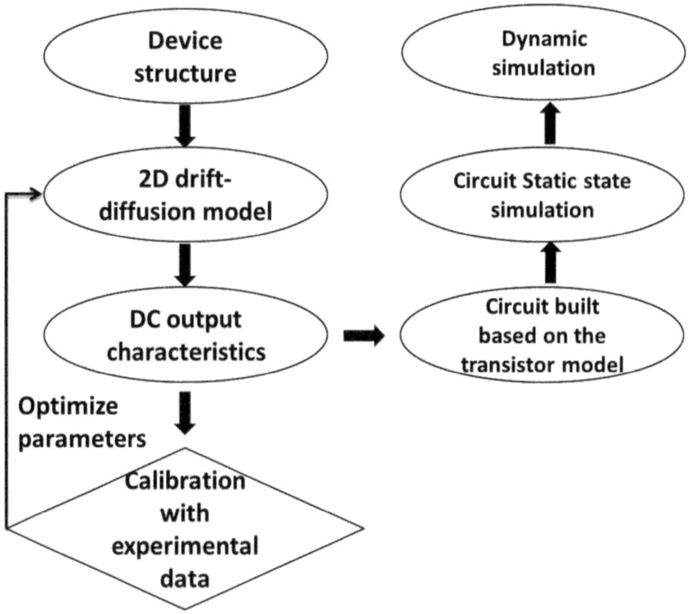

Fig. 3.13 Simulation methodology flowchart [6]

its circuit implementation to evaluate its switching performance. The hybrid model gives an inexpensive and accurate way to project and benchmark the performance that can be extended to any GaN-based power transistors.

Since the CAVET is normally on the device, a cascoded CAVET approach was adopted to achieve normally off operation. The cascoded CAVET was subjected to switching operation and detailed performance matrix was generated.

Figures 3.14 and 3.15 show the turn-on and turn-off switching characteristics of a 1.2 kV cascoded CAVET, with a $R_{\mathrm{ON,\ TOT}}$ of 80mΩ. From the waveforms shown in Fig. 3.14, the turn-on delay time $T_{\mathrm{on\text{-}delay}}$ is 2 ns, and rise time T_r is 16 ns, showing a turn-on $\Delta v/\Delta t$ of 40 kV/μs. The total gate charge of the CAVET is 88nC. The turn-off delay time $T_{\mathrm{off\text{-}delay}}$ is 29.5 ns, and fall time T_f is 18 ns, giving a turn-off $\Delta v/\Delta t = 35.6$ kV/μs.

Figure 3.16 shows the switching losses comparison of different power transistors under the same test current of $I_D = 20$A. Compared to the commercial SiC MOSFET (Cree CMF20120D [16]), the total switching energies at $I_D = 20$A of which are 610 μJ, the cascoded CAVET has more than 3 × lower switching energy loss.

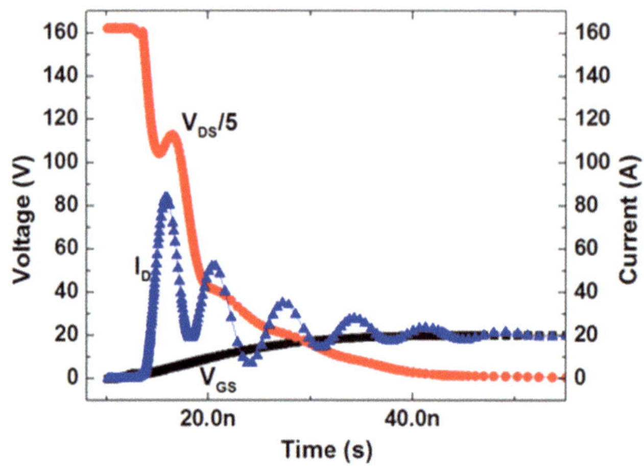

Fig. 3.14 Simulated turn-on switching waveforms of the cascoded CAVET [6]

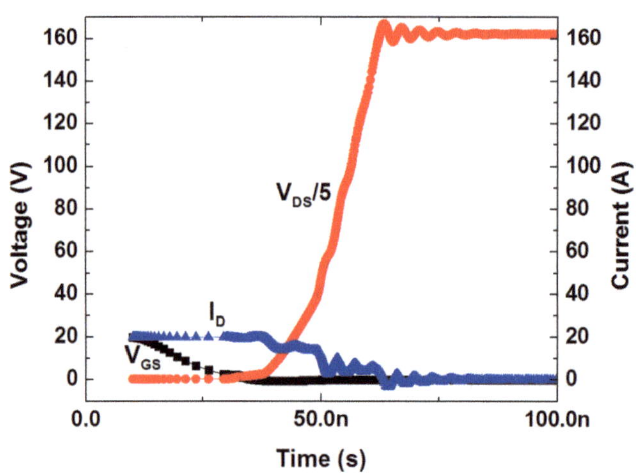

Fig. 3.15 Simulated turn-off switching waveforms of the cascoded CAVET [6]

3.3.1 Discussion of the Fabrication Process

Compared to the HEMT, the fabrication process of the GaN CAVET is more complex. The key challenge of the CAVET fabrication is the CBL. The CBL is used to create an electrostatic barrier for electrons to block current, originated from the drain and flowing to the source, flow through any other path other than the designed aperture. The current blocking layer can be formed by either Mg-implanted or Mg-doped p-GaN. In this section, four approaches of fabrication will be discussed.

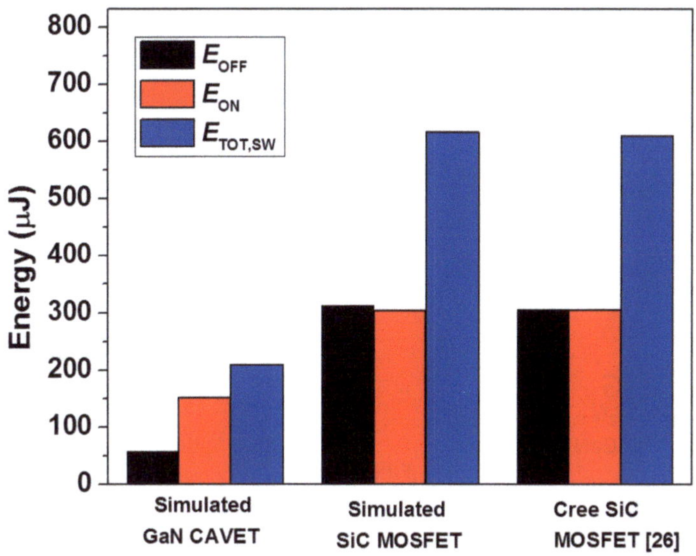

Fig. 3.16 Comparison of switching losses of different transistors at an operating voltage of 800 V and load current of 20 A. The model we developed works well for SiC MOSFET as evident from the close agreement of the simulated and experimental SiC MOSFET results [6]

Planar CAVET with Mg-Implanted CBL

The process flow of the CAVET with Mg-implanted CBLs is shown in Fig. 3.17. The fabrication begins with a lightly doped n-GaN grown on a n + GaN substrate. First of all, a metal or a hard mask is deposited on the designed aperture region. Then, the Mg ion implantation is conducted to achieve a CBL. Due to the physical mask blocking the aperture from getting implanted, Mg-implanted CBL is formed around aperture. Subsequently, a 130-nm-thick UID GaN and a 25-nm-thick AlGaN structure, followed by a 30-nm-thick Si_3N_4, are grown on top of the implanted sample using MBE or MOCVD. The full device is then fabricated with gate, source, and drain electrode deposition.

Planar CAVET with Mg-doped CBL

The process flow of the CAVET with CBLs formed by selective area p-GaN regrowth is shown in Fig. 3.18. First, a Silicon Dioxide (SiO_2) mask is used to protect the aperture region against GaN etching. A 400-nm-deep etching of the top GaN layer (the same layer where the aperture is defined) is conducted followed by selective growth of p-GaN layers to define the CBL. Afterward, the AlGaN/GaN layers are regrown by MBE or MOCVD, and the full device fabrication is carried out in the same manner as the one described for a planar Mg-implanted CAVET.

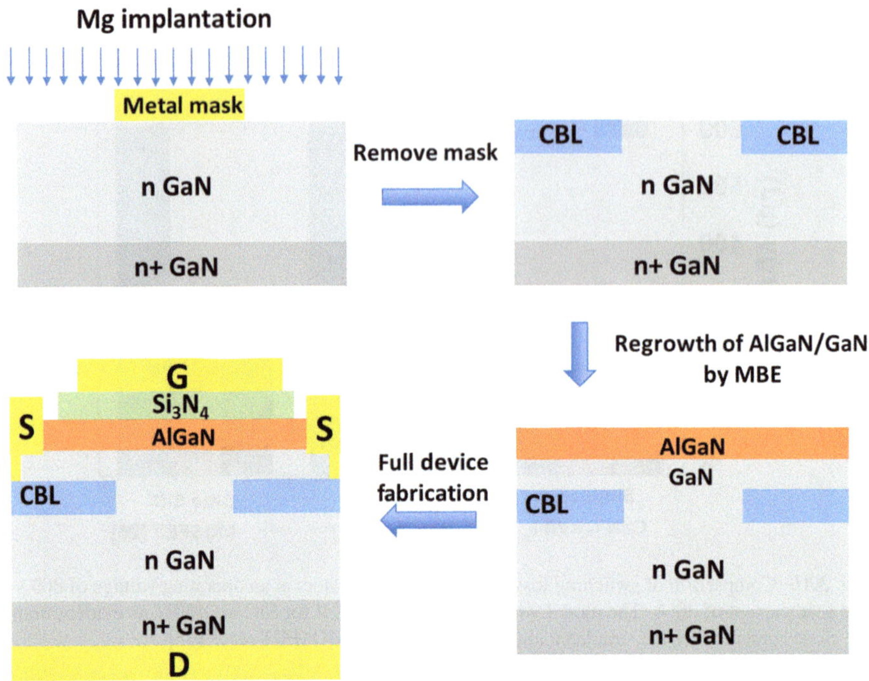

Fig. 3.17 A schematic of a CAVET process flow with the CBL formed by Mg implantation [4]

Planar CAVETs with Regrown Aperture Region

The process flow of the CAVET with aperture region formed by n-GaN regrowth is shown in Fig. 3.19. The epitaxial layers include a n-GaN as the drift region and a heavily doped p-GaN. First, the p-GaN in the aperture region is etched away. After removing the mask, the etched aperture region is refilled by regrowth of GaN. The channel regrowth process is identical to the one described in the previous sections. After the regrowth, the buried p-GaN is activated by etching vias and subsequently annealing in the nitrogen ambient at 700 °C to diffuse out the hydrogen. Electrodes are deposited to complete the device.

CAVET with Aperture Region Formed by Si Ion Implantation

The process flow of the CAVET with aperture region formed by Si ion implantation is shown in Fig. 3.20. The device fabrication begins with an epitaxial p-n structure. The sample is subjected to Si ion implantation with a photoresist mask covering CBL region. The implantation, therefore, takes place in the aperture region only. After implantation, the SiO_2 protection layer was removed using hydrofluoric acid followed by a postimplantation annealing at 1280 °C to activate the implanted Si and heal crystal damages. The rest of the process is similar to those described in earlier sections.

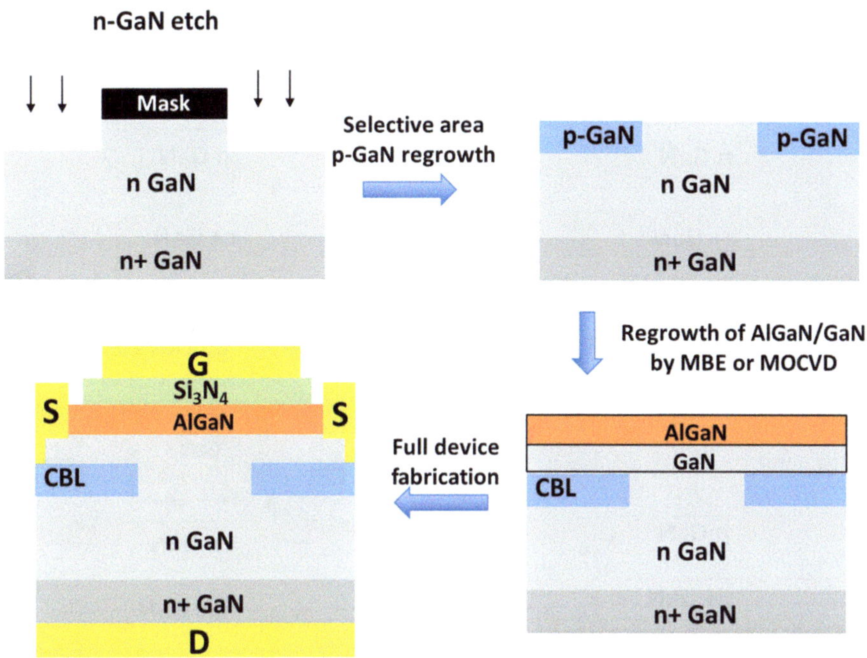

Fig. 3.18 A schematic of a CAVET process flow with the CBL formed by selective area p-GaN regrowth [13]

3.3.2 *Trench CAVET*

The conventional GaN CAVET, schematically shown in Fig. 3.3, typically uses Mg-implanted *p*-GaN for the CBL. A major fabrication challenge is posed by Mg out-diffusion during the regrowth process. First, the channel resistance can go very high due to the diffusion of Mg species into the channel, thereby depleting 2DEG. To prevent the diffusion of Mg into the regrown channel, the regrowth temperature of the channel in a CAVET is restricted, making it rely on low-temperature growth processes. Low-temperature MOCVD regrowth process does not typically result into a high-quality material. MBE regrowth although has proven to be successful in arresting Mg diffusion, there are concerns on the formation of vertical highly conductive paths under metal rich growth conduction, causing shorts in the device. An MBE-regrown channel requires exposure of the sample to the atmosphere for prolonged interval, and process limitations make it challenging to remove environmental contaminants from the regrowth interface. Low-temperature flow modulation epitaxy by MOCVD may be considered as an alternative, but it is not yet understood whether the material quality produced in such a fashion can support high-current and high-voltage devices.

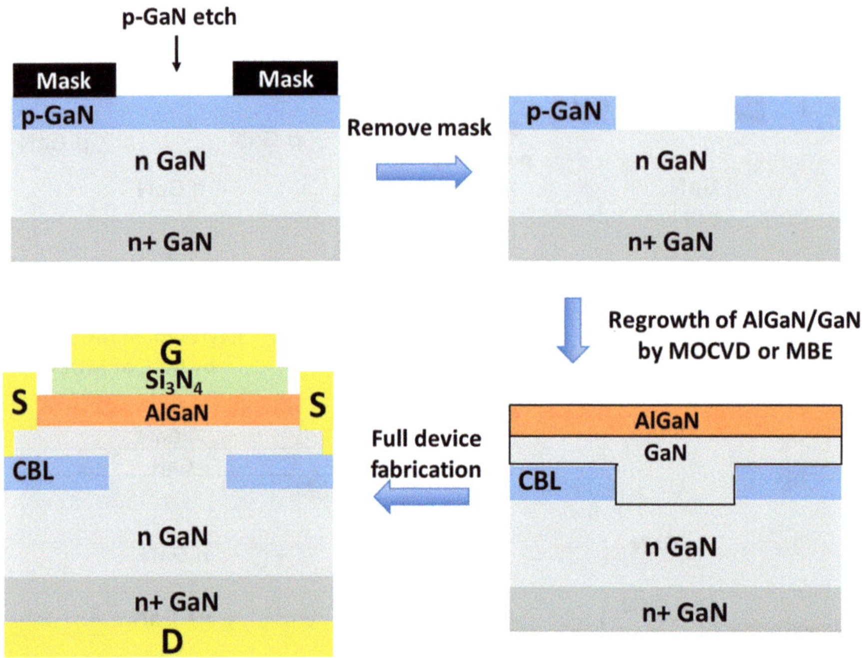

Fig. 3.19 A schematic of a CAVET process flow with the aperture region formed by regrowth [3]

A solution to the aforementioned problem was proposed by introducing a trench gate structure into the conventional CAVET, as shown in Fig. 3.21 [9]. Instead of using Mg-implanted *p*-GaN for the CBL, the trench CAVET adopts MOCVD-grown Mg-doped *p*-GaN as the CBL material. The trench CAVET utilizes the regrown AlGaN/GaN on the trench sidewall as the channel. The trench sidewall angle determines the polarization scale of the channel: a 90-degree angle indicates a nonpolar plane, and a 45-degree angle indicates a semipolar plane. The threshold voltage in a trench CAVET can be adjusted by the trench sidewall angle. The functioning of a trench CAVET is similar to a trench MOSFET, except for the AlGaN/GaN channel. In a conventional trench MOSFET, the channel is formed by inversion layer between the oxide and the *p*-type semiconductor. The electron mobility in the inversion layer is typically below 50 cm^2/Vs, limited by the interface roughness scattering. However, in a trench CAVET, the channel is formed by the 2DEG, which takes the advantage of the high mobility in the HEMT structure. Ideally, the channel mobility in a trench CAVET can be as high as 1690 cm^2/Vs [8].

In GaN, a robust oxide technology is not available yet. However, thanks to the development of GaN HEMT in the last decade, the in situ MOCVD grown Silicon Nitride (grown Si_3N_4) on AlGaN/GaN structure has low interface traps, which is essential to the device reliability. A trench CAVET takes the advantage of the matured HEMT gate dielectric technology to suppress gate leakage.

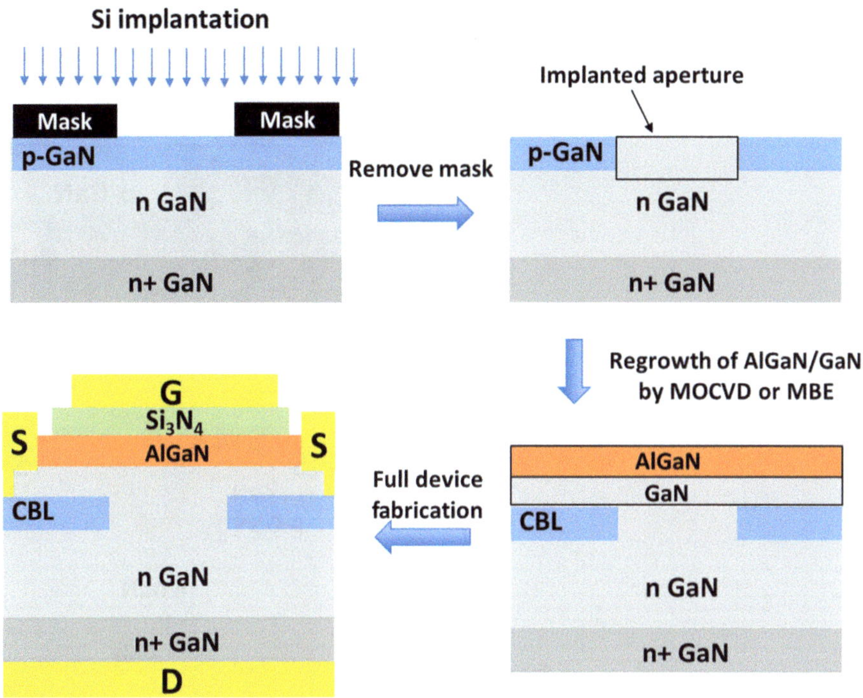

Fig. 3.20 A schematic of a CAVET process flow with the aperture region formed by Si ion implantation [14]

Fig. 3.21 Schematic illustrations of the trench CAVET

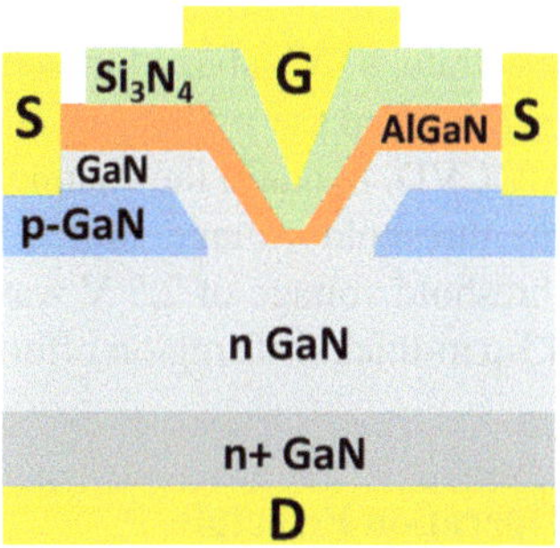

The first MIS gate trench CAVET was reported by Ji et al. in 2016 on bulk GaN substrates. The SEM cross-section picture of the device is shown in Fig. 3.22. The device had a breakdown voltage of 225 V limited by the gate-to-drain breakdown. Further improvement led to over 800 V in blocking voltage with below 3 Ωcm^2.

The p-GaN gate structure is widely used in normally off lateral GaN HEMTs. Due to the high electron density of 2DEG induced by the polarization charge, the threshold voltage of p-GaN gated HEMT is typically less than 2 V. However, a more positive threshold voltage can be realized in a p-GaN gated trench CAVET

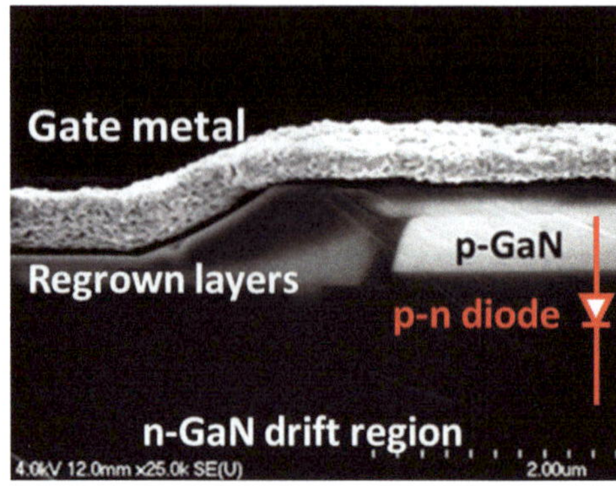

Fig. 3.22 Side profile of the MIS gate trench CAVET from SEM [9]

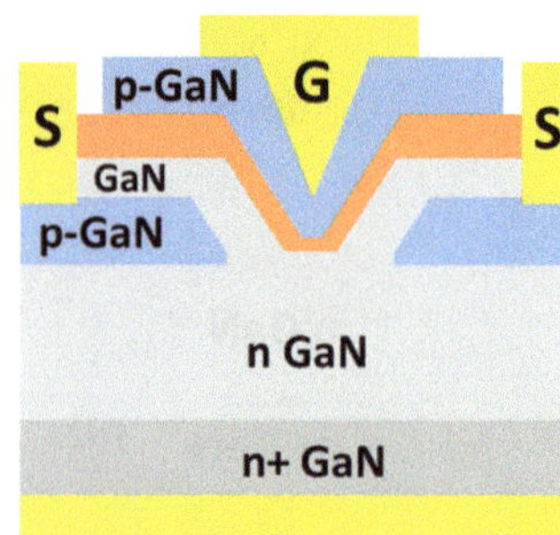

Fig. 3.23 Trench CAVET with a p-GaN gate layer [8]

with the channel formed at the semipolar plane. This was first demonstrated by Shibata et al. in 2016 (Fig. 3.23) [8]. Based on a p-n epitaxial structure on bulk GaN substrate, a "V"-shaped trench was formed using inductively coupled plasma (ICP) etching, and the p-GaN/AlGaN/GaN triple layers were regrown over the trench by MOCVD. Because the channel locates at the semipolar plane instead of the c-plane, the threshold voltage shifted toward the positive side by 1.5 V. A large positive threshold voltage of 2.5 V was demonstrated with 1700 V blocking capability on a 13-μm-thick drift region offering a low specific on-resistance of 1 $m\Omega\cdot cm^2$ [8].

Operation Principle

The cross section of the MIS gate trench CAVET is shown in Fig. 3.24. The channel of the trench CAVET is located at the semipolar plane of AlGaN/GaN heterostructure. Under the zero-gate bias, due to the polarization charge in the semipolar AlGaN/GaN heterostructure, the electrons are accumulated in the triangular quantum well, forming a conductive 2DEG channel. The conduction band profile is shown in Fig. 3.14. In order to turn the device off, a negative gate bias is required.

Figure 3.25 shows the cross-sectional structure of the p-GaN gate trench CAVET and the conduction band profile. Due to the conductivity modulation of the p-GaN

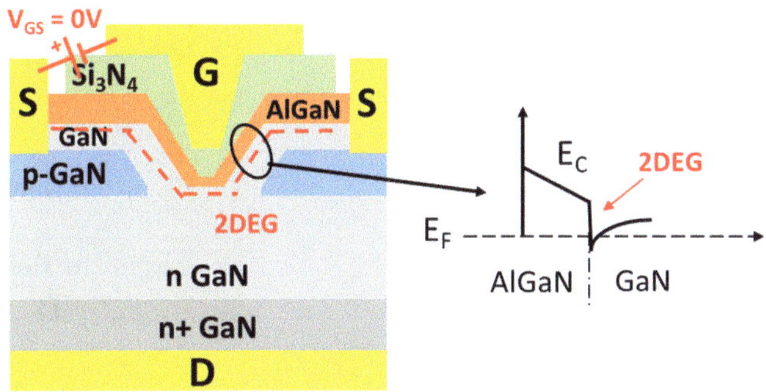

Fig. 3.24 Cross section of the on-state MIS gate trench CAVET and the conduction band profile

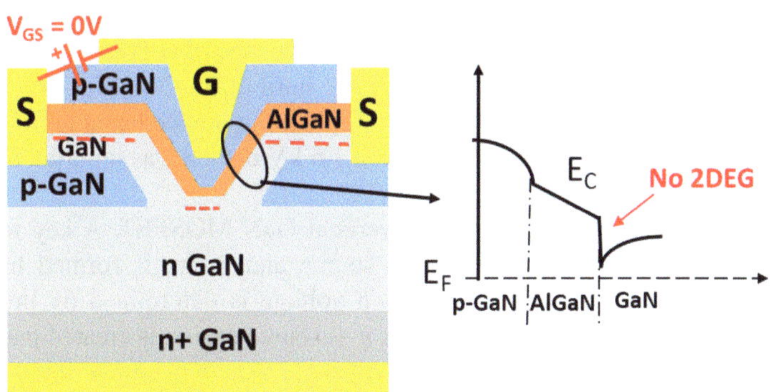

Fig. 3.25 Cross section of the on-state p-GaN gate trench CAVET and the conduction band profile

gate layer, the electrons are depleted completely under a zero-gate bias. In order to turn on the device, a positive gate bias is required. However, the maximum gate bias on the p-GaN gate trench CAVET is limited to 4 V; otherwise, gate leakage is increased due to the forward biasing of the p-n junction between the gate and the source.

3.4 MOSFETs

GaN MOSFETs are the other branch of devices that are showing promising performance offering a normally off solution which is a significant drawback of any GaN HEMT-based design.

There are two types of MOSFET reported so far: (1) nonregrowth-based MOSFET and (2) regrowth-based MOSFET, Oxide, GaN interlayer FET (OGFET). In this section, the nonregrowth-based MOSFET will be presented, while the regrowth-based MOSFET (OGFET) will be introduced in the next section.

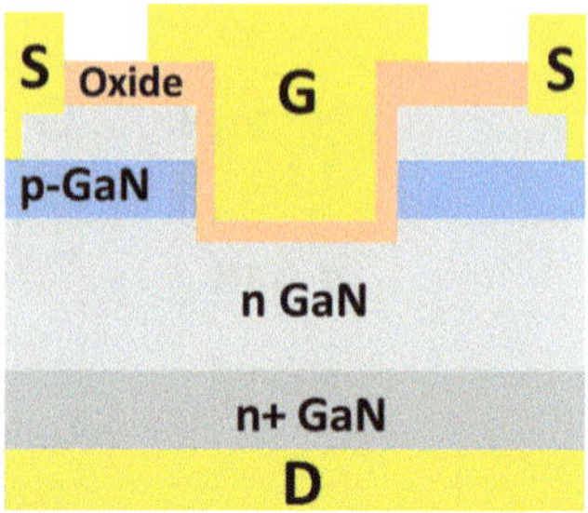

Fig. 3.26 Structure of GaN MOSFET

Since the development of dry etching technology in the early 1990s, the trench MOSFET has been a dominant device structure for power electronics [17]. To date, both Si- and SiC-based trench MOSFETs have been commercialized and shown excellent performance. However, because of the absence of bulk GaN substrate, the GaN-based trench MOSFET was not available until 2004–2005. In 2007, the first GaN MOSFET was reported by Otake et al. [18]; the device has a high threshold voltage of 5.1 V. After 7 years of research, a 1.6 kV device was reported by Oka et al. in 2014 [10].

Figure 3.26 shows the structure of a vertical GaN MOSFET. A key feature of these devices, the *p-n* junction between source and drain, is formed by *p*-base and *n*-drift regions. The device breakdown voltage is determined by the reverse characteristics of the main *p-n* junction. A $n+$ source region is created partially on top of the *p*-base region, while the junction between the $n+$ source region and the *p*-base region is connected to the source contact to improve the breakdown voltage by eliminating the *n-p-n* open base effect. The channel located at the etched sidewall is formed by the inversion layer of the MOS structure.

Compared to the CAVET structure, there are two basic advantages of the MOSFET: (1) the MOSFET is a reliable normally off device with a high threshold voltage over 2 V; (2) the absence of the regrowth makes the process less challenging, reducing the cost and turnaround time. Such advantages of the MOSFET make it an attractive design for vertical GaN transistors. However, for GaN MOSFET, the biggest challenge lies in the channel electron mobility of the device. During on-state, the electrons flow through the inversion layer of the sidewall MOS structure, and the channel electron mobility is limited by the surface roughness and impurity scattering. Another issue along with the poor channel property is the device reliability. The GaN MOSFET cannot be widely recognized without a strong reliability track.

3.4.1 Regrowth-Based MOSFET (OGFET)

GaN OGFET is a modified structure based on the conventional trench MOSFET. Compared to the conventional trench MOSFET, the OGFET has two features: (1)

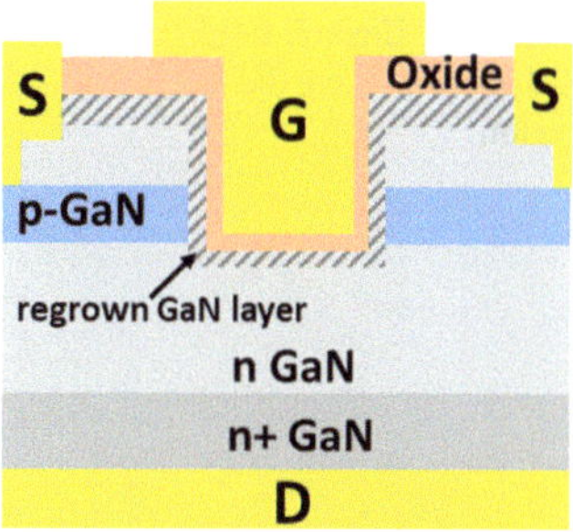

Fig. 3.27 Structure of GaN OGFET [19]

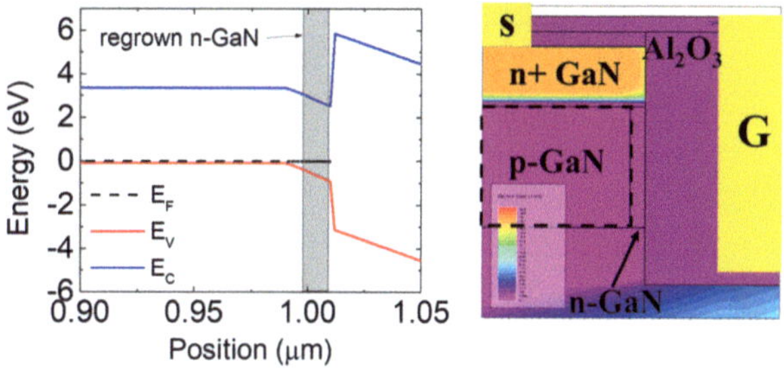

Fig. 3.28 Energy band diagram and the electron distribution of the OGFET in off-state

an unintentional doped (UID) GaN interlayer is used as the channel region, which enhances the channel electron mobility to reduce the Coulomb scattering by the dopants; (2) the oxide is in situ grown by MOCVD, which reduces the interface states and improves the gate oxide reliability. The novelty of the OGFET lies in enhancing the channel electron mobility without sacrificing the normally off behavior (Fig. 3.27).

Operation Principle of the OGFET

The working principle of the OGFET is similar to the MOSFET. Under a zero V_{GS}, the electrons in the GaN insert layer (shaded region in Fig. 3.28) are depleted by the p-GaN base region making the OGFET go to its off-state. The energy band diagram and the simulated electron concentration contour are shown in Fig. 3.28. The p-n diode formed by the p-GaN base region and the n-GaN drift region is used to hold the high off-state blocking voltage. The electric field distribution along the p-GaN base region and the n-GaN drift region is shown in Fig. 3.29.

Under a positive V_{GS} (15 V), the electrons are accumulated in the UID GaN insert layer, and the transistor is in its on-state. The energy band diagram and the electron concentration are shown in Fig. 3.30. Because of the enhanced channel electron mobility, the OGFET has smaller $R_{ON,TOT}$ compared to the conventional trench MOSFET.

In 2016, Gupta et al. reported the first OGFET results based on the sapphire substrates; the device showed a 60% $R_{ON,\,TOT}$ reduction while maintaining the

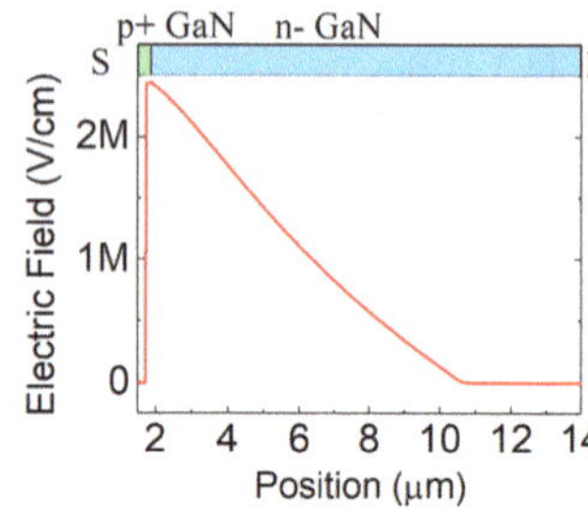

Fig. 3.29 Electric field distribution along the drift region

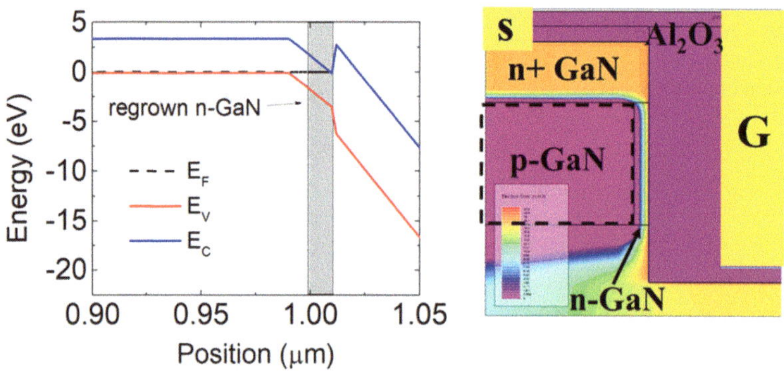

Fig. 3.30 Energy band diagram and the electron distribution of the OGFET in on-state

threshold voltage >2 V [19]. In 2017, Gupta et al. demonstrated an OGFET on bulk GaN substrates with a breakdown voltage of 990 V and a low $R_{on,sp}$ of 2.6 mΩ·cm^2 [20]. In the same year, Ji et al. demonstrated a high-performance OGFET with a breakdown voltage over 1.43 kV. By using a 10 nm unintentional GaN interlayer as the channel, a low $R_{ON,\,TOT}$ of 2.2 mΩ·cm^2 was achieved, which demonstrated an excellent on-state performance [11]. The *I*-*V* characteristics of the fabricated OGFET are shown in Fig. 3.31. Figure 3.32 shows the transfer I_D-V_{GS} characteristics and the gate leakage. The threshold voltage, V_{TH}, defined at a current level of 10^{-4} A/cm^2 (I_{on} / I_{off} = 10^6), obtained was 4.7 V (when V_{GS} sweeps up). A clockwise hysteresis of ΔV_{TH} of 0.3 V was observed. A subthreshold slope of 283 mV/decade was measured from $I_D = 10^{-5}$ A/cm^2 to 10^{-2} A/cm^2. Figure 3.33 shows the off-state measurement of a unit cell device with an under a V_{GS} of -10 V. It is shown that a breakdown voltage of 1435 V was obtained at a current level of 50 mA/cm^2.

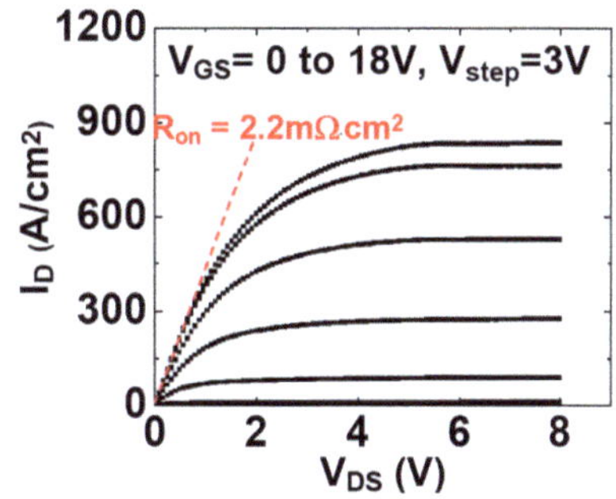

Fig. 3.31 *I-V* characteristics of fabricated OGFET with saturation current density of 850 A/cm^2 and $R_{on,sp}$ of 2.2 mΩ·cm^2 [11]

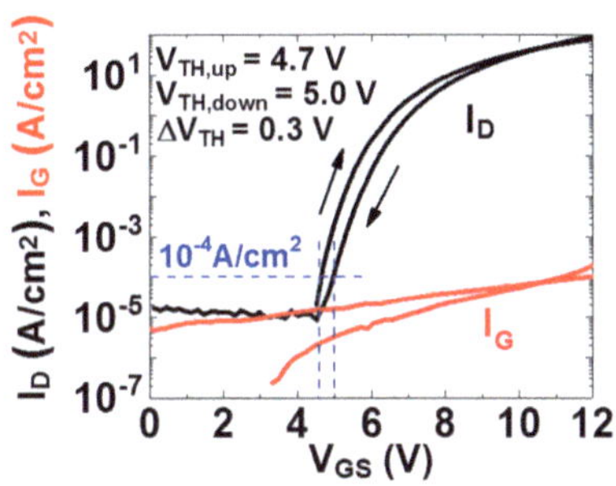

Fig. 3.32 Transfer characteristics of the fabricated OGFET (black curves) and gate leakage (red curves). The threshold voltages of sweep up and down were 4.7 V and 5 V (defined at a current level of 10^{-4} A/cm^2). The subthreshold slope was 283 mV/decade measured from 10^{-5} A/cm^2 to 10^{-2} A/cm^2 [11]

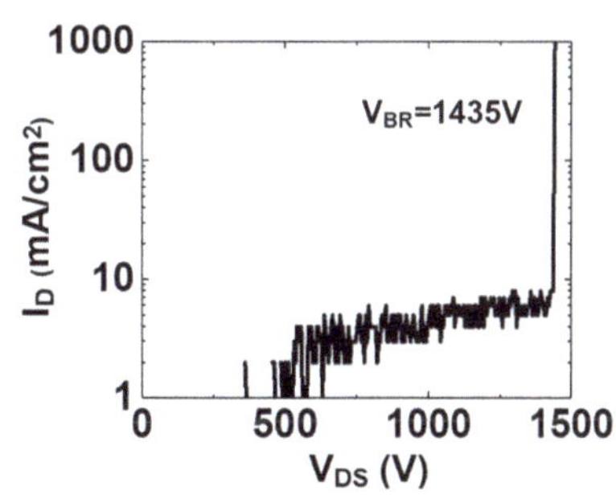

Fig. 3.33 Off-state characteristics of the fabricated OGFET. Breakdown voltage (V_{BR}) was 1435 V defined at off-state leakage of 50 mA/cm^2 [11]

3.4.2 OGFET Switching Performance

Based on the reported 1.4 kV OGFET fabricated on the bulk GaN substrate [11], a physics-based device model was developed and then integrated with a circuit model to study the dynamic characteristics and power losses.

The switching waveforms of the transistor during the turn-off and turn-on transients in a double-pulse test circuit are shown in Fig. 3.34 and 3.35. The total turn-off time is ~30 ns, while the turn-on time is 47 ns. The total gate charge is ~89 nC, which includes a Q_{GD} of ~17 nC.

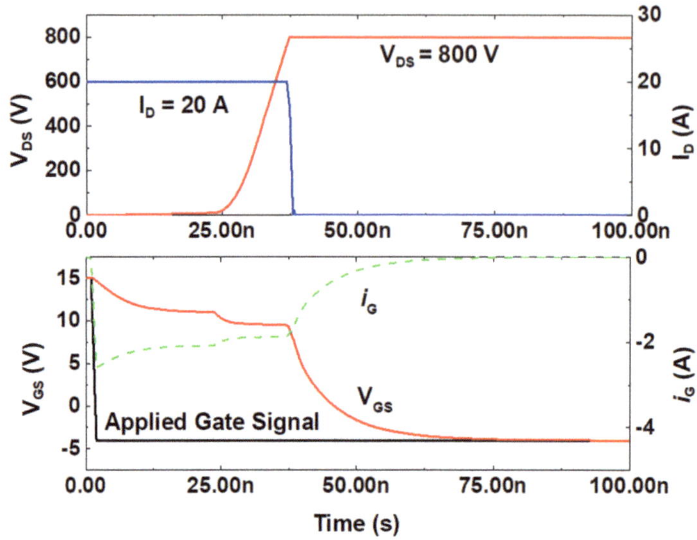

Fig. 3.34 Waveforms during the turn-off transient. The $t_{d(off)}$ is 21.3 ns, t_f is 9 ns, and the $\Delta v/\Delta t$ is 71 kV/μs. The gate charge is 89 nC. The gate resistance is 7.5 Ω, and the frequency is 100 kHz [21]

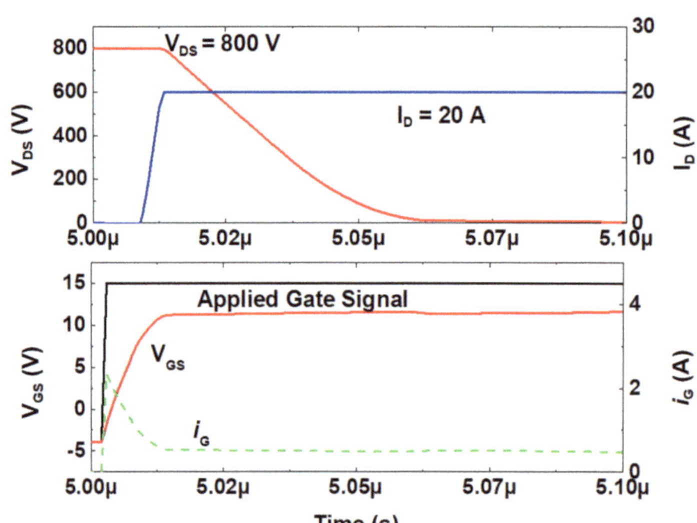

Fig. 3.35 Waveforms during the turn-on transient. The $t_{d(on)}$ is 14 ns, the t_r is 33 ns, and the $\Delta v/\Delta t$ is 19.4 kV/μs [21]

Table 3.1 shows the comparison of CREE SiC MOSFET [16], simulated SiC MOSFET, simulated GaN CAVET, and simulated GaN OGFET with the same voltage and current rating. Owing to the high channel electron mobility as well as the bulk electron mobility, GaN vertical devices show faster switching speed and remarkably lower energy loss.

Table 3.1 Switching performance comparison of SiC MOSFET, GaN OGFET, and GaN CAVET

Parameters	Cree SiC MOSFET [16]	Simulated SiC MOSFET [6]	Simulated GaN OGFET	Simulated GaN CAVET [6]
V_{BR} (V)	1.2	1.2	1.4	1.3
R_{on} (Ω)	80	80	75	80
Q_G (nC)	40	55	21.3	29.5
$t_{on\text{-}delay}$ (ns)	38	25	9	18
$t_f(ns)$	13	10	14	2
$t_{off\text{-}delay}$ (ns)	24	27	33	16
$t_r(ns)$	90.8	157	89	88
E_{on} (μJ)	305	312	348	57
E_{off} (μJ)	305	304	92	152
E_{ts} (μJ)	610	616	440	209

3.5 Conclusion

This chapter discussed the two types of GaN vertical transistors citing various reported work to describe their development. Both CAVETs and MOSFETs are proving to be very promising device technologies enabling high-efficiency power switches. While they share a lot in common in terms of design space, a distinctive feature of CAVET is realized in the use of AlGaN/GaN to induce 2DEG channel unlike relying low-mobility inversion layer channel like a MOSFET. Although very different from each other in their channel implementation, MOSFETs and CAVETs both rely on high-quality low-defect density drift region for high-voltage operation. Maintaining high electron mobility in the drift region is of great importance to both designs, particularly with higher (>2 KV) blocking voltages.

References

1. Y.-F. Wu, J. Gritters, L. Shen, R.P. Smith, B. Swenson, kV-class GaN-on-Si HEMTs enabling 99% efficiency converter at 800 V and 100 kHz. IEEE Trans. Power Electron. **29**(6), 2634–2637 (2014)
2. J. Honea, J. Kang, High-speed GaN switches for motor drives. Power Electron. Europe **3**, 38–41 (2012)
3. I. Ben-Yaacov, Y.-K. Seck, U.K. Mishra, S.P. DenBaars, AlGaN/GaN current aperture vertical electron transistors with regrown channels. J. Appl. Phys. **95**(4), 2073 (2004). https://doi.org/10.1063/1.1641520
4. S. Chowdhury, B.L. Swenson, U.K. Mishra, Enhancement and depletion mode AlGaN/GaN CAVET with mg-ion-implanted GaN as current blocking layer. IEEE Electron Device Lett **29**(6), 543–545 (2008). https://doi.org/10.1109/LED.2008.922982
5. S. Chowdhury, M.H. Wong, B.L. Swenson, U.K. Mishra, CAVET on bulk GaN substrates achieved with MBE-regrown AlGaN/GaN layers to suppress dispersion. IEEE Electron Device Lett **33**(1), 41–43 (2012)

6. D. Ji, Y. Yue, J. Gao, S. Chowdhury, Dynamic modeling and power loss analysis of high-frequency power switches based on GaN CAVET. IEEE Trans. Electron Devices **63**(10), 4011–4017 (2016)
7. H. Nie, Q. Diduck, B. Alvarez, A.P. Edwards, B.M. Kayes, M. Zhang, G. Ye, T. Prunty, D. Bour, I.C. Kizilyalli, 1.5-kV and 2.2-mΩ-cm2 vertical GaN transistors on bulkGaN substrates. IEEE Electron Device Lett **35**(9), 939–941 (2014). https://doi.org/10.1109/LED.2014.2339197
8. D. Shibata, R. Kajitani, M. Ogawa, K. Tanaka, S. Tamura, T. Hatsuda, M. Ishida, T. Ueda, 1.7kV/1.0 mΩ·cm2 normally-off vertical GaN transistor on GaN substrate with regrown p-GaN/AlGaN/GaN semipolar gate structure, in *Proceedings of IEEE Electron Devices Meeting (IEDM)*, (2016), pp. 248–251. https://doi.org/10.1109/IEDM.2016.7838385
9. D. Ji, M.A. Laurent, A. Agarwal, W. Li, S. Mandal, S. Keller, S. Chowdhury, Normally OFF trench CAVET with active mg-doped GaN as current blocking layer. IEEE Trans. Electron Devices **64**(3), 805–808 (2016). https://doi.org/10.1109/TED.2016.2632150
10. T. Oka, Y. Ueno, T. Ina, K. Hasegawa, Vertical GaN-based trench metal oxide semiconductor field-effect transistors on a free-standing GaN substrate with blocking voltage over 1.6 kV. Appl. Phys. Express **7**(2), 021002 (2014). https://doi.org/10.7567/APEX.7.021002
11. D. Ji, C. Gupta, S.H. Chan, A. Agarwal, W. Li, S. Keller, U.K. Mishra, S. Chowdhury, Demonstrating > 1.4 kV OG-FET performance with a novel double field-plated geometry and the successful scaling of large-area devices, in *Proceedings of IEEE Electron Devices Meeting (IEDM)*, (2017), pp. 223–226
12. Y. Zhang, M. Sun, D. Piedra, J. Hu, Z. Lin, X. Gao, K. Shepard, T. Palacios, 1200 V GaN vertical fin power field-effect transistors, in *Proceedings of IEEE Electron Devices Meeting (IEDM)*, (2017), pp. 215–218
13. R. Yeluri et al., Design, fabrication, and performance analysis of GaN vertical electron transistors with a buried p/n junction. Appl. Phys. Lett. **106**(18), 183502 (2015)
14. D. Ji, A. Agarwal, W. Li, S. Keller, S. Chowdhury, Demonstration of GaN current aperture vertical electron transistors with aperture region formed by ion implantation. IEEE Trans. Electron Devices **65**(2), 483–487 (2018)
15. D. Ji, S. Chowdhury, A discussion on the DC and switching performance of a gallium nitride CAVET for 1.2kV application, in *Proc. IEEE 3rd Workshop on Wide Bandgap Power Devices and Applications (WiPDA)*, vol. 2-4, (2015), pp. 174–179
16. CMF20120D datasheet, Available: http://www.cree.com/~/media/Files/Cree/Power/Data%20Sheets/CMF20120D.pdf
17. B.J. Baliga, *"Fundamentals of Power Semiconductor Devices," Chapter 6* (Springer U.S., New York, 2008)
18. H. Otake et al., GaN-based trench gate metal oxide semiconductor field effect transistors with over 100 cm2/(V-s) channel mobility. Jpn. J. Appl. Phys. **46**(7L), L599–L601 (2007)
19. C. Gupta, S.H. Chan, Y. Enatsu, A. Agarwal, S. Keller, U.K. Mishra, OG-FET: An in-situ oxide, GaN interlayer based vertical trench MOSFET. IEEE Electron Device Lett **37**(12), 1601–1604 (2016). https://doi.org/10.1109/LED.2016.2616508
20. C. Gupta, C. Lund, S.H. Chan, A. Agarwal, J. Liu, Y. Enatsu, S. Keller, U.K. Mishra, In-situ oxide, GaN interlayer based vertical trench MOSFET (OG-FET) on bulk GaN substrates. IEEE Electron Device Lett **38**(3), 353–355 (2017). https://doi.org/10.1109/LED.2017.2649599
21. D. Ji, W. Li, S. Chowdhury, Switching performance analysis of GaN OG-FET using TCAD device-circuit-integrated model, IEEE International Symposium on Power Semiconductor Device & ICs (ISPSD), May 2018

Chapter 4
Reliability of GaN-Based Power Devices

Gaudenzio Meneghesso, Enrico Zanoni, Matteo Meneghini, Maria Ruzzarin, and Isabella Rossetto

No new product is possible without reliability: this is especially true for new and emerging technology, such as gallium nitride-based devices. For GaN power transistors, breakdown mechanisms play a significant role. The reduction of the robustness and of the long-term reliability still represents a serious issue that must be taken into consideration. The first part of the chapter deals with the above mentioned aspects and mainly focuses on the permanent degradation induced in GaN-based devices by off-state time-dependent mechanisms.

The second part of the chapter analyzes the degradation mechanisms which affect HEMTs with a p-type gate submitted to a high positive bias. An overview of the main results reported in the literature concerning the origin of the permanent degradation is discussed. Results on the recoverable trapping mechanisms are furthermore provided.

The third part analyzes the instabilities in MISHEMT structures. A detailed analysis of negative bias threshold voltage instabilities (NBTI) is discussed in terms of dependence on the applied temperature and performance worsening induced by a cascode configuration.

4.1 Off-State Time-Dependent Degradation Mechanisms

The reliability of AlGaN/GaN HEMTs can be limited by several breakdown mechanisms, which can be mainly summarized in:

G. Meneghesso · E. Zanoni · M. Meneghini · M. Ruzzarin · I. Rossetto (✉)
Department of Information Engineering, University of Padova - DEI, Padova, Padova, Italy
e-mail: gaudenzio.meneghesso@unipd.it; enrico.zanoni@unipd.it; matteo.meneghini@unipd.it; ruzzarin@dei.unipd.it

G. Meneghesso et al. (eds.), *Gallium Nitride-enabled High Frequency and High Efficiency Power Conversion*, Integrated Circuits and Systems,
https://doi.org/10.1007/978-3-319-77994-2_4

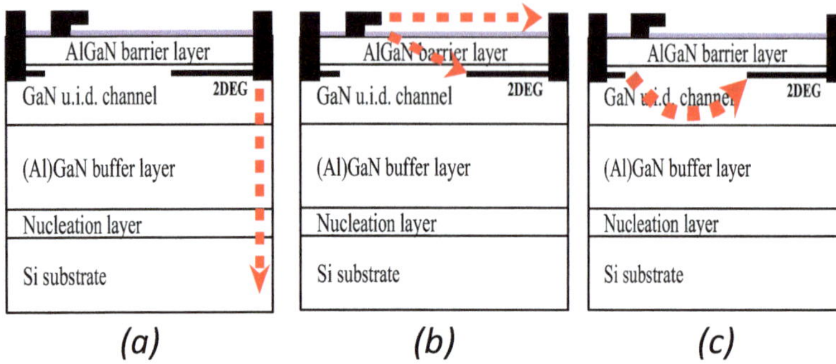

Fig. 4.1 Schematic depicting different typologies of breakdown mechanisms, namely, (**a**) vertical drain-to-substrate breakdown, (**b**) off-state lateral breakdown of the gate-drain region, (**c**) off-state drain-source lateral breakdown

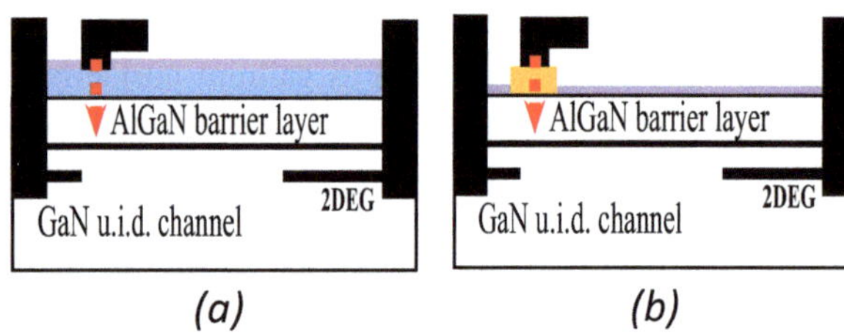

Fig. 4.2 Schematic depicting different typologies of breakdown mechanisms under forward gate bias in (**a**) MIS structures and (**b**) devices with a p-type gate

1. Vertical drain-to-substrate breakdown of the (Al)GaN buffer layer (Fig. 4.1a) [1, 2].
2. Off-state lateral breakdown of the gate-drain region, induced by the high electric field (Fig. 4.1b). Previous papers ascribe this mechanism to the degradation of the Schottky junction [3–5] (inverse piezoelectric effect, defect generation/percolation processes, electrochemical reaction at the surface of the transistors) and to the failure of the passivation layer (described in detail in this section).
3. Off-state drain-source lateral breakdown induced by an increase in subthreshold leakage current and by punch-through effects (Fig. 4.1c) [6, 7].
4. Forward breakdown of the gate junction [8–12]. This aspect, as described in detail in the following section, represents an issue for e-mode devices aimed at power applications such as structures with a p-type gate (Fig. 4.2a) and MIS/MOSHEMT devices (Fig. 4.2b).

Recent papers [13–15] demonstrated that the reliability of GaN HEMTs can be severely limited by time-dependent breakdown processes similarly to what is observed in gate oxides in standard CMOS devices. The robustness of power devices, showing a breakdown voltage >1000 V when evaluated by fast DC sweeps, may be reduced by several hundreds of volts under long-term stress.

In some cases, the situation is substantially different from the case of silicon CMOS devices: in silicon transistors, an oxide is present, while in GaN devices failure may be due to the time-dependent breakdown of the depleted semiconductor itself.

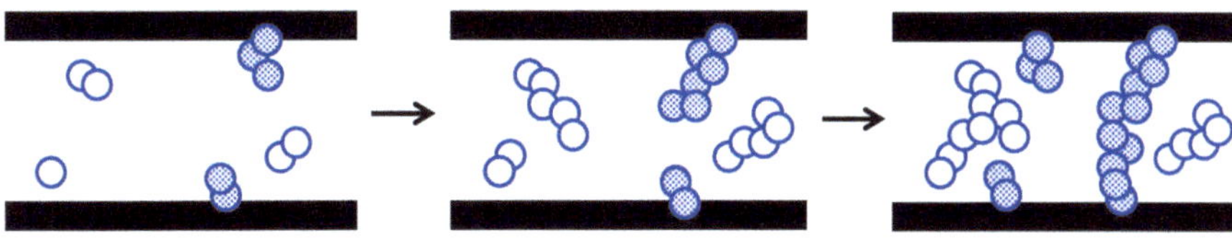

Fig. 4.3 Schematic representation of the model suggested to describe the TDDB in oxide films

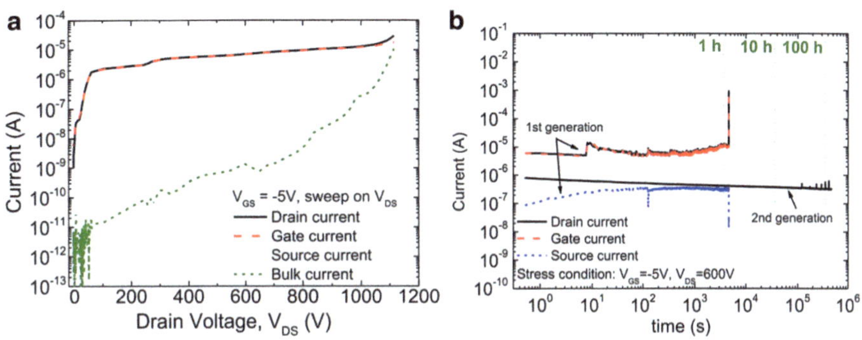

Fig. 4.4 (**a**) Leakage current contribution monitored during an off-state DC voltage sweep (non-catastrophic); (**b**) representative example of an off-state DC stability test discussing the signature of the degradation mechanisms. (© 2015 IEEE. Reprinted, with permission, from [14])

An extensive discussion was provided for the time-dependent dielectric breakdown (TDDB) in the oxide films, due to their importance in CMOS integrated circuits. Degraeve et al. [16, 17] suggested that the breakdown originates from traps located in a random position (in the oxide) even in a fresh device. Under high-voltage stress, new traps are generated until a conducting path is created from one interface to another, thus leading to the breakdown condition (Fig. 4.3).

Moens et al. [13] suggests that, under off-state long-term stress tests, d-mode devices fail as a consequence of a TDDB-like breakdown mechanism. The authors suggest that, when the device is reverse-biased under a high drain voltage, the 2DEG is depleted and the buffer may behave as a defective dielectric. The time-dependent mechanism is found to be Weibull-distributed in ten A-rated devices tested at 200 °C for drain voltage values higher than 900 V.

The evidence of time-dependent degradation processes in GaN-based HEMTs submitted to off-state stress was furthermore discussed by Meneghini et al. [14]. Under off-state long-term stress, failure was observed at voltage levels significantly lower than the breakdown voltage determined by a DC sweep (Fig. 4.4a). The authors ascribe the origin of the degradation to the hard failure of the SiN passivation layer at the edge of the gate on the drain side.

DC stability tests performed in off-state at $V_D = 600$ V demonstrate the following signature (see Fig. 4.4b) [14]: (i) the gate-drain leakage shows a gradual and steplike increase, presumably due to the creation of defect-related leakage paths next to the gate. These latter can be in principle ascribed to the degradation of the gate Schottky

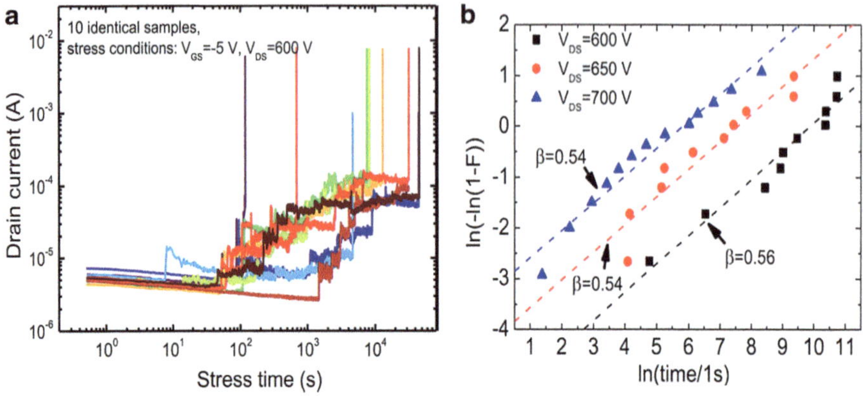

Fig. 4.5 (**a**) Constant voltage stress performed at (V_G, V_D) = (−5 V, 600 V) on a set of ten identical devices. (**b**) Weibull distribution for three set of devices stressed in off-state ($V_G = -5$ V) at room temperature with a V_D ranging from 600 V to 700 V. (© 2015 IEEE. Reprinted, with permission, from [14])

junction and/or of the insulating properties of the dielectric under the gate terminal. (ii) Long stress times lead to a permanent degradation which mainly results in a sudden non-recoverable increase of the off-state drain current.

Statistical analysis confirms that the failure can be described by a Weibull distribution (refer to Degraeve et al. [16] for detailed information) with a shape factor lower than 1 (Fig. 4.5), confirming the high device-to-device variability and indicating the presence of an extrinsic breakdown mechanism. The breakdown process is found to be time- and field-dependent with a time to failure exponentially dependent on the drain bias level applied.

The role of the SiN passivation layer in the catastrophic failure is validated by 2D numerical simulations (Fig. 4.6). The simulations indicate that, under off-state conditions, the maximum electric field is in the edge of the gate head on the drain sides. Under high drain bias levels, the corresponding simulated field peak is comparable with the breakdown electric strength of the SiN (6 MV/cm). Conversely, the simulated electric field in the AlGaN is much lower than the corresponding breakdown value.

Comparable results were provided by Rossetto et al. [15] on similar devices. The authors demonstrated, by means of microscopical investigation, that the SiN passivation layer may play a significant role in the hard failure of GaN-based HEMTs. The creation of defect-related leakage paths next to the gate is suggested by the detection of hot spots by means of a cooled CCD camera. The intensity of the EL signal (in the detected hot spots) increases, up to the hard failure, with the increase of the noise superimposed to the leakage current and/or with the steplike behavior. Evidence of the breakdown of the SiN passivation layer is provided by TEM analysis performed on a set of devices after hard failure (Fig. 4.7).

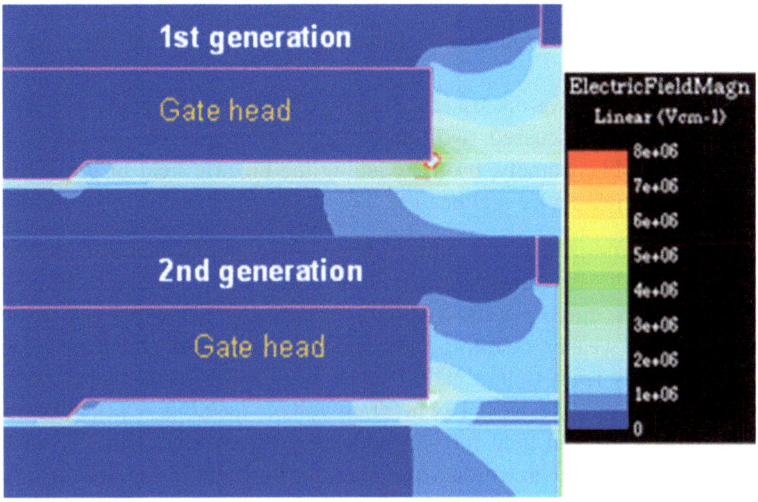

Fig. 4.6 Simulated electric field distribution close to the gate edge for two generations of devices under test. (© 2015 IEEE. Reprinted, with permission, from [14])

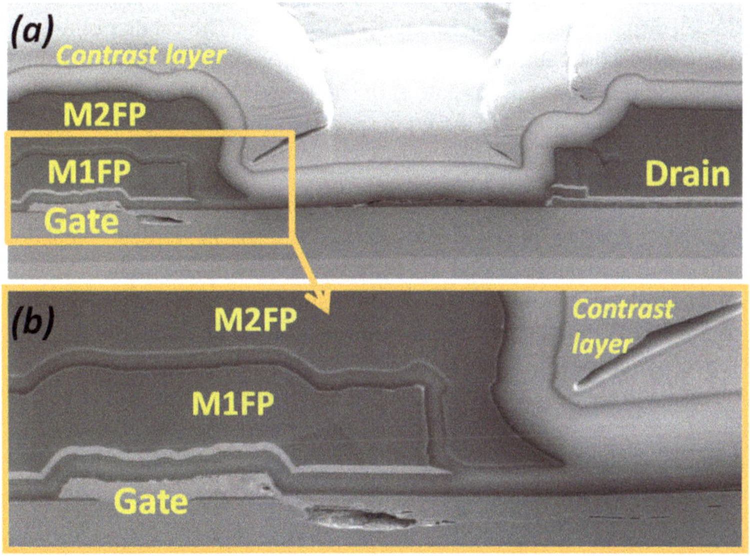

Fig. 4.7 (**a**) Transmission electron microscope (TEM) performed on a cross-section of a device after hard failure induced by a constant voltage test, showing a severe degradation in the gate side/edge. (**b**) Enlargement of the damaged portion of the device. (© 2017 IEEE. Reprinted, with permission, from [15])

A short circuit path is detected across the SiN passivation layer and the AlGaN barrier layer near the gate edge on the drain side, thus confirming that time-dependent breakdown mechanisms may origin from the robustness of the passivation layer at the edge of the gate, in correspondence of the peak of the electric field.

According to the above-discussed failure mechanism, Meneghini et al. [14] and Rossetto et al. [15] discussed two approaches to improve the reliability of the AlGaN/GaN HEMTs, namely, by (i) reducing the electric field in the 2DEG and (ii) by optimizing the device architecture.

(i) A significant increase in the time to breakdown can be obtained through the reduction of the maximum electric field under the gate overhead, achieved by using GaN epitaxy with a different electrical behavior. Gen-2 devices exhibit much weaker coupling between drain potential and body potential. This effectively leads to less 2DEG retraction (depletion) for Gen-1 devices as compared to Gen-2 devices at same drain voltage (see Fig. 4.8a), accompanied by a higher electrical field for Gen-1 devices across SiN. The TTF of Gen-2 samples is more than three orders of magnitude longer than that of Gen-1 devices (Fig. 4.8b) [14].
(ii) A second approach for the improvement of the robustness of the SiN layer was discussed by Rossetto et al. [15]. The deposition of an extra layer of silicon nitride (with PECVD technique in the case under analysis) induces a reduction of the electric field across the SiN passivation layer. The validity of the approach is demonstrated by an increase of the failure voltage by more than 200 V in devices submitted to an off-state step-stress and by an increase by more than two orders of magnitude observed in samples submitted to an off-state constant voltage stress (Fig. 4.9).

The time-dependence of breakdown mechanisms, already discussed in off-state conditions, is found to play a significant role even in vertical breakdown. A systematic analysis was provided by Borga et al. [18]. The authors demonstrate that, when submitted to two terminal (drain-to-substrate) stress, degradation in GaN-based transistors is time-dependent. Catastrophic failure is Weibull-distributed and characterized by an exponential dependence on the applied electric field; oppositely, a weak thermal dependence is shown in time to failure.

Drain-to-substrate current conduction can be described by the space charge limited model. At high drain voltage values, the current behavior is no more ohmic, as for low bias levels until not all the traps are ionized. Under high drain bias levels the significant depletion of the buffer and the ionization of the traps leads to a significant increase of the current, due to the trap-filled limited model.

The current behavior for long stress times and high drain bias is consistent to defect percolation theory, which explains the time-dependent breakdown of dielectrics under high electric fields. Under these conditions the depleted buffer can be considered as a leaky dielectric. Occurrence of dislocations, which differently from insulators can be observed in wide band gap semiconductors, can enhance vertical leakage and accelerate defect generation processes, thus facilitating premature breakdown. In the case under analysis, long stress times lead to an increase of the

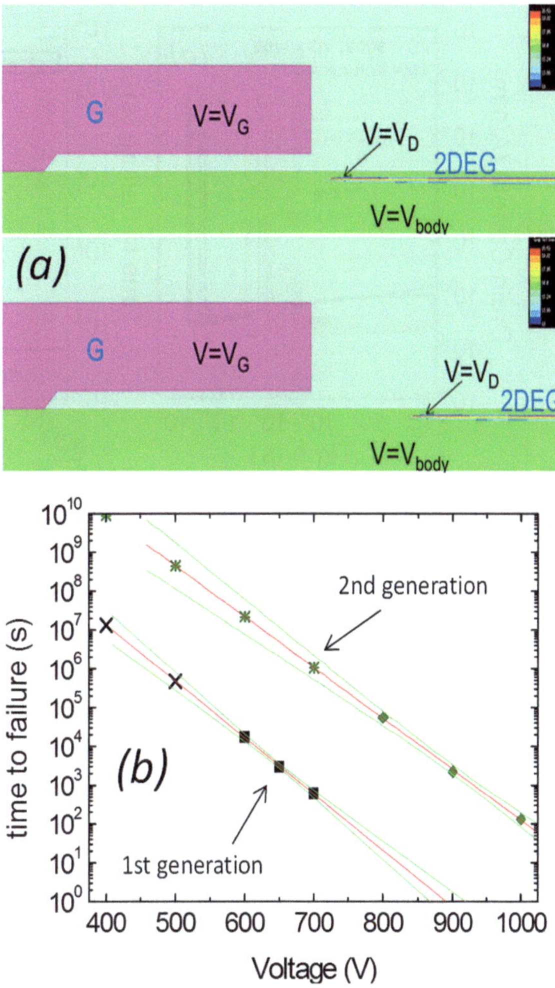

Fig. 4.8 (**a**) Visual representation of the difference between devices of first (top) and second generation (bottom) in terms of 2DEG depletion and body potential (false color superimposed to device structure). (**b**) Corresponding time to failure as a function of applied drain voltage. (© 2015 IEEE. Reprinted, with permission, from [14])

noise superimposed to the leakage current, indicating a defect generation process, until the occurrence of a significant increase of the current in correspondence of the hard failure.

4.2 Time-Dependent Failure of Structures with P-Type Gate

Devices aimed at power applications such as power-switching converters must fulfill several requirements in terms of device performance. A large gate bias swing, together with a good robustness toward forward gate bias, is an essential aspect to be considered. Although commercial devices usually operate at a gate voltage between

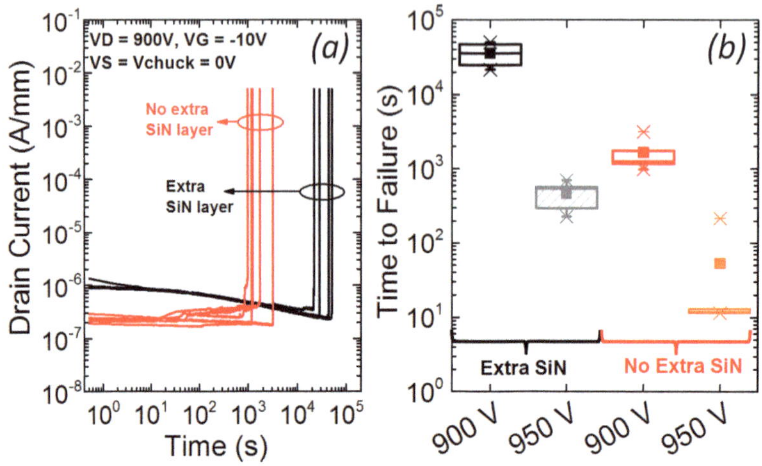

Fig. 4.9 (**a**) Drain current monitored during a constant voltage stress in devices with and without an additional SiN layer. (**b**) Corresponding time to failure. (© 2017 IEEE. Reprinted, with permission, from [15])

0 V and 7 V, devices are able to withstand a much higher gate bias level. According to the device main applications, the devices must guarantee a high robustness toward off-state (operating voltage is usually 600–650 V) and reduced parasitic losses in order to guarantee high switching frequencies. In commercial devices, parasitic effects are usually defined by an on-resistance in the range of 50–70 mΩ and a gate charge of approximately 6 nC [19].

Devices used in high-power applications must switch from an off-state high-voltage condition to a low-voltage high-current state. The normally off condition is therefore an essential point in order to assure safe operation. In case of failure of the gate driver, the device will be indeed forced to an off-state condition.

In this scenario, two main approaches are used to fulfill the enhancement-mode operation and to improve the device robustness: (i) the use of a p-type stack (hereafter called devices with a p-type gate or p-type structures); solutions which use a gate metal deposited on the p-type layer with a ohmic or Schottky contact were both discussed in the literature; (ii) the insertion of a thin insulator layer below the gate metal with the aim of reducing the parasitic effects and, with proper recess, of enabling e-mode operation (hereafter called MISHEMT structures).

Both solutions are limited by two major aspects in terms of reliability and dynamic performance:

(i) Devices with a p-type gate and MISHEMTs are found to be extremely sensitive to time-dependent phenomena. The long-term reliability is therefore significantly undermined since the failure, being time-dependent, can be reached at lower bias levels for longer stress times. As discussed in the previous section, TDDB was extensively investigated in oxide films; on the basis of the parameters which describe the Weibull distribution, an estimation of the time to failure and information concerning the typology of failure can be provided [16, 17].

(ii) Devices with a p-type gate and MISHEMTs suffer from threshold voltage instabilities, worsened by the increase of the temperature. Instabilities, although recoverable, strongly undermine the performance of the device and, under limit conditions, may lead to premature failure of the devices. A more detailed discussion is reported in this section.

Several results were discussed in the literature in order to formulate hypotheses concerning the hard failure in p-type structures under forward gate bias. The analyses are mainly aimed at studying the time-dependence of the breakdown mechanism and at understanding the origin of the failure mechanism.

The time-dependence of the degradation mechanisms was studied by several authors on different device technologies. Rossetto et al. [12, 20] demonstrated that devices with a p-type gate submitted to a high gate bias overstress undergo a time-dependent failure; the time to failure is exponentially dependent on the bias level applied and can be described by a Weibull distribution (Fig. 4.10a). A good correlation was furthermore found between the gate leakage current and the time to failure. This aspect, observed by several authors in different technologies, confirms the role of pre-existing defect states in the acceleration/definition of the failure mechanism, consistently with the theory of the TDDB in oxides described by Degraeve et al. [16] (see previous section for further details). The correlation between the initial gate leakage current and the time to failure is confirmed by Tallarico et al. [11]. The authors modeled the abovementioned correlation with an exponential law for different gate bias levels. According to this model, the authors were able to predict the maximum allowed initial gate current to reach a mean time to failure of 10 years at the operating voltage. Analogous results were reported by Tapajna et al. in terms of time-dependence of the failure process and correlation between the time to failure and the initial gate leakage current [8, 9].

Tapajna et al. [9] and Rossetto et al. [12] finally studied the dependence of the temperature on the time to failure (Fig. 4.10b). The results suggest that the failure mechanism is thermally activated. It is worth mentioning that the different activation energy (0.1 eV in [9] and 0.5 eV in [12]) depends on the different technologies and/or on the different degradation mechanisms involved (Fig. 4.10b).

The origin of the failure mechanism was extensively discussed in the literature. Although several hypotheses were formulated, a clear definition of the dominant degradation process is still under debate.

Wu et al. [10] firstly studied the origin of the failure mechanism on a p-type structure having a Schottky contact between the TiN gate metal and the p-GaN layer. The authors suggested that, under high gate bias overstress, the Schottky metal/p-GaN diode is in reverse bias (Fig. 4.11); the electrons in the channel, after being emitted over the AlGaN layer and being injected into the p-GaN, have enough energy (owing to the high electric field) to promote avalanche breakdown. This hypothesis is supported by the positive temperature dependence of the forward

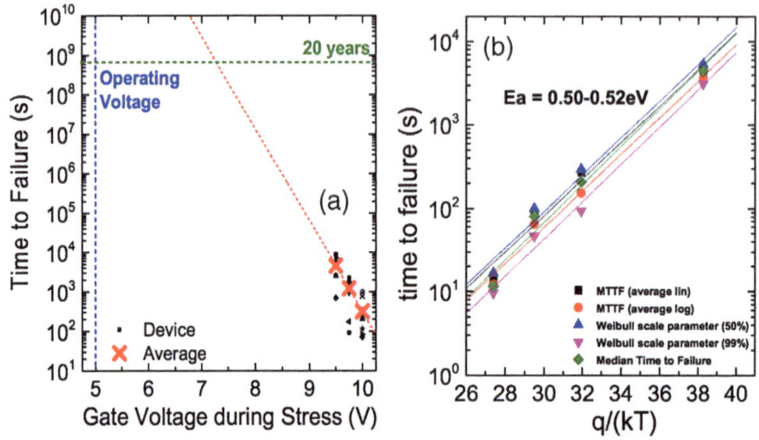

Fig. 4.10 (**a**) Dependence of TTF on stress voltage. Ten identical devices were stressed for each voltage level. (**b**) Results of stress tests carried out at different temperature levels. TTF was found to be dependent on temperature, with activation energy in the range 0.50–0.52 eV. (Reprinted from [20] with permission from Elsevier)

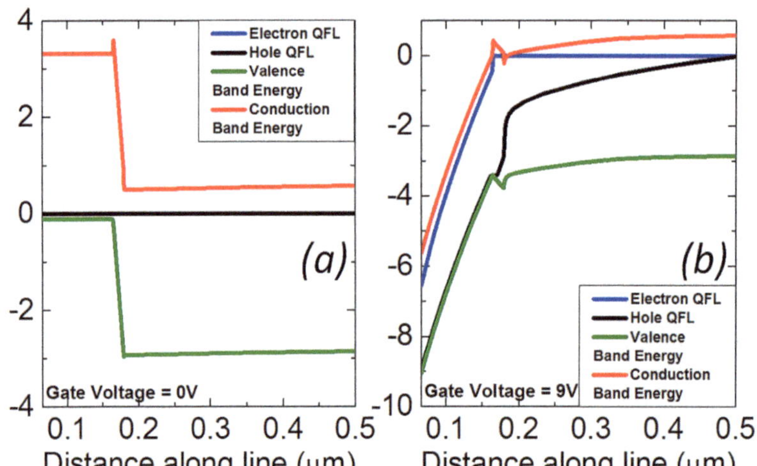

Fig. 4.11 Band diagram of a GaN HEMT with a p-GaN gate (**a**) under equilibrium ($V_G = 0$ V) and (**b**) under gate bias overstress ($V_G = 9$ V). (© 2016 IEEE. Reprinted, with permission, from [12])

gate breakdown voltage evaluated at different ambient temperatures. Furthermore, at high gate bias levels, the authors detected a weak electroluminescence, presumably ascribed to the recombination of the generated electron-hole pairs before the hard failure.

The origin of the degradation process under forward gate bias was furthermore discussed by Rossetto et al. [12] by means of 2D simulations, electrical

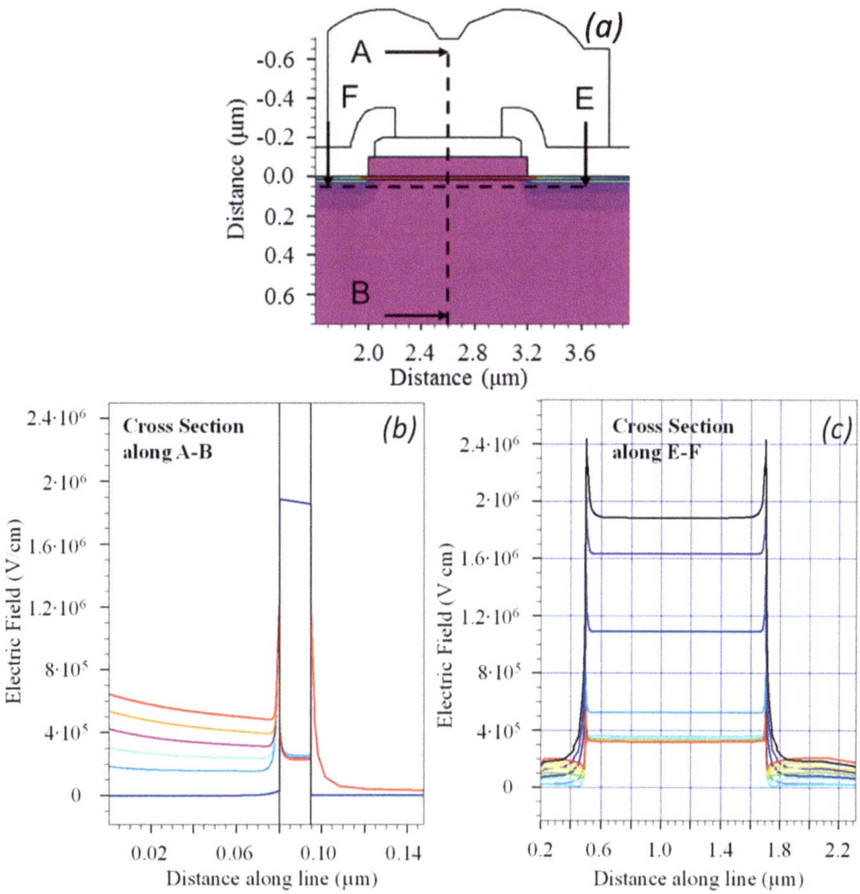

Fig. 4.12 (**a**) Schematic of the simulated structure. Simulation (ATLAS) of the electric field (**b**) across the p-GaN and AlGaN layers and (**c**) along the AlGaN layer. (© 2016 IEEE. Reprinted, with permission, from [12])

measurements, and emission microscopy. At high gate bias levels, the Schottky junction between the gate metal and the p-type layer is reverse-biased. Under this condition, the creation of a depleted region may favor the generation of defect-related percolative paths which, under long-term stress, lead to the hard failure of the devices. Under this condition, a contribution of the avalanche mechanisms is also possible.

The creation of defect-related percolative paths, assisted by possible nonuniformities and by the high-current flowing, is promoted by the high electric field in the p-GaN. Conversely to the electric field in the AlGaN, the electric field in the p-GaN layer increases with the gate bias level applied (Fig. 4.12).

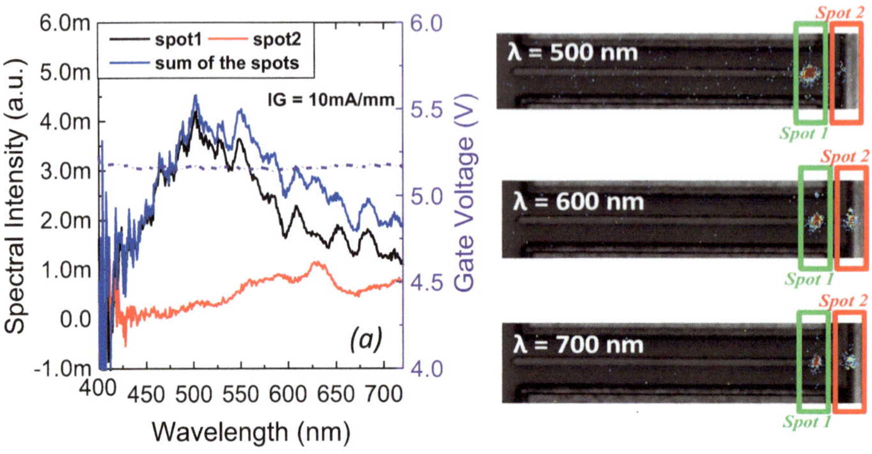

Fig. 4.13 (**a**) Spectra detected with a CCD camera. (Right) corresponding emission microscopy is reported for $\lambda = 500$ nm, $\lambda = 600$ nm, $\lambda = 700$ nm. (Reprinted from [20] with permission from Elsevier)

The electric field is found to increase also in the SiN layer on the edge of the gate. Although the electric field is much lower than the breakdown field of the SiN, possible spikes and nonuniformities can promote the catastrophic failure.

Emission microscopy, performed on already damaged devices belonging to different technologies, indicates that the EL signal is mainly due to bremsstrahlung radiation of the hot electrons and yellow luminescence (Fig. 4.13) [20]. The latter results from the contribution of trap states related to gallium vacancies or carbon-related defects.

Tallarico et al. [11] confirms that the p-GaN layer plays a major role in the failure of p-type structures under forward gate bias overstress and ascribe the breakdown event to the creation of a percolative path in the depleted region of the p-GaN layer, near to the metal/p-GaN interface, where the simulated electric field peaks under forward gate bias conditions. The suggested model indicates that the equivalent circuit of the p-GaN gate is formed by two back-to-back junctions, corresponding to the Schottky metal/p-GaN junction and the PiN p-GaN/AlGaN/GaN diode. The two junctions are, respectively, reversely and directly biased at high gate bias levels. The authors suggest that, for high V_G levels, the voltage drop across the AlGaN is saturated; the additional applied voltage falls entirely in the p-GaN region. The p-GaN layer, which behaves as a depleted region, undergoes high electric fields which may lead to the formation of percolative paths.

Another mechanism that can contribute to the degradation was proposed by Tapajna et al. on devices with a p-type gate formed via Ni/Au metallization [8, 9]. The authors suggest that, as a consequence of the high electric field and current flowing in the p-GaN, a conductive path is generated due to trap generation and/or a percolation mechanism, leading to the catastrophic failure. The authors furthermore

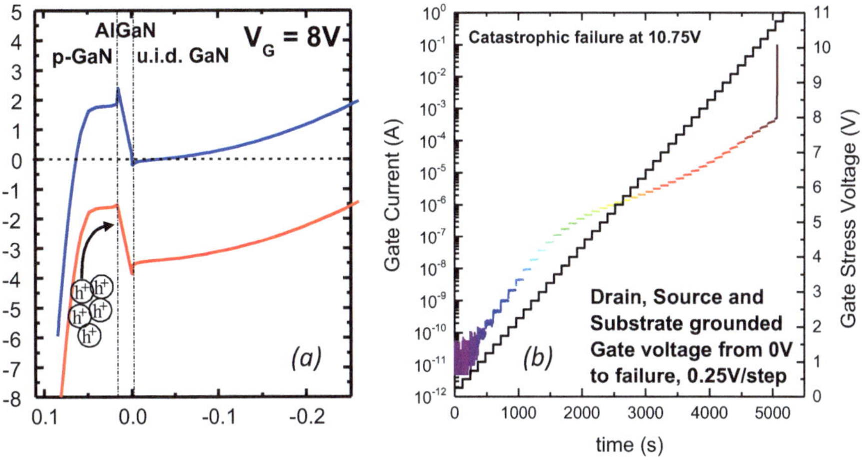

Fig. 4.14 (**a**) Schematic depicting the mechanism responsible for the NBTI recoverable phenomenon. (**b**) Gate current monitored during a forward gate bias step-stress (null drain and source voltage)

suggest that the creation of donor-like traps at the p-GaN/AlGaN interface plays a key role in the formation of percolative paths and, thus, in the origin of the hard failure.

Permanent degradation is a key aspect in the definition of the performance of p-type structures under forward gate bias. Nevertheless, under similar conditions, the device performance may be significantly undermined by trapping and recoverable effects. A little discussion is reported in literature on this topic [21], and the recoverable effects of the trap states under forward gate bias are still under discussion.

Figure 4.14 reports the impact of recoverable and permanent degradation measured on a commercial device. The device under test, submitted to a forward gate bias step-stress, is rated for operating at $V_G = 5$ V. Figure 4.14b demonstrates that hard failure occurs at $V_G > 10$ V; no soft degradation (permanent effects) is demonstrated at lower bias levels. Considerations about permanent degradation (occurring at $V_G > 10$ V) were already discussed.

After each step of the stress, a full DC characterization is performed in order to detect the possible presence of recoverable mechanisms. Figure 4.15 demonstrates that stress induces a negative shift (NBTI) of threshold voltage for bias levels lower than the device robustness. A possible explanation for the NBTI mechanism detected is the following. A high gate bias overstress allows the injection of holes toward the p-GaN/AlGaN interface, where holes can be accumulated due to the discontinuity in the valence band (Fig. 4.10a). The positive charge temporarily accumulated at the p-GaN/AlGaN interface and/or in the AlGaN barrier leads to a negative threshold voltage shift. The correlation between the negative V_{TH} shift and the gate current monitored during the stress indicates that the gate current

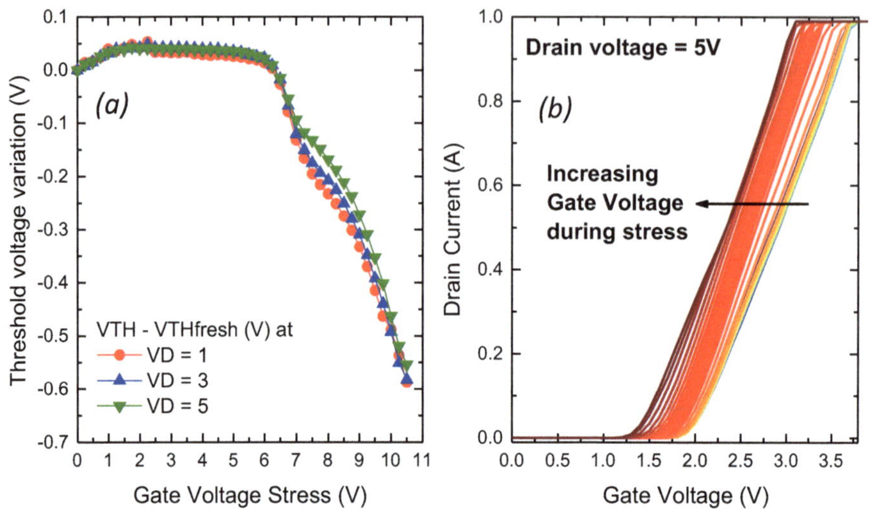

Fig. 4.15 Evaluation of NBTI mechanism on a HEMT with p-type structure submitted to forward gate bias overstress. (**a**) Negative variation of the V_{TH}; (**b**) transfer characteristics monitored after each step of the stress at $V_D = 5$ V (saturation regime)

injected during the stress significantly impacts the NBTI mechanism. In principle, additional trapping mechanisms may occur at the p-GaN/AlGaN interface (e.g., accumulation of electrons in the conduction band) and contribute to the instabilities of the threshold voltage.

It is worth mentioning that the on-resistance measured at operating voltage faces no variation until the occurrence of the hard failure, indicating that the recoverable trapping mechanisms do not affect the on-resistance in real operating conditions.

4.3 Positive and Negative Bias Threshold Voltage Instabilities in MISHEMT Structures

MISHEMT (metal-insulator-semiconductor high-electron-mobility transistor) structures are commonly used to fulfill the requirements imposed by high-power applications. The insertion of a thin dielectric layer under the gate significantly improves device performance by reducing the parasitic effects, enhancing the gate bias swing and increasing the device robustness. Furthermore, by means of proper recess, it allows the implementation of e-mode device.

Nevertheless, the performance of these structures is affected by two major limits, namely, the high sensitivity toward time-dependent degradation (not discussed in this chapter) and the influence of threshold voltage instabilities.

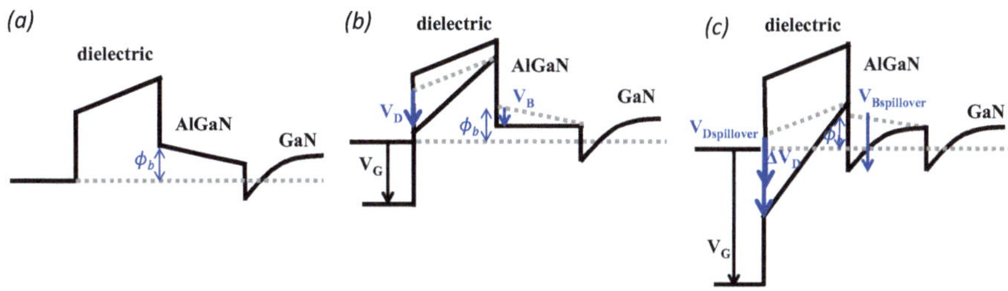

Fig. 4.16 Band diagram variation with different positive gate bias levels, namely, (**a**) thermal equilibrium, (**b**) low positive gate bias levels, (**c**) high positive gate bias levels. In condition (**c**) a second channel is formed between the dielectric and the AlGaN layer

Although several dielectrics and deposition techniques were studied [22–24], threshold voltage instabilities still limit significantly the device performance. Instabilities are induced by charge trapping under the gate located in the dielectric or at the dielectric/IIIN interface. The importance of this aspect is strengthened by the influence of the temperature and of the bias level applied, both positive (PBTI) and negative (NBTI), at the gate. This section is divided into three main subsections: (i) PBTI, (ii) NBTI, and (iii) instabilities in cascode configuration.

4.3.1 Positive Bias Threshold Voltage Instabilities in MISHEMTs

The origin of the PBTI mechanism was explained by Lagger et al. [22]. A schematic principle of mechanism is depicted in Fig. 4.16. (a) At thermal equilibrium, there is negligible electron flow between the channel and the interface; (ii) at positive low gate bias levels, electrons may flow toward the III-N interface through the barrier layer; (iii) at very high gate bias levels, called spill-over regime in [22], a second channel is formed at the dielectric/AlGaN interface. Due to the absence in the voltage drop V_D-V_B, electrons are provided by the second channel and can get trapped in the dielectric or at the interface between the dielectric and the AlGaN barrier layer, inducing an increase of the ΔN_{it} (namely, the number of trapped electron density at the interface).

The charge trapping responsible for PBTI mechanisms is significantly influenced by the material used as dielectric layer [22, 25–27]. An explanation was tentatively suggested by Wu et al. [27], by comparing the effects of charge trapping in devices with a ALD Al_2O_3 and PEALD SiN insulator layer under the gate. Experimental results demonstrate that PEALD SiN devices are characterized by a wide distribution of defect levels near the GaN conduction band; conversely in Al_2O_3 devices trap states are due to a narrower distribution far from the GaN conduction band. The authors suggest that, in the case of SiN, a low dependence of ΔV_{TH} on gate voltage applied corresponds to a higher distribution of defects around the Fermi level, where the electrons can be trapped even at low-voltage values. In the case of Al_2O_3, the distribution of defects is narrower and far from the Fermi level, leading to less accessible defects at low gate bias levels.

The charge trapping is furthermore significantly influenced by the deposition technique. Rossetto et al. [24] and Meneghesso et al. [23] demonstrated that the deposition method used significantly impacts the threshold voltage positive shift. In the case under analysis, SiN is used as insulator under the gate; a PEALD deposition significantly lowers the V_{TH} dynamic shift ($\approx$ 2 V) with respect to a RTCVD SiN deposition technique. Furthermore, the authors demonstrate that, independently from the insulator and/or from the deposition used, the V_{TH} dynamic shift is promoted by the injection of electrons in the gate insulator under gate forward bias. This aspect is confirmed by the strong correlation between the dynamic V_{TH} shift and the forward gate current in the corresponding bias point.

4.3.2 Negative Bias-Induced Threshold Voltage Instability

This section describes the instabilities of main MIS-HEMTs parameters, when the devices are exposed to negative gate voltages under high temperature. Several analyses report that GaN MIS-HEMTs do not show any V_{TH} shift at room temperature [28]; however, it is demonstrated that they can suffer from a negative V_{TH} shift when operating at low stress voltages and high temperatures (more realistic conditions) [29]. One approach to study this kind of instabilities is to stress the devices with HTRB (high temperature reverse bias) tests. A study of the negative threshold voltage instability based on combined electrical and temperature-dependent investigation is reported by [29]. GaN-on-Si HEMTs with partially recessed AlGaN were tested with a set of stress/recovery experiments, and the variation in the main parameters (V_{TH} and R_{ON}) was monitored with fast $I_D V_G$ and $I_D V_D$ measurements at 90 °C. The results reported in Fig. 4.17a show that a negative gate bias of -10 V induces a negative shift of the threshold voltage (-3.2 V after 5000 s of stress), and during the recovery phase (when the device is left unbiased), the shift does not recover completely (Fig. 4.17b). The stress causes also a significant decrease in on-resistance (-33%) which is correlated to the decrease in V_{TH} as it is shown in Fig. 4.17c. The strong correlation between the two parameters demonstrates that the two instabilities are induced by the same physical mechanism. Moreover, the process responsible of the instability of V_{TH} and R_{ON} is activated by temperature. Figure 4.18 shows the V_{TH} shift monitored for the same stress performed at different temperature: at room temperature, the variation is negligible (blue curve in Fig. 4.18a); at higher temperatures, the shift increases and is recoverable (Fig. 4.18b). The V_{TH} shift during stress was found to have a stretched exponential trend [30], while during the recovery phase, it follows a logarithmic time-dependence [31]. The exponential fit of the stress is reported in Fig. 4.18a, and the logarithmic fit for the recovery phase is reported in Fig. 4.18b for the curve at 90 °C.

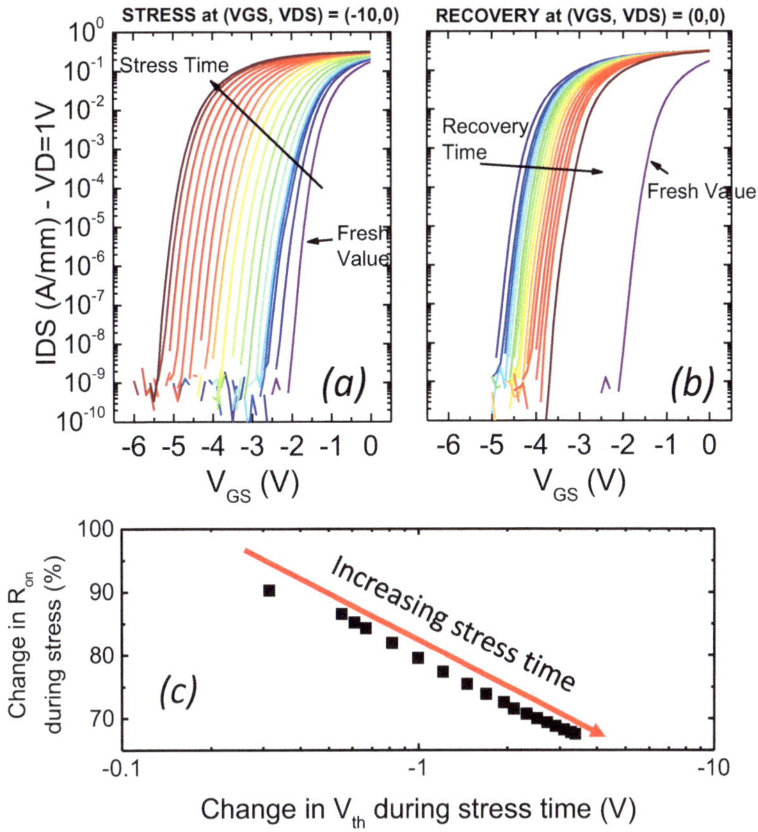

Fig. 4.17 (**a**) Negative V_{TH} shift induced by exposing the devices to off-state stress ($V_{GS} = -10$ V, $V_{DS} = 0$ V, $T = 90$ °C): the V_{TH} shift was characterized by repeatedly interrupting the stress experiment for the execution of a fast I_D-V_D and I_D-V_G (drain current vs drain voltage and drain current vs gate voltage, respectively) measurements. After 5000 s of stress, a rest phase (**b**) was performed with all terminals at zero voltage at 90 °C. In this phase, the V_{TH} shift partially recovered. Correlation between the variation in on-resistance and the variation in threshold voltage measured for increasing stress times at 90 °C (**c**). (© 2016 IEEE. Reprinted, with permission, from [29])

On the basis of the experimental results, a physical interpretation has been developed and schematically represented in Fig. 4.19. The dielectric layer or the SiN/AlGaN interface contains acceptor-like defects that may be responsible of the threshold voltage instabilities in reverse bias. At equilibrium the traps are neutral when they are above the Fermi level and negatively charged when they are below it. When subjected to negative bias at high temperature, the electrons trapped in the trap states are released, and they can pass through the 3.7 nm AlGaN barrier by trap assisted tunneling toward the GaN layer [32]. Moreover, the defects depletion results in a net positive charge responsible for a more negative shift of V_{TH} and for

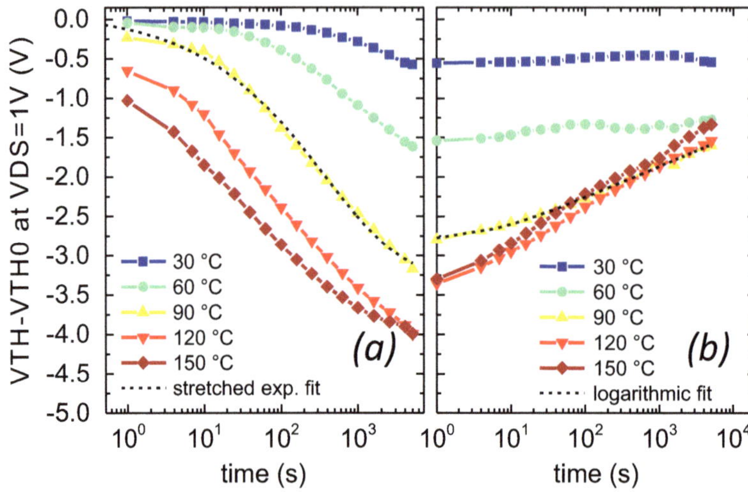

Fig. 4.18 Variation in threshold voltage measured during (**a**) stress and (**b**) recovery at several temperature levels. Bias conditions are the following: $V_{GS} = -10$ V, $V_{DS} = 0$ V. The V_{TH} shift was found to follow a stretched exponential trend $\sim\exp^{(-t/\tau)\beta}$, with β in the range 0.2–0.7 depending on temperature (**a**) (the exponential fit of the stress phase is reported for the curve at 90 °C). The V_{TH} shift was found to be recoverable at temperatures higher than 90 °C, and the recovery was found to have a logarithmic time-dependence (**b**). (© 2016 IEEE. Reprinted, with permission, from [29])

a correlated decrease of R_{ON}. The activation energy calculated for this process is 0.37 eV. During the recovery phase, the device is biased at 0 V, and some traps go below the Fermi level, refilled by electrons. Electrons are supposed to come from the accumulation layer or the metal by means of defect-assisted conduction through the AlGaN barrier or the SiN layer, respectively. The mechanism activated during the recovery phase is slower (logarithmic time-dependence) than the process induced by the stress (exponential trend). This can be explained by the fact that the recovery process is limited by two factors:

1. The electrons that have already fill traps induce a repulsive action on other electrons [28, 32, 33].
2. The amount of defects available for conduction through the barrier may be low [33].

In conclusions, an analysis of instabilities induced by a reverse bias in GaN-based MIS-HEMTs is presented: standard RT measurements are not sufficient to detect negative V_{TH} shift; thus, a comprehensive characterization was performed by carrying out high temperature reverse bias (HTRB) stresses. A model to explain the measured V_{TH} drift is reported on the basis of the experimental evidence.

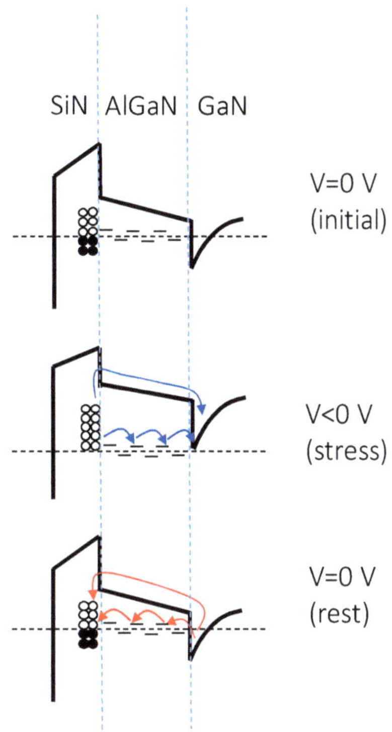

Fig. 4.19 Schematic representation of the process responsible for the negative V_{TH} shift. The surface donors are not indicated for simplicity. (© 2016 IEEE. Reprinted, with permission, from [29])

4.3.3 Constant Source Current-Induced Degradation

In several cases, normally-on GaN MIS-HEMTs are used in cascode configuration, in combination with a silicon MOSFET in order to achieve normally off operation as reported in Fig. 4.20c [34]. The leakage current of the MOSFET in off-state is equal to a constant source current IS for the HEMT subjected to high drain-source voltage. Thus, in order to evaluate the reliability of GaN MIS-HEMTs is important to consider two different conditions: high temperature reverse bias (HTRB) stress, as reported in the previous section and in Fig. 4.20a, and also high temperature source current (HTSC) stress (see Fig. 4.20b) to assess the impact *of subthreshold drain-source leakage and high field [35]. In the HTSC test,* the leakage through the Si MOSFET in off condition is emulated by drawing a constant current out of the source of the HEMT while applying a high drain bias (150 V). Figure 4.21 shows the variation of V_{TH} and of R_{on} when the device is submitted to HTRB stress with $V_{GS} = -10$ V, $V_{DS} = 150$ V (violet curve), and HTSC stress with different I_S (100 nA, 1 μA, 2 μA, 10 μA) at 150 °C. The results of the HTSC stress demonstrate a stronger increase of the on-resistance for higher source current imposed (Fig. 4.21b), without significant variation of the threshold voltage (Fig. 4.21a). R_{on} shows a very fast recovery by setting all the terminals grounded.

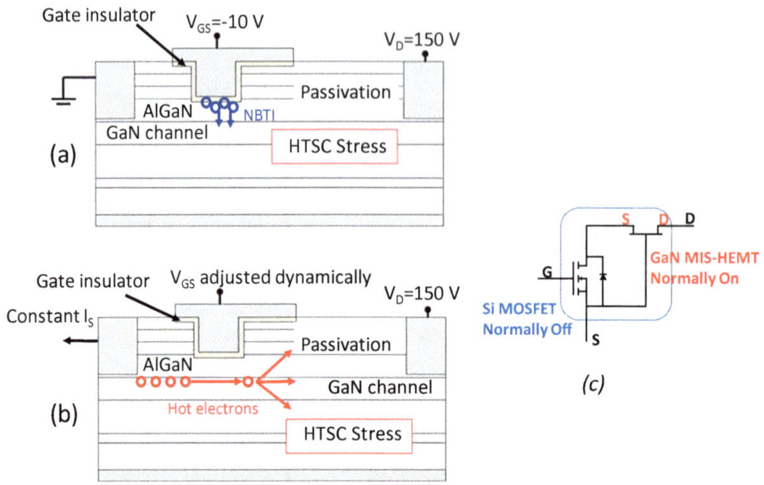

Fig. 4.20 Structure of a MIS-HEMT in two different stress conditions: (**a**) HTRB and (**b**) HTSC. The HTSC stress protocol proposed in [35] is representative of HEMTs used in cascode configuration with a silicon MOSFET (**c**) in order to achieve normally OFF operations. When the MOSFET is in the off-state, its leakage current flows to the HEMT while a high drain-source bias is applied, and it corresponds to a constant source current I_S for the HEMT. (© 2016 IEEE. Reprinted, with permission, from [35])

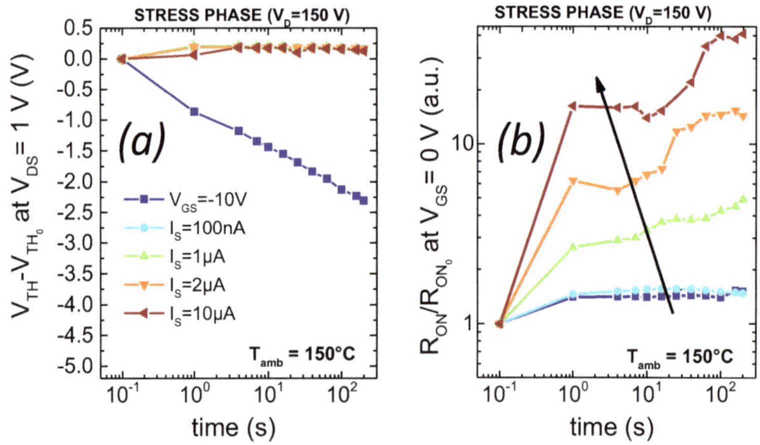

Fig. 4.21 Variation of the threshold voltage (**a**) and the on-resistance (**b**) measured during 200 s of stress. Two stress protocols are compared: HTRB stress with $V_{GS} = -10$ V, $V_{DS} = 150$ V (violet curve) and HTSC stress with different I_S (100 nA, 1 µA, 2 µA, 10 µA) at 150 °C. (© 2016 IEEE. Reprinted, with permission, from [35])

The effect of the temperature on the degradation process was evaluated by considering 1 µA of source current and by varying the temperature (from 130 °C to 180 °C) as reported in Fig. 4.22. The variation of the on-resistance is found to be

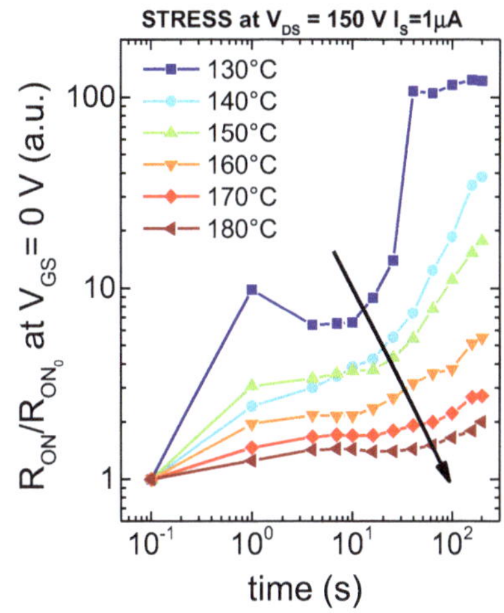

Fig. 4.22 Variation of the on-resistance measured during 200 s of stress for different temperatures. Stress conditions: V_{GS} adjusted dynamically, $V_{DS} = 150$ V, $I_s = 1$ µA. (© 2016 IEEE. Reprinted, with permission, from [35])

stronger at low temperatures. The monotonic dependence of the degradation process on I_S (Fig. 4.21b), correlated to the negative temperature dependence (Fig. 4.22), leads to suppose that the positive variation of R_{on} can be ascribed to the injection of hot electrons toward the gate-drain access region as depicted in Fig. 4.20b [35]. The fast recovery process suggests that trapping occurs in the semiconductor material, because in the dielectric detrapping process should be slower due to the hopping conduction [36].

The activation energy of the degradation mechanism is equal to −1.3 eV (indicating the negative temperature dependence) which decreases with increasing temperatures when the probability of hot electrons injection reduces [37]. In conclusion the leakage current of the cascoded Si MOSFET, combined with the high off-state field, may result in an increase in on-resistance due to hot electrons injected toward the gate-drain access region; these results underline the importance of analyzing the MIS-HEMTs under HTSC tests.

4.4 Conclusions

Gallium nitride-based HEMTs are very promising devices for power electronic applications due to the high breakdown voltage, improved on-resistance, and low parasitic effects. Recently, a significant effort has been spent in order to optimize the technology starting from the material properties to reduce trapping effects. However, together with good performance, the greatest challenge for a new technology is achieving a high level of reliability and stability.

The first part of the chapter discusses the time-dependence of hard failure in case of vertical and off-state lateral breakdown. At high drain bias levels, the buffer may behave as a leaky dielectric and premature failure can consequently occur at voltage levels much lower than the breakdown voltage determined by a DC sweep. Consistently to percolation theory the time to failure is Weibull-distributed and exponentially correlated to the applied field; moreover degradation is preluded by an increase of the leakage current and of the noise to it superimposed.

This section furthermore discusses a case study, describing the role of the silicon nitride in the degradation. Failure may occur in the passivation layer near the edge of the gate head where, according to simulations, the electric field peaks. Severe degradation is demonstrated by optical investigation, such as intensity of the electroluminescence signal and TEM evaluation. Two approaches to significantly improve the robustness are finally suggested, namely, the reduction of the electric field in the 2DEG by using a GaN epitaxy with a different electrical behavior and the optimization of the device architecture by the deposition of an extra layer of silicon nitride.

In the second part of the chapter, we reported an overview concerning the permanent degradation and the recoverable trapping effects in p-type devices subjected to forward gate bias. The time-dependence observed in permanent degradation is found to be well correlated to the initial current value, confirming the role of pre-existing defects in the failure mechanisms. The role of temperature and the dependence on the applied field are moreover defined. A summary of the main theories concerning the origin of the failure mechanism is included, namely, the impact of avalanche breakdown and the generation of defect-related leakage paths as a consequence of the depletion region formed in the p-type layer under high gate bias overstress. The electric field simulated in the AlGaN layer, conversely to the one reported in the pGaN and in the SiN layer, is found to have a negligible role in forward gate bias degradation.

Despite the high robustness achieved under forward gate bias, p-type devices still suffer from recoverable trapping mechanisms. Although it has negligible impact on on-resistance in real-life operating conditions, NBTI is considered one of the major limits in the performance of p-type structures, especially as a consequence of the strong correlation between instabilities and gate current during the stress. A possible explanation suggests that NBTI can be mainly ascribed to holes injected toward the p-GaN/AlGaN interface at high gate bias.

Finally, in the last section we reported about the main mechanisms that cause the threshold voltage instabilities in MIS-HEMTs. MIS-HEMT performance is strongly affected by threshold voltage and on-resistance instabilities. However, standard measurements at room temperature are not sufficient to investigate the phenomena because these latter are influenced by high-voltage and high temperature conditions. Thus, different stresses were carried out studying PBTI, NBTI, and instabilities in cascode configuration.

When a MIS-HEMT is submitted to high positive bias, electrons may flow toward the III-N interface creating a second channel at the interface between the insulator and the AlGaN layer and inducing a positive shift of the threshold voltage (PBTI). The positive V_{TH} shift is promoted by electron injection and influenced by the material and/or the deposition method used for the insulator.

On the contrary, the reverse bias and high temperatures (HTRB) induce a negative non-completely recoverable threshold voltage (NBTI) on MIS-HEMTs, correlated with a decrease of R_{ON} and strengthened by temperature. The trapping mechanism responsible of the ΔV_{TH} can be ascribed to the presence of a net positive charge (due to defects on insulator/AlGaN interface), and it has an exponential trend. The recovery process is slower (logarithmic time-dependence).

In order to evaluate the reliability of GaN MIS-HEMTs in cascode configuration with a silicon MOSFET, it is important to perform high temperature source current (HTSC) stress to evaluate the impact of the leakage current of the MOSFET on the device. The results reveal a recoverable increase of the on-resistance for higher source current imposed, without ΔV_{TH}. The phenomenon is stronger at low temperatures, and it can be ascribed to the injection of hot electrons toward the gate-drain access region.

References

1. G. Meneghesso, M. Meneghini, E. Zanoni, Breakdown mechanisms in AlGaN / GaN HEMTs : An overview. Jpn. J. Appl. Phys. **53**, 100211–1001/8 (2014)
2. I.B. Rowena, S.L. Selvaraj, T. Egawa, Buffer thickness contribution to suppress vertical leakage current with high breakdown field (2.3 MV/cm) for GaN on Si. IEEE Electron Device Lett. **32**(11), 1534–1536 (2011)
3. E. Zanoni, M. Meneghini, A. Chini, D. Marcon, G. Meneghesso, AlGaN/GaN-based HEMTs failure physics and reliability: Mechanisms affecting gate edge and Schottky junction. IEEE Trans. Electron Devices **60**(10), 3119–3131 (2013)
4. J. Joh, J.A. del Alamo, Critical voltage for electrical degradation of GaN high-electron mobility transistors. Electron Device Lett. IEEE **29**(4), 287–289 (2008)
5. F. Gao et al., Role of oxygen in the OFF-state degradation of AlGaN / GaN high electron mobility transistors. Appl. Phys. Lett. **99**, 223506–2231/3 (2011)
6. M. Meneghini et al., OFF -state degradation of AlGaN / GaN power HEMTs : Experimental demonstration of time-dependent drain-source breakdown. IEEE Trans. Electron Devices **61**(6), 1987–1992 (2014)
7. S.R. Bahl, M. Van Hove, X. Kang, D. Marcon, M. Zahid, S. Decoutere, New Source-side Breakdown Mechanism in AlGaN/GaN Insulated-Gate HEMTs. Proceedings of 25th international symposium on power semicond. devices ICs, pp. 419–422, 2015
8. M. Tapajna, O. Hilt, E. Bahat-Treidel, J. Würfl, J. Kuzmík, Investigation of gate-diode degradation in normally-off p-GaN/AlGaN/GaN high-electron-mobility transistors. Appl. Phys. Lett. **107**, 193506–1931/4 (2015)
9. M. Tapajna, O. Hilt, J. Würfl, J. Kuzmík, Gate reliability investigation in normally-off p-type-GaN cap/AlGaN/GaN HEMTs under forward bias stress. Electron Device Lett. IEEE **37**(4), 385–388 (2016)
10. T.-L. Wu et al., Forward bias gate breakdown mechanism in enhancement-mode p-GaN gate AlGaN/GaN high-electron mobility transistors. IEEE Electron Device Lett. **36**(10), 1001–1003 (2015)
11. A.N. Tallarico et al., Investigation of the p-GaN gate breakdown in forward-biased GaN-based power HEMTs. IEEE Electron Device Lett. **38**(1), 99–102 (2017)
12. I. Rossetto et al., Time-dependent failure of GaN-on-Si power HEMTs with p-GaN gate. IEEE Trans. Electron Devices **63**(6), 2334–2339 (2016)

13. P. Moens et al., Technology and design of GaN power devices. Proceedings of 45th European solid-state device Res. conference, pp. 64–67, 2015
14. M. Meneghini et al., Extensive investigation of time-dependent breakdown of GaN-HEMTs submitted to OFF-state stress. IEEE Trans. Electron Devices **62**(8), 2549–2554 (2015)
15. I. Rossetto et al., Field-related failure of GaN-on-Si HEMTs : Dependence on device geometry and passivation. IEEE Trans. Electron Devices **64**(1), 73–77 (2017)
16. R. Degraeve et al., New insights in the relation between electron trap generation and the statistical properties of oxide breakdown. IEEE Trans. Electron Devices **45**(4), 904–911 (1998)
17. R. Degraeve, B. Kaczer, G. Groeseneken, Degradation and breakdown in thin oxide layers: Mechanisms, models and reliability prediction. Microelectron. Reliab. **39**(10), 1445–1460 (1999)
18. M. Borga et al., Evidence of time-dependent vertical breakdown in GaN-on-Si HEMTs. IEEE Trans. Electron device **64**(9), 3616–3621 (2017)
19. http://www.gansystems.com/gs66508t.php, 2017.
20. I. Rossetto et al., Study of the stability of e-mode GaN HEMTs with p-GaN gate based on combined DC and optical analysis. Microelectron. Reliab. **64**, 547–551 (2016)
21. M. Meneghini, O. Hilt, J. Wuerfl, G. Meneghesso, Technology and reliability of normally-off GaN HEMTs with p-type gate. Energies **1**(10), 153 (2017)
22. P. Lagger et al., Role of the dielectric for the charging dynamics of the dielectric/barrier interface in AlGaN/GaN based metal-insulator-semiconductor structures under forward gate bias stress. Appl. Phys. Lett. **105**, 33512 (2014)
23. G. Meneghesso et al., Trapping and reliability issues in GaN-based MIS HEMTs with partially recessed gate. Microelectron. Reliab. **58**, 151–157 (2016)
24. I. Rossetto et al., Impact of gate insulator on the dc and dynamic performance of AlGaN/GaN MIS-HEMTs. Microelectron. Reliab. **55**, 1692–1696 (2015)
25. T. Wu et al., Time dependent dielectric breakdown (TDDB) evaluation of PE-ALD SiN gate dielectrics on AlGaN / GaN recessed gate D-mode MIS-HEMTs and E-mode MIS-FETs. IEEE international reliability physics symposium, pp. 4–9, 2015
26. T. Wu et al., The impact of the gate dielectric quality in developing Au-free D-mode and E-mode recessed gate AlGaN / GaN transistors on a 200mm Si substrate. Proc. 27th international symposium on power semiconductor devices IC's, pp. 225–228, 2015
27. T. Wu et al., Toward understanding positive bias temperature instability in fully recessed-gate GaN MISFETs. IEEE Trans. Electron Devices **63**(5), 1853–1860 (2016)
28. M. Van Hove et al., Fabrication and performance of au-free AlGaN/GaN-on-silicon power devices with Al_2O_3 and Si_3N_4/Al_2O_3 gate dielectrics. IEEE Trans. Electron Devices **60**(10), 3071–3078 (2013)
29. M. Meneghini et al., Negative bias-induced threshold voltage instability in GaN-on-Si power HEMTs. Electron Device Lett. IEEE **37**(4), 474–477 (2016)
30. O. Mitrofanov, M. Manfra, Mechanisms of gate lag in GaN / AlGaN / GaN high electron mobility transistors. Microelectron. Reliab. **34**, 33–53, 2004 (2003)
31. W. Choi, H. Ryu, N. Jeon, M. Lee, H.-Y. Cha, K.-S. Seo, Improvement of Vth instability in normally-off GaN MIS-HEMTs employing PEALD-SiN_x as an interfacial layer. Electron Device Lett. IEEE **35**(1), 30–32 (2014)
32. P. Lagger, C. Ostermaier, G. Pobegen, D. Pogany, Towards understanding the origin of threshold voltage instability of AlGaN/GaN MIS-HEMTs. Tech. Dig. - Int. Electron Devices Meet. IEDM, pp. 299–302, 2012
33. P. Lagger, M. Reiner, D. Pogany, C. Ostermaier, Comprehensive study of the complex dynamics of forward bias-induced threshold voltage drifts in GaN based MIS-HEMTs by stress / recovery experiments. IEEE Trans. Electron Devices **61**(4), 1022–1030 (2014)
34. X. Huang, Z. Liu, Q. Li, F.C. Lee, Evaluation and application of 600 V GaN HEMT in Cascode structure. IEEE Trans. Electron Devices **29**(5), 2453–2461 (2014)

35. M. Ruzzarin et al., Evidence of hot-electron degradation in GaN- based MIS-HEMTs submitted to high temperature constant source current stress. Electron Device Lett. IEEE **37**(11), 1415–1417 (2016)
36. B.S. Eller, J. Yang, R.J. Nemanich, Electronic surface and dielectric interface states on GaN and AlGaN. J. Vac. Sci. Technol. A **31**(5), 50807 (2013)
37. E. Zanoni, G. Meneghesso, Impact ionization in compound semiconductor devices, in *Handbook of Advanced Electronic and Photonic Materials and Devices*, ed. by H. S. Nalwa (Ed), (Academic press, US), US, p. 2001

Chapter 5
Validating GaN Robustness

Kenichiro Tanaka, Ayanori Ikoshi, and Tetsuzo Ueda

5.1 Introduction

5.1.1 Reliability Issues Specific to GaN Power Transistors

How we validate the robustness of GaN power transistors? As they are booming in the market recently, this issue becomes more and more important. The reliability of GaN power transistors are investigated first by the standardized tests for Si power transistors [1–9]. However, the standardized tests for Si are not sufficient to guarantee the robustness of GaN transistors when they are employed in power converter applications, because the reliability of GaN under switching operation is different from that of Si transistors.

It is known for GaN power transistors that the ON-state resistance under switching operation, which we call hereafter dynamic Rdson, can be much larger than their direct current (DC) ON-state resistance, Rdson. This phenomenon is called current collapse, where the drain current is reduced and thus ON-state resistance is increased once a high drain bias is applied to a GaN transistor [10, 11]. Current collapse is caused by the negative charge trapping in the structure under high-voltage stress. The increase in dynamic Rdson elevates the device temperature, which in turn can result in a thermal instability and finally device destruction.

K. Tanaka (✉)
Panasonic Corporation, Moriguchi-shi, Japan
e-mail: tanaka.kenichiro@jp.panasonic.com

A. Ikoshi · T. Ueda
Panasonic Corporation, Kadoma-shi, Japan
e-mail: ikoshi.ayanori@jp.panasonic.com; ueda.tetsuzo@jp.panasonic.com

G. Meneghesso et al. (eds.), *Gallium Nitride-enabled High Frequency and High Efficiency Power Conversion*, Integrated Circuits and Systems,
https://doi.org/10.1007/978-3-319-77994-2_5

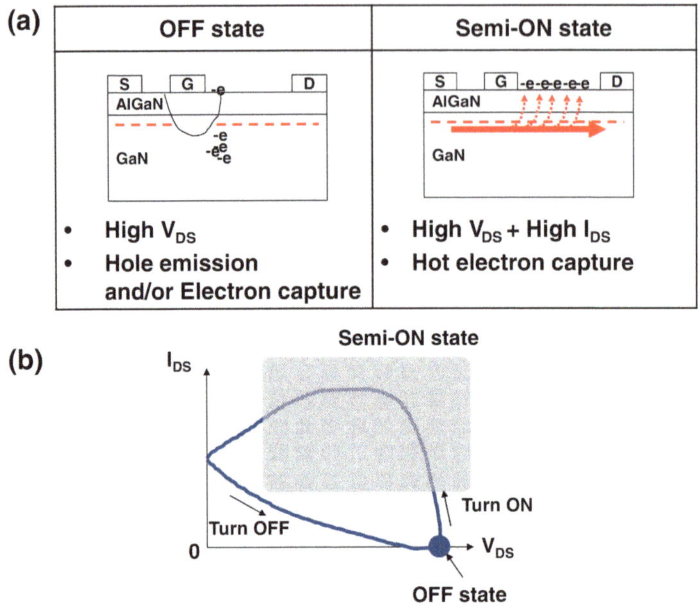

Fig. 5.1 (**a**) Mechanism for current collapse in OFF and semi-ON states. (**b**) Schematic switching locus in an inductive-load switching, where OFF state and semi-ON state are indicated

Since current collapse has been the critical issue for GaN transistors, it has been intensively investigated. There are mainly two mechanisms suggested for current collapse, which are schematically shown in Fig. 5.1a.

Firstly, current collapse is driven by a high drain bias in the OFF state as shown in Fig. 5.1a. When a high V_{DS} is applied to the device in the OFF state, hole emission and/or electron capture is induced at the aluminum gallium nitride (AlGaN) surface [12] or the epilayer [13], which partially depletes the channel, leading to the current collapse. There are numerous reports on the traps that are responsible for current collapse [10–14]. For instance, we reported that there are hole traps in the GaN epitaxial layer, and the hole emission in the OFF state plays a crucial role in the current collapse phenomena [15].

Secondly, what we call "semi-ON" state also induces current collapse as shown in Fig. 5.1a. Here the semi-ON state means that a high V_{DS} and a high I_{DS} are applied simultaneously to the device, which corresponds to the shaded area in Fig. 5.1b. When the device is operated under the so-called hard switching, in which the temporal V_{DS} and I_{DS} are overlapped in the switching event, the device is momentarily subject to the "semi-ON" state. In the semi-ON state, large amount of hot electrons are induced, and they gain sufficient energy to overcome the AlGaN barrier or the buffer layer beneath the two-dimensional electron gas (2DEG) channel, some of which are trapped at the surface or the (Al)GaN epitaxial layer and will partially deplete the channel [16–20]. As a result, current collapse is observed (the dynamic Rdson is increased). Meneghini et al. points out that the charge trapping in the semi-ON state is field-dependent and is a very fast phenomenon [17], which indicates that this mechanism can happen even if the time duration when the device is under the semi-ON state lasts only a fraction of time.

Since these two mechanisms are superimposed during the switching event, current collapse is more severely observed as the switching voltage and/or switching current increases. In other words, current collapse depends on the shape of I_{DS}-V_{DS} switching locus [17, 21–26].

5.1.2 *Switching Safe Operating Area (SSOA)*

Since whether or not GaN power transistor can be switched safely is determined by whether or not current collapse is induced, the safety switching is determined primarily by the switching locus. Therefore, the concept of "switching safe operating area (SSOA)," wherein a device can be switched safely, is proposed for GaN power transistors [27]. Seemingly similar, the proposed SSOA is different from the conventional safe operating area (SOA) in that the SSOA for GaN is determined primarily by current collapse, while the conventional SOA for Si is determined mainly by thermal issue.

Figure 5.2 exemplifies the conventional SOA defined for Si power transistors [28], which is defined by the ON-state resistance, maximum current, breakdown voltage, maximum power limitation, and thermal instability limitation. Here the color lines shown at the right-upper corner are limited mainly by the thermal issue or maximum power rating. The right-upper corner is produced by repeatedly imposing short pulses to the break point of the device [28].

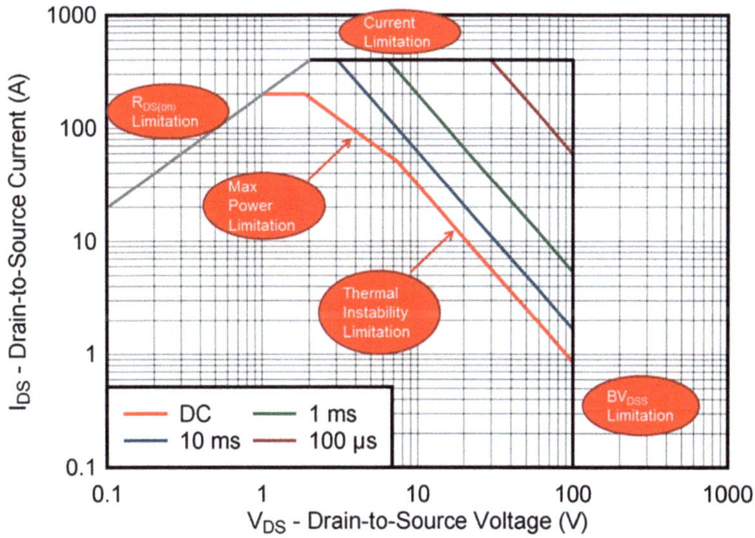

Fig. 5.2 Conventional safe operating area (SOA) for Si power transistors [28]

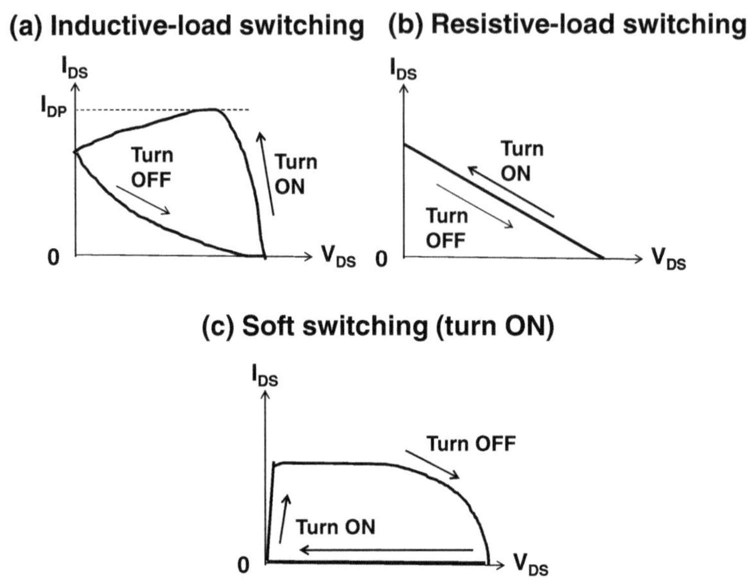

Fig. 5.3 Schematic locus curves during (**a**) inductive-load switching, (**b**) resistive-load switching, and (**c**) soft switching under typical switching applications

However, conventional SOA for Si is not sufficient for validating the robustness of GaN transistors as employed in switching power applications. Since GaN lateral power transistors have small input and output capacitances, their switching event is typically as fast as on the order of ns. In such a fast switching, heat elevation is considered to be extremely small; thus, it is expected that a GaN transistor can be switched safely in all of the rectangular area determined by the maximum current and the rated voltage. In reality, however, if higher V_{DS} and I_{DS} are applied simultaneously to GaN power transistors in the switching event, they can experience instability due to current collapse.

The shape of switching locus is strongly dependent on the load and the switching condition. Figures 5.3a–c show the typical locus curves for inductive-load (LR-load), resistive-load (R-load), and soft switching (at turn ON) in power converter applications, respectively. In our experience, current collapse is more severely observed for LR-load switching, because its locus is typically larger than that for R-load or soft switching, and thus the charge trapping under semi-ON state increases the dynamic Rdson.

To take an example, Fig. 5.4 summarizes the circuit schematics and the corresponding loci under the LR-load or R-load switching event obtained for our early-stage current-collapse-ridden gate injection transistor (GIT). We perform continuous switching with increasing V_{DD} to obtain the relative dynamic Rdson as a function of V_{DD}, which is shown in Fig. 5.5. Here the relative dynamic Rdson is the

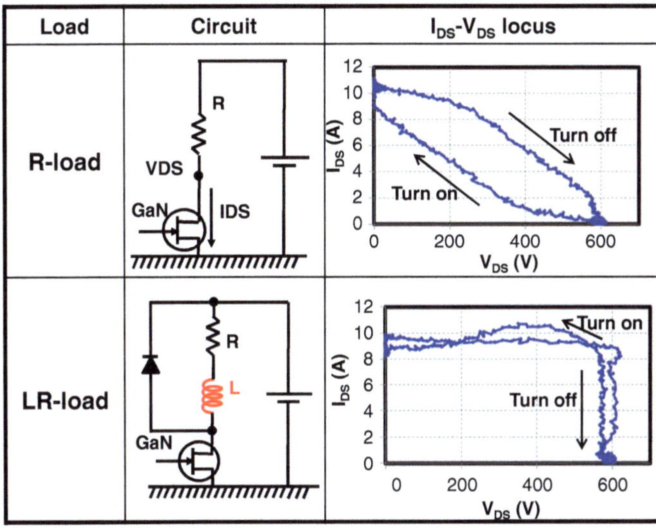

Fig. 5.4 Electric circuit for resistive-load (R-load) and inductive-load (LR-load) switching and the corresponding loci measured for a current-collapse-ridden GIT in the switching event

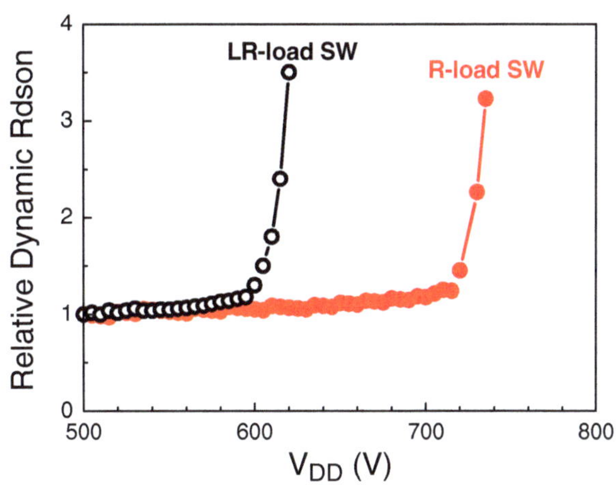

Fig. 5.5 Relative dynamic Rdson as a function of V_{DD} measured for a current-collapse-ridden GIT in a resistive-load (R-load) and inductive-load (LR-load) switching operation

dynamic Rdson normalized by its DC ON-state resistance. The dynamic Rdson for the LR-load switching is increased under the lower V_{DD} by roughly 20% compared with that for the R-load switching.

Thus, LR-load switching circuit can be used as a means to determine the SSOA limit, as shown in Fig. 5.6a, because the LR-load switching imposes more harsher stress on the device in one switching cycle, which enables us to evaluate the lifetime in a much shorter test time. There are some variations of the switching circuits of this kind, such as a boost converter or a boost converter with output tied to the

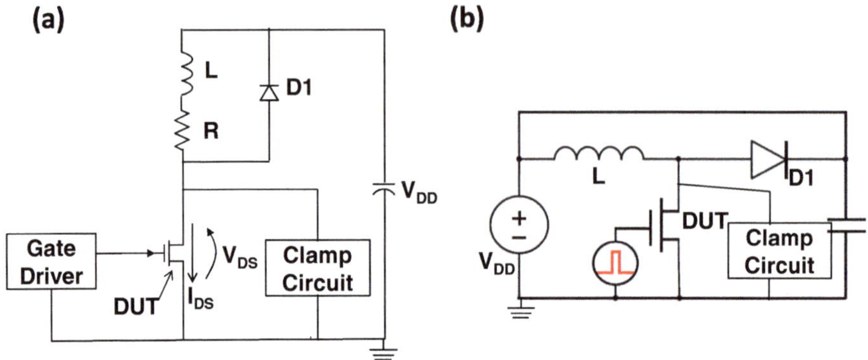

Fig. 5.6 Electric circuit for SSOA measurement: (**a**) inductive-load switching and (**b**) boost converter with output tied to input voltage source [23]

input voltage source as shown in Fig. 5.6b [23]. In the circuits, dynamic Rdson is measured with a clamping circuit, in which V_{DS} is clamped at a certain value in the OFF state so that the ON-state V_{DS} is measured with accuracy [27, 29, 30].

However, LR switching is not the only option to define the SSOA. LR-load switching is generally carried out in a low-duty cycle (<5%) with a low frequency (kHz range, for instance [31]), which is not in accordance with a typical switching applications for GaN power transistors. For example, GaN transistors can be utilized for MHz region high-frequency power converter application, in which they are switched with a duty cycle of around 50% combined with the soft-switching technique [32]. Although the stress during the one switching cycle under soft-switching operation is much smaller than the LR-load switching, we cannot rule out the possibility that the degradation under higher-frequency soft-switching operation is comparable to that under the LR-switching operation. To our knowledge, there is not comprehensive approach for the robustness of GaN power transistors under the soft-switching operation. The degradation mechanism under high-frequency soft-switching operation awaits further investigation to accelerate the widespread use of GaN power transistors.

SSOA is defined as the area inside which the device's dynamic Rdson is not increased and its junction temperature T_J is below its rated value. Figure 5.7 summarizes a flowchart for obtaining the SSOA that is standardized for GaN power transistors [27]. In order to determine the SSOA for GaN power transistors, the devices are switched with varying the switching conditions to find the limit where both the dynamic Rdson and junction temperature remain within the normal values.

There is one more thing to be added here. Although SSOA defines the area inside which the device can be switched safely, it does not necessarily guarantee the robustness under long-period continuous-switching operation. When the device is exposed to a continuous-switching operation for a long period of time, the device would be worn out. Here again, the long-time degradation is also dependent on

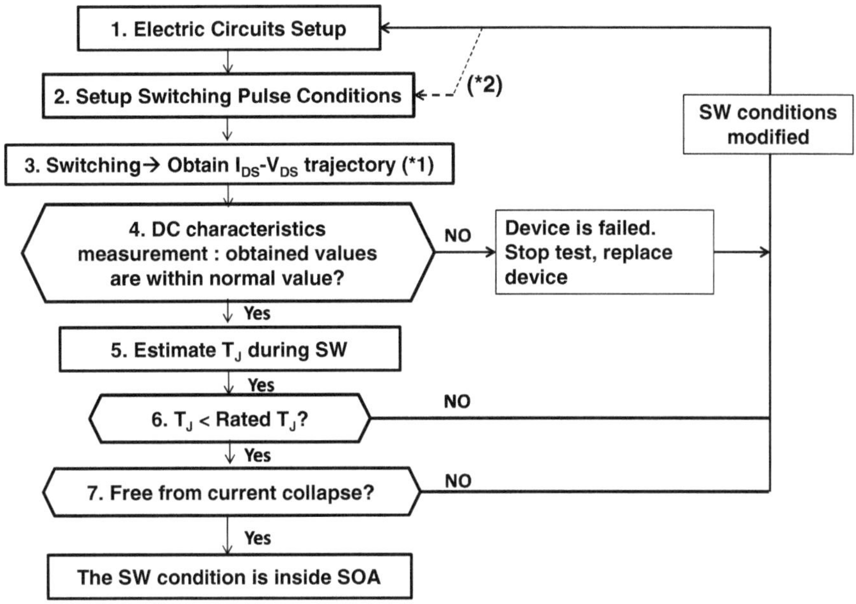

Fig. 5.7 A flowchart for determining SSOA. (*1) In double-pulse test, device is switched only twice. In the continuous-pulse switching, the switching test is stopped intermittently, and the following procedure is carried out. (*2) If modification of experimental setup is not needed, only switching pulse condition is modified [27]

the shape of switching locus. This is because the charge trapping or the wear-out occurs accumulatively under a continuous-switching operation and they are severely accelerated under semi-ON state conditions.

We thus propose "long-time switching safe operating area (lSSOA)" inside, which the device can be switched for a long period of time. On the other hand, the SSOA that guarantees the switching for a short period of time is called hereafter sSSOA. We suppose that the method for determining the lSSOA is rather complex and varies from device to device. As far as we know, there is neither generalized method to define the lSSOA of GaN transistors nor the report on lSSOA. In Sect. 5.2.5 [42], we exemplify how we determine the lSSOA for HD-GIT.

The sSSOA is obtained by the so-called double-pulse test (DPT), where the device is switched only twice to capture the waveforms. Figure 5.8 shows the schematic waveforms in DPT [27]. Since the device is switched only twice, the temperature rise of the device is limited or negligible in most cases. Thus the DPT is suitable to measure the sSSOA.

On the other hand, lSSOA is determined by dynamic high-temperature operating lifetime (D-HTOL) test in which continuous-pulse switching is carried out for a long period of time. Figure 5.9 shows the schematic waveforms under continuous LR-load switching. In this case, the device's T_J, its DC characteristics, and dynamic Rdson are intermittently checked to verify that they are within their normal values.

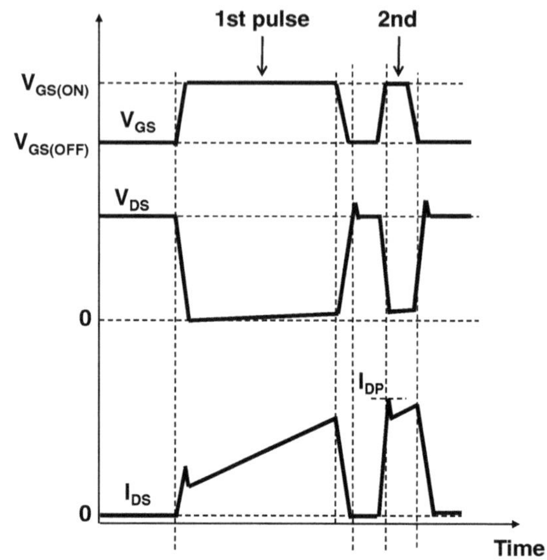

Fig. 5.8 Waveforms of V_{GS}, V_{DS}, and I_{DS} in double-pulse test for inductive-load switching [27]

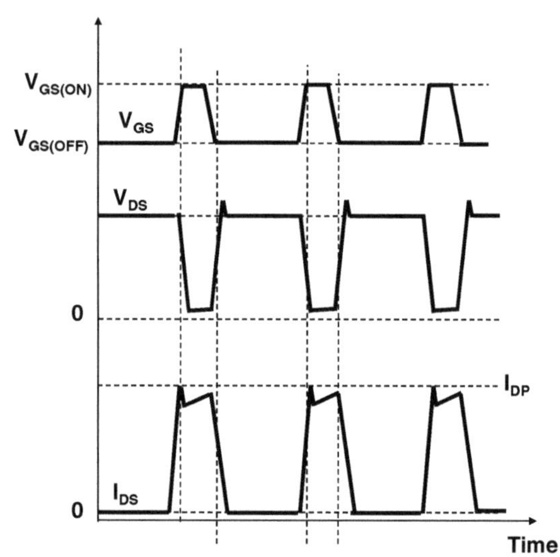

Fig. 5.9 Waveforms of V_{GS}, V_{DS} , and I_{DS} in continuous switching for inductive-load switching

Figure 5.10 shows the schematic SSOA for GaN power transistors. Here the lines (i)–(iii) are determined by the ON-state resistance, maximum current, and rated voltage, respectively. The line (iv) is sSSOA, which is determined by DPT. The line (v) is lSSOA determined by D-HTOL test, which is obtained by the acceleration factors of V_{DD}, I_{DP}, and T_J on lifetime.

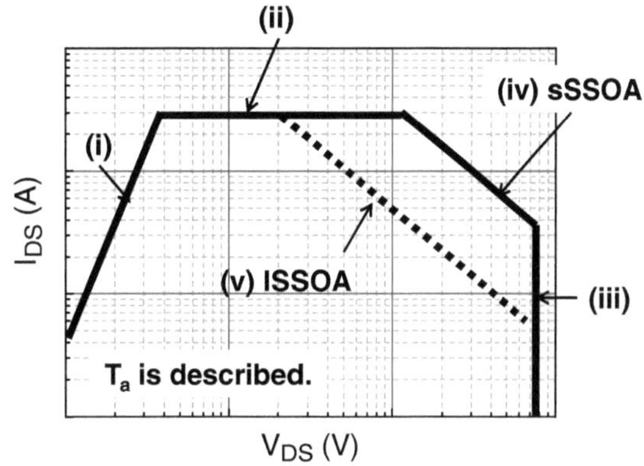

Fig. 5.10 Schematic switching safe operating area (SSOA) for GaN transistor. (i)–(v) are explained in the text. lSSOA and sSSOA are long-time and short-time switching safe operating area, respectively

5.2 Reliability Validation on Hybrid-Drain-Embedded Gate Injection Transistor (HD-GIT)

In this section, we present how we validate the robustness of our commercially available GaN power transistors. We suppose the method presented here can be applied to other GaN transistors at least with some modifications.

5.2.1 Device Structure

Figure 5.11a shows the schematic cross section of our normally OFF GaN-based transistor commercially available, which we call as hybrid-drain-embedded gate injection transistor (HD-GIT), hereafter [31, 33, 34]. The device structure of HD-GIT is based on a normally OFF gate-injection transistor (GIT) whose device structure is presented schematically in Fig. 5.11b [35]. A p-type GaN gate is formed over the AlGaN/GaN channel to serve normally OFF operation in both devices, and AlGaN/GaN heteroepitaxial structure is grown on the buffer layer over a cost-effective 6-inch Si substrate. A notable structure of HD-GIT structure is that p-GaN (pdrain) connected to the drain is formed over a thick AlGaN barrier layer to suppress the current collapse. Since the AlGaN barrier layer underneath the pdrain is thick enough, the electrons underneath the pdrain are not depleted, and thus the ON-state resistance is not increased. Holes are injected from the pdrain only when high voltage is applied to the pdrain, and the injected holes prevent the gate-to-drain region negatively charged. As a result, current collapse is not observed in an LR-load switching under the V_{DS} of 850 V which is far above the rated voltage (600 V) [31, 33]. In addition, since the hole injection from the pdrain alleviates the internal electric field [33], we find drastic improvement in the reliability of HD-GIT than GIT [36].

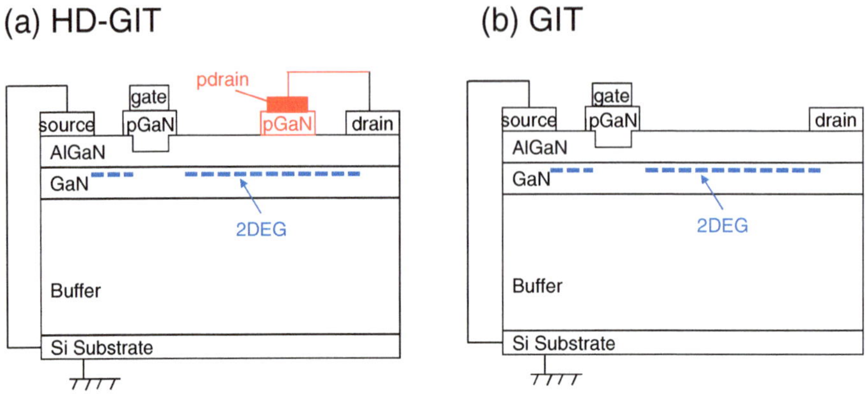

Fig. 5.11 Schematic cross section of (**a**) HD-GIT [31, 33, 34] and (**b**) GIT [35]

All devices employed in the reliability tests are packaged in an $8 \times 8\,\text{mm}^2$ surface-mount-type dual-flat no-leads (DFN) form wherein parasitic inductance is minimized so as to be utilized in high-frequency switching applications. The specific ON-state resistance, rated voltage, breakdown voltage, and threshold voltage of HD-GIT are 70 mΩ, 600, 1000, and 1.2 V, respectively.

5.2.2 *Fundamental Reliability Tests*

As a first step toward validating the robustness of HD-GIT, we perform fundamental reliability tests on HD-GIT conforming to the JEDEC standard utilized for validating the reliability of Si power transistors [36]. The tests focus primarily on the reliability for the DC characteristics of the device. Table 5.1 summarizes the test items, conditions, compliances, quantities, and results for HD-GIT. Here the alternating current (AC) gate bias test (item 11) is beyond JEDEC standard for Si power transistor, while others are in compliance with the standard.

1. High-Temperature Reverse Bias (HTRB): HD-GITs are maintained in the OFF state under an ambient temperature of T_a and a bias voltage of V_{DS}. The reliability under HTRB test is investigated with scrunity under various OFF state conditions. It is found that the OFF-state degradation of HD-GIT is explained by time-dependent dielectric breakdown (TDDB) mechanism [37] and crucially depends on the OFF-state voltage and temperature. It is also found that the lifetime under HTRB strongly depends heavily on the leakage current before the HTRB test, which means we can predict the HTRB lifetime with accuracy. The acceleration factor of V_{DS} and the activation energy on the lifetime are obtained to be 0.037 and 0.62 eV, respectively. Based on the

Table 5.1 Basic reliability test results for HD-GIT

Test	Conditions	Compliance	Quantity	Result
1. HTRB	$T_a = 150\,°C$, $V_{DS} = 480\,V$, 1000 h	JESD22-A108 [27]	45 pcs × 3 lots	PASS
2. H3TRB	$T_a = 85\,°C$, humidity = 85%, $V_{DS} = 480\,V$, 1000 h	JESD22-A101 [27]	45 pcs × 3 lots	PASS
3. DC HTGS (+)	$T_J = 150\,°C$, $V_{GS} = 4.0\,V$, 1000 h	JESD22-A108 [27]	45 pcs × 3 lots	PASS
4. DC HTGS (−)	$T_J = 150\,°C$, $V_{GS} = -12\,V$, 1000 h	JESD22-A108 [27]	45 pcs × 3 lots	PASS
5. HTS	$T_a = 150\,°C$	JESD22-A103 [27]	45 pcs × 3 lots	PASS
6. LTS	$T_a = -65\,°C$	JESD22-A119 [27]	45 pcs × 3 lots	PASS
7. TC	$T_a = -55/150\,°C$, Dwell time = 30 min each, 1000 cycles	JESD22-A104 [27]	45 pcs × 3 lots	PASS
8. IOL	$T_a = 25\,°C$, $\Delta T_J = 100\,°C$, 15000 cycles (1 cycle: ON / OFF = 2/2 min)	JESD22-A105 [27]	6 pcs × 1 lot	PASS
9. ESD, HBM	$T_a = 25\,°C$, ±2 kV	JS-001 [2]	3 pcs × 1 lot	PASS
10. ESD, CDM	$T_a = 25\,°C$, ±500 V	JS-002 [2]	3 pcs × 1 lot	PASS
11. AC, HTGS	Continuous pulses (7.7/ − 7.2 V) are applied to V_{GS}.	Beyond JEDEC	5 pcs × 1 lot	PASS

obtained acceleration factors, we estimate the lifetime of 0.1% failure under $V_{DS} = 480\,V$ (80% derated) at 80 °C is longer than 1000 years, which is long enough for most conventional power converter applications [36].

2. High-Humidity High-Temperature Reverse Bias (H3TRB): Reverse bias is applied to HD-GIT under humidity condition.
3. High-Temperature DC Gate Bias, Positive [DC HTGS (pos.)]: A DC positive voltage V_{GS} is applied to the gate at a high temperature with the estimated junction temperature T_J of 150 °C.
4. High-Temperature DC Gate Bias, Negative [DC HTGS (neg.)]: A DC negative voltage V_{GS} is applied to the gate at a high temperature.
5. High-Temperature Storage (HTS): HD-GITs are stored at a high temperature.
6. Low-Temperature Storage (LTS): HD-GITs are stored at a low temperature.
7. Temperature Cycling (TC): The temperature of HD-GITs is switched periodically between a low and a high temperature.
8. Intermittent Operational Life (IOL): DC power is switched ON and OFF periodically to HD-GITs so that T_J is increased by 100 °C during the turn ON.
9. Electrostatic discharge with human body model (ESD, HBM): ESD test is performed based on human body model.

Table 5.2 Pass criteria for the reliability tests

Test items	Conditions for pass
$I_{GSS}@V_{GS} = -10\,V$	<1 mA
$V_{GSF}@I_{GS} = 100\,\mu A/mm$	2.5~4.5 V
$V_{TH}@I_{DS} = 10\,\mu A/mm$	0.7~1.6 V
Rdson@$I_{GS} = 100\,\mu A/mm$	<100 mΩ
$I_{DSS\_D}@V_{DS} = 600\,V$	≤10 µA
$I_{DSS\_G}@V_{DS} = 600\,V$	≤10 µA

10. Electrostatic discharge with charge device model (ESD, CDM): ESD test is performed based on charge device model.
11. AC Gate Bias [AC HTGS]: Continuous pulses with the maximum and the minimum peak voltages of 7.7 and −7.2 V are applied to the gate for 1000 h. The switching frequency is 100 kHz, and the on time is 2 µs. Here the estimated T_J is 150 °C.

After we conduct the above tests, we check the DC characteristics of HD-GITs. The pass/failure criteria are summarized in Table 5.2. We confirm that the HD-GITs pass under all the above reliability tests.

5.2.3 Short-Time Switching Safe Operating Area (sSSOA)

Let us move on the reliability of HD-GIT under switching operation. The black line in Fig. 5.12 shows the sSSOA for HD-GITs, which is determined by DPT. Here the DPT is performed on purpose above the rated voltage of the devices (600 V) to investigate the switching capability limit of the devices. Since the HD-GIT is free from current collapse, the sSSOA forms the rectangular area limited by the rated voltage and the maximum current. On the other hand, the sSSOA for GIT shown as the red lines is much smaller than that for HD-GIT, because current collapse occurs in the semi-ON state. The striking improvement of HD-GIT in the sSSOA area is originated from the hole injection from the pdrain [33], which is discussed briefly in Sect. 5.3.

5.2.4 Dynamic High-Temperature Operation Life (D-HTOL) Test

We conduct D-HTOL test on HD-GIT where we employ the electric circuit shown as Fig. 5.6. Here we employ an inductive with a resistive load with a SiC diode (D1) in parallel and set the switching frequency to be 50 kHz with a duty ratio of 3% [42].

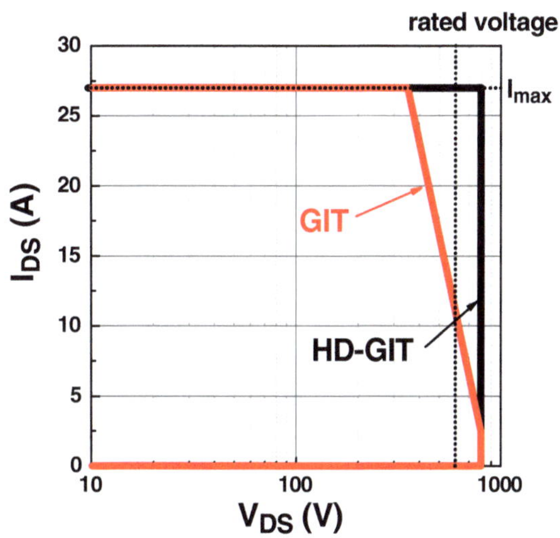

Fig. 5.12 Short-time switching safe operating area (sSSOA) of HD-GIT and GIT determined by double-pulse test

We find that the long-time switching lifetime L_{sw} strongly depends on I_{DP} and the input voltage V_{DD}, while it relies weakly on the temperature. Their acceleration factors on the lifetime L_{sw} under long period of time are discussed here.

Firstly, we carry out continuous switching under $V_{DD} = 640$ V with $I_{DP} = 27$ A at the ambient temperature T_a of 50 and 95 °C. Figure 5.14a is the obtained Weibull plots. The obtained m factor is 1.7, being larger than 1, which indicates that the degradation mode is wear-out. The inset in Fig. 5.14a is the Arrhenius plot for the median time to failure (MTTF), from which the activation energy is determined to be 0.1 eV. The obtained activation energy is very small, which suggests L_{sw} is weakly dependent on T_J. This is presumably because hot-electron-trapping effect in GaN transistor is less prominent at high temperature because hot-electron energy at high temperature is generally reduced due to activated phonon scattering [38].

Secondly, we investigate the acceleration factor of V_{DD} on L_{sw} with keeping I_{DP} the same, in which we resolder appropriate loads in the electric circuit of Fig. 5.6 to keep the I_{DP} constant. Figure 5.13a shows the I_{DS} -V_{DS} loci during the switching under several V_{DD}. Figure 5.14b presents the Weibull plots for L_{sw} under several V_{DD} with a constant I_{DP} of 27 A. L_{sw} is shortened with the increase in V_{DD}. The inset in Fig. 5.14b shows MTTF as a function of V_{DD}, from which the voltage acceleration factor β_v is obtained to be 0.039. Although T_J is supposed to be higher for larger V_{DD}, since L_{sw} is only weakly dependent on T_J in HD-GIT, L_{sw} seems to depend only on V_{DD}.

Thirdly, we investigate the acceleration factor of I_{DP} on L_{sw}. Figure 5.13b shows the I_{DS} -V_{DS} loci during the switching event with the variation of I_{DP}. Here, applying constant voltage of 640 V to V_{DD}, we perform D-HTOL test with varying I_{DP}. Figure 5.14c shows the Weibull plots for the several I_{DP}. L_{sw} is shortened as I_{DP} increases. The inset in Fig. 5.14a shows the MTTF as a function of I_{DP}, from which the voltage acceleration factor β_c is obtained to be 0.47. Based on the

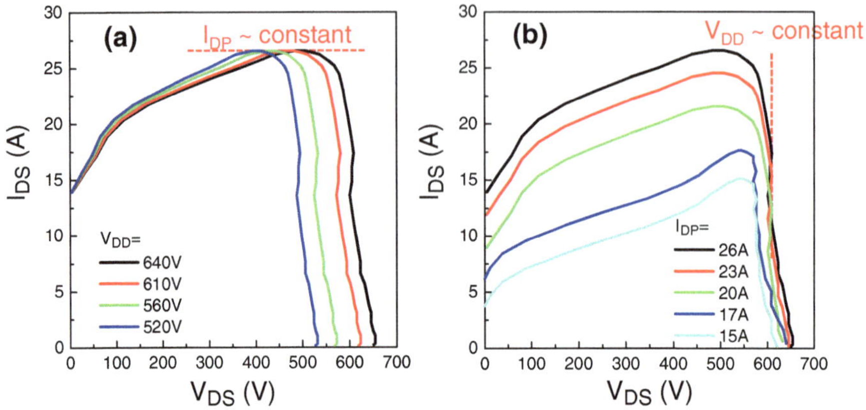

Fig. 5.13 I_{DS} -V_{DS} loci under switching reliability tests, in which (**a**) stress input voltage V_{DD} is varied with peak-switching current constant, and (**b**) peak-switching I_{DP} current is varied with input stress voltage constant [42]

acceleration factors obtained here, we estimate L_{sw} for the LR-load switching under a condition of $V_{\mathrm{DD}} = 400\,\mathrm{V}$, $I_{\mathrm{DP}} = 27\,\mathrm{A}$, and $T_{\mathrm{J}} = 95\,°\mathrm{C}$ to be 6000 h.

Furthermore, we roughly estimate the switching lifetime for a 3 kW totem-pole power factor correction (PFC) circuit to be roughly 24 years, which is obtained based on the assumption that each switching pulses shortens the lifetime by the amount estimated by the instantaneous I_{DP} and V_{DD} values combined with the obtained acceleration factors. We suppose the expected lifetime is long enough for most of the PFC applications.

In fact, the application-level reliability on GaN transistors is still controversial, and more comprehensive research is needed because there are only limited number of lifetime data when employed in some real applications so far [6, 39, 40]. However, since the lifetime estimated in this report is sufficiently long, we believe we can step forward to employ GaN transistors in power converter applications. We expect further reliability investigation is performed in the application level in the foreseeable future.

5.2.5 Long-Time Switching Safe Operating Area (lSSOA)

Based on the obtained acceleration factors for the D-HTOL test, let us define the lSSOA for HD-GIT in this section [41, 42]. If we assume that the acceleration of I_{DP} and V_{DD} are independent, L_{sw} can be described as

$$L_{\mathrm{sw}} = A \cdot \exp\{-(\beta_{\mathrm{v}} V_{\mathrm{DD}} + \beta_{\mathrm{c}} I_{\mathrm{DP}})\}, \tag{5.1}$$

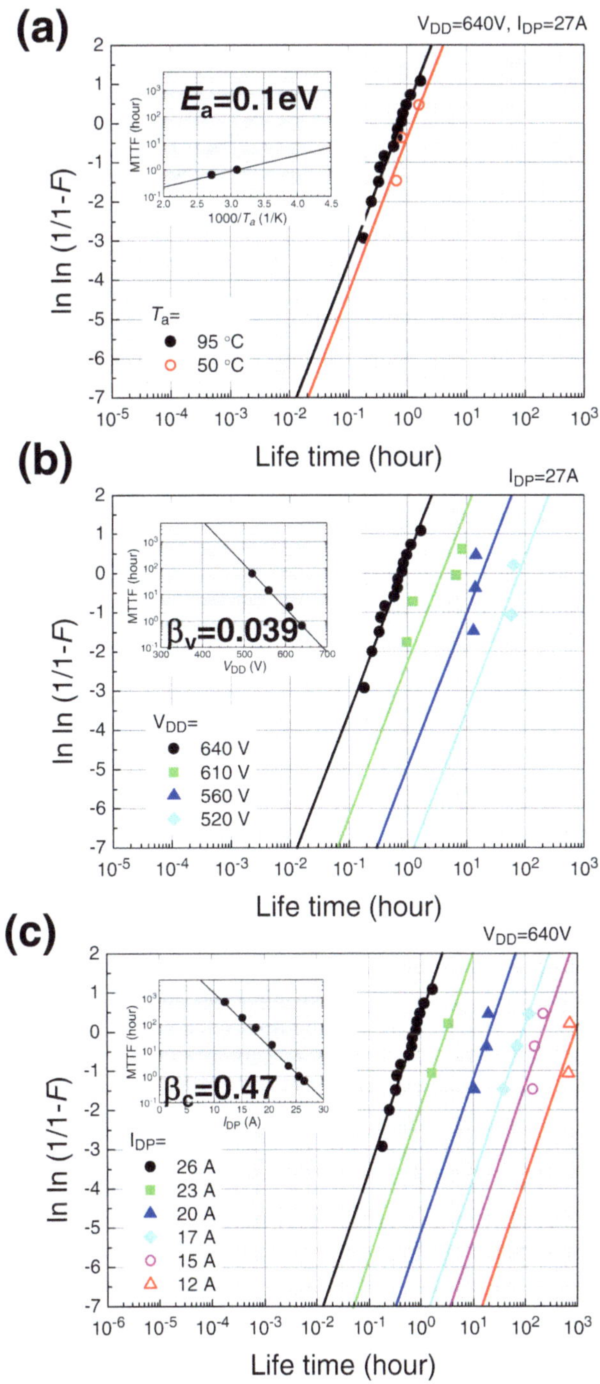

Fig. 5.14 Weibull plots with the variation of (**a**) ambient temperature T_a, (**b**) input voltage V_{DD}, and (**c**) peak-switching current I_{DP}. The insets in each figures show corresponding acceleration factors [42]

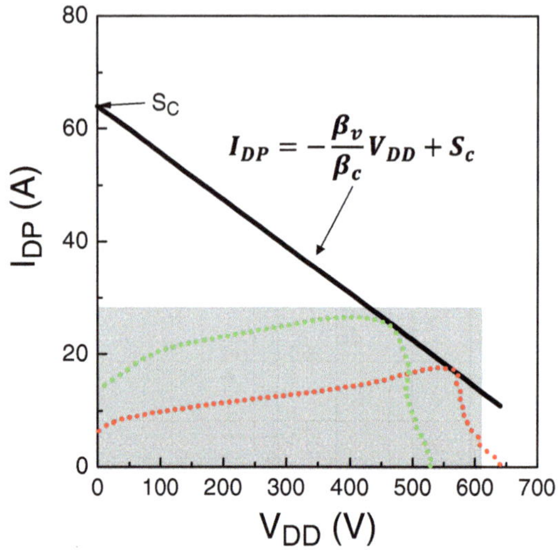

Fig. 5.15 The switching conditions which intersect with the solid line have the same switching lifetime. For example, the dotted green and red switching loci are expected to have the same switching lifetime. Shaded area is the area limited by the rated voltage and the maximum current of the device under test [42]

where A is a constant. Here the temperature acceleration is neglected for the sake of simplicity because its dependence is small. This equation means that the set of (I_{DP}, V_{DD}) that gives the same value of ($\beta_v V_{DD} + \beta_c I_{DP}$) have the same L_{sw}. Henceforth, the line

$$I_{DP} = -(\beta_v/\beta_c)V_{DD} + S_c \tag{5.2}$$

have the same L_{sw} for a given S_c.

The solid line in Fig. 5.15 describes the Eq. (5.2) with S_c of 64 A. The green and red dotted lines in Fig. 5.15 show the switching loci of I_{DS} -V_{DS} under the conditions of $V_{DD} = 530$ V with $I_{DP} = 26$ A and $V_{DD} = 640$ V with $I_{DP} = 17$ A, respectively. Since the solid line presents the (I_{DP}, V_{DD}) giving the same L_{sw} as shown in Fig. 5.15, the minimum S_c with which the line of Eq. (5.2) intersects at one point with the switching locus determines the corresponding L_{sw}. Thus L_{sw} under the loci of the green line and the red line is speculated to be the same.

Based on this speculation, we plot the measured MTTF as a function of S_c, which is shown in Fig. 5.16. The MTTF is shorter for the larger S_c. The solid line is the linear fit for the experimental data, where we find the clear relation between MTTF and S_c. From this relation, we extract the S_c values for 1-, 5-, and 10-year switching operation. By drawing the line with a slope of $-\beta_v/\beta_c$ from (0, S_c), we extract the lSSOA for 1-, 5-, and 10-year switching operation, which are shown in Fig. 5.17. Here the obtained area is cut by the rated voltage and the maximum current of the device. To our knowledge, this is the first demonstration of the lSSOA of GaN power transistors, which is beneficial to the power converter designers planning to use GaN power transistors.

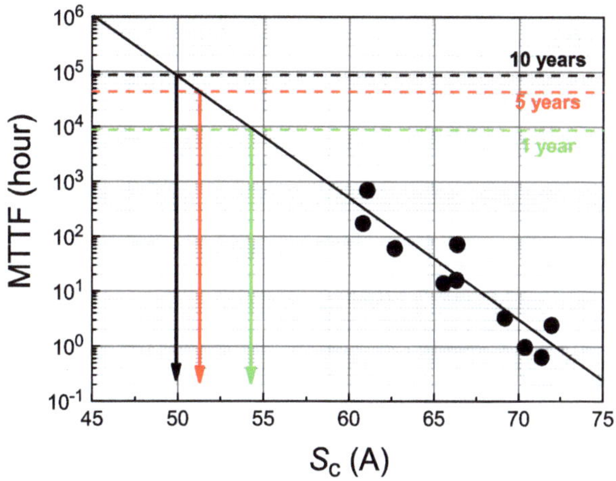

Fig. 5.16 MTTF as a function of S_c [42]. Here S_c is explained in the text

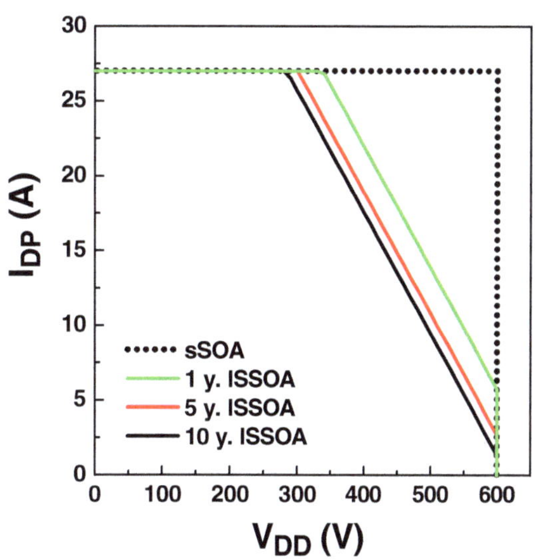

Fig. 5.17 Long-time switching SOA (lSSOA) for HD-GIT under 1-, 5-, and 10-year operation. The dotted line indicates short-time switching safety operating area (sSSOA)

5.2.6 *Short-Circuit Capability (SCC) Test*

Short-circuit events can happen in power converter applications. It is required that the device should withstand for several μs before the protection circuit starts working, and there have been some reports on the SCC test on GaN transistors [43–45]. SCC test on HD-GIT is carried out employing the electric circuit as shown in Fig. 5.18a. For comparison, SCC test on GIT is also performed. In each cycle, the device is maintained in the OFF state of $V_{DS} = 400$ V with $V_{GS} = 0$ V for several seconds before it is switched to ON state for 4 μs, which is again followed by a transition to the OFF state. Between each cycles, we wait for 3 min to measure

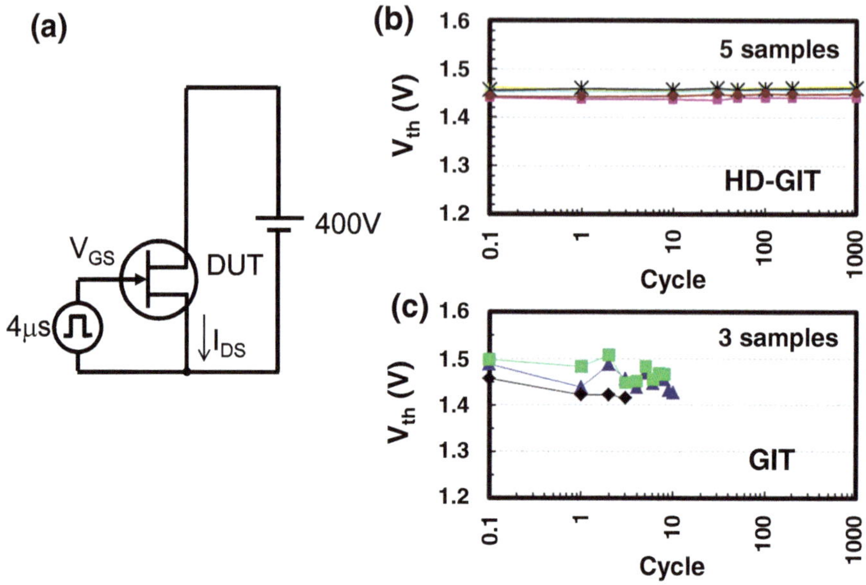

Fig. 5.18 (**a**) Electric circuit for short-circuit capability test. DUT: device under test. The variation of V_{th} as a function of cycle numbers N for (**b**) HD-GIT and (**c**) GIT under short-circuit capability test. The data on the left ends are the V_{th} value before the test [36]

the threshold voltage V_{th} at room temperature to check whether the device is not seriously degraded. The cycle is performed 1000 times.

Figure 5.18b shows the behavior of V_{th} as a function of the cycle number N obtained for HD-GIT. The V_{th} variation is negligible, indicating that HD-GIT is stable under the SCC test. We confirm that the other DC characteristics are also within their normal values.

On the other hand, Fig. 5.18c shows the behavior of V_{th} as a function of N obtained for GIT. GIT is broken before N reaches 10. Furthermore, the variation of V_{th} for GIT is much larger than that for HD-GIT. The fact that HD-GIT endures the SCC test while GIT does not suggests that the heat effect is not the main reason for the breakdown of GIT in the SCC test, and the main reason is considered to be current collapse. The drastic improvement of HD-GIT than GIT for SCC test is due to the unique device structure of HD-GIT, where hole injection from pdrain suppresses current collapse effectively [36].

5.3 Physical Mechanism for High Reliability of HD-GIT

In this chapter, we briefly present why HD-GIT exhibits prominent robustness. We believe that the high robustness of HD-GIT is originated from the hole injection

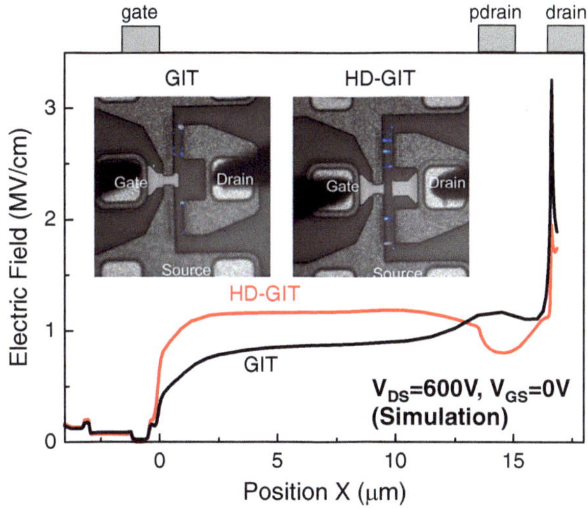

Fig. 5.19 Simulated absolute electric-field distribution at the two-dimensional electron gas (2DEG) channel. Inset shows the experimental electroluminescence images of GIT and HD-GIT taken for the OFF condition ($V_{DS} = 600$ V and $V_{GS} = 0$ V) [33]

from the pdrain, which suppresses the current collapse and thus alleviates the internal electric field not only in the OFF state but also in the semi-ON state based on the experimental and simulation study.

In the OFF state, the hole injection from the pdrain prevents the negative charge trapping in the structure (see Fig. 5.1a). Figure 5.19 shows the simulated absolute electric-field distribution at the two-dimensional electron gas (2DEG) channel of HD-GIT and GIT. The maximum electric field in HD-GIT is reduced compared with that in GIT by the hole injection from the pdrain. The insets in Fig. 5.19 show the experimental EL images of GIT and HD-GIT obtained for the OFF state ($V_{DS} = 600$ V and $V_{GS} = 0$ V), respectively. The electroluminescence (EL) position is located mainly at the drain side in GIT, while it is uniformly distributed between the gate-pdrain region in HD-GIT, which supports the simulation in the OFF state [33]. The reduction in the maximum electric field in the HD-GIT at the drain-side edge helps improve the reliability under HTRB test [36].

As well as in the OFF state, the hole injection from the pdrain in HD-GIT is quite effective to suppress hot-electron trapping in the semi-ON state (see Fig. 5.1). Figure 5.20 summarizes the EL images of the semi-ON states at room temperature taken for GIT and HD-GIT at several V_{GS} under $V_{DS} = 150$ V conditions. In GIT, the EL signal is located at the gate side when $V_{GS} = 2.0$ V; however, it shifts to the drain side as V_{GS} increases to 3.6 V. At the intermediate $V_{GS} = 3.0$ V, the EL signal is observed both at the gate and the drain sides. The result implies that the hot-electron trapping is much induced as the V_{GS} and thus I_{DS} increases, which leads to the increase in the trapped electron density near 2DEG channel. As a result, the maximum electric field is shifted from the gate to the drain side as V_{GS} increases in GIT [17].

In HD-GIT, on the other hand, the ELs remain at the gate side at any V_{GS} values, which indicates that the hot-electron trapping is suppressed effectively in HD-GIT.

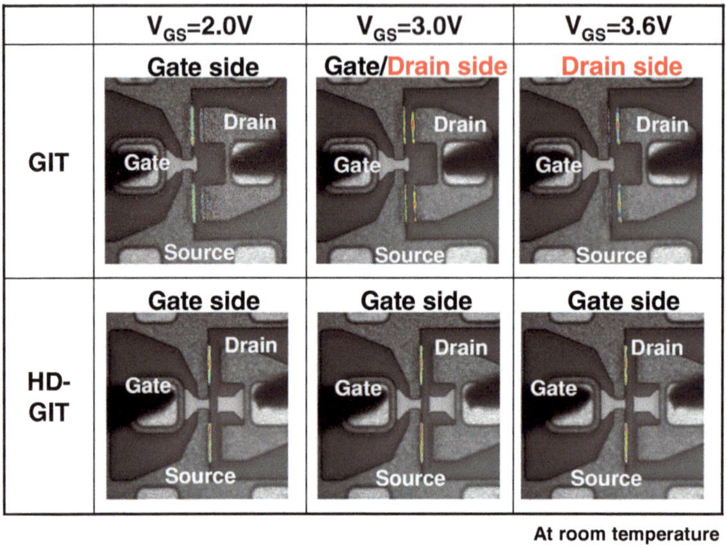

Fig. 5.20 Electroluminescence of GIT and HD-GIT taken at several V_{GS} under $V_{DS} = 150$ V. The electroluminescence signal positions are indicated [33]

In HD-GIT, the hole injection from the pdrain prevents the gate-pdrain access region from being negatively charged due to the hot electrons. The results indicate that the current collapse is improved in the semi-ON state, and thus HD-GIT maintain exhibits higher reliability than GIT [33, 36].

5.4 Summary

We review how we validate the robustness of our hybrid-drain-embedded gate injection transistor (HD-GIT) commercially available. In addition to the standardized reliability test for Si power transistor, we perform double-pulse test (DPT) and dynamic high-temperature operation life (D-HTOL) test to define the short-time switching safe operating area (sSSOA) and the long-time switching safe operating area (lSSOA). We believe that the method described here can be applied to other GaN power transistors. We expect that the GaN transistors whose robustness is guaranteed are employed more widely to realize more efficient and compact power-switching applications.

Acknowledgements The authors would like to gratefully acknowledge the members of Panasonic Corporation for their technical advices and helps throughout the work. Our sincere thanks also go to Dr. Daisuke Ueda, Kyoto Institute of Technology, for his technical advices.

References

1. JEDEC standard, JESD22 (2016)
2. ANSI/ESDA/JEDEC joint standard, JS-001/002
3. T. Kikkawa, T. Hosoda, K. Imanishi, K. Shono, K. Itabashi, T. Ogino, Y. Miyazaki, A. Mochizuki, K. Kiuchi, M. Kanamura, M. Kamiyama, S. Akiyama, S. Kawasaki, T. Maeda, Y. Asai, Y. Wu, K. Smith, J. Gritters, P. Smith, S. Chowdhury, D. Dunn, M. Aguilera, B. Swenson, R. Birkhahn, L. McCarthy, L. Shen, J. McKay, H. Clement, J. Honea, S. Yea, D. Thor, R. Lal, U. Mishra, P. Parikh, in *IEEE International Electron Devices Meeting (IEDM)*, San Francisco, Dec 2014, 2.6
4. Y.-F. Wu, J. Gritters, L. Shen, R.P. Smith, J. McKay, R. Barr, R. Birkhahn, in *IEEE Workshop on Wide Bandgap Power Devices and Applications (WiPDA)*, Columbus, Oct 2013, 6
5. T. Kikkawa, T. Hosoda, K. Shono, K. Imanishi, Y. Asai, Y. Wu, L. Shen, K. Smith, D. Dunn, S. Chowdhury, P. Smith, J. Gritters, L. McCarthy, R. Barr, R. Lal, U. Mishra, P. Parikh, in *IEEE International Reliability Physics Symposium (IRPS)*, Monterey, Apr 2015, 6C-1
6. B.K. Rhea, L.L. Jenkins, F.T. Werner, W.E. Abell, R.N. Dean, in *IEEE Workshop on Wide Bandgap Power Devices and Applications (WiPDA)*, Blacksburg, Oct 2015, 206
7. P. Moens, A. Banerjee, P. Coppens, A. Constant, P. Vanmeerbeek, Z. Li, F. Declercq, L. de Schepper, H. de Vleeschouwer, C. Liu, B. Padmananbhan, W. Jeon, J. Guo, A. Salib, M. Tack, in *European Solid State Device Research Conference (ESSDERC)*, Graz, Sept 2015, 64
8. EPC (2015) [Online]. http://epc-co.com/epc/DesignSupport/eGaNFETReliability.aspx
9. I. Rosetto, M. Meneghini, O. Hilt, E. Bahat-Treidel, C. de Santi, S. Dalcanale, J. Wuerfl, E. Zanoni, G. Meneghesso, IEEE Trans. Electron Dev. **63**, 2334 (2016)
10. S.C. Binari, K. Ikossi, J.A. Roussos, W. Kruppa, D. Park, H.B. Dietrich, D.D. Koleske, A.E. Wickenden, R.L. Henry, IEEE Trans. Electron Dev. **48**, 465 (2001)
11. A.F.M. Anwar, S.S. Islam, R.T. Webster, Appl. Phys. Lett. **84**, 1970 (2004)
12. H. Hasegawa, T. Inagaki, S. Ootomo, T. Hashizume, J. Vac. Sci. Technol. **21**, 1844 (2003)
13. D. Bisi, M. Meneghini, C. de Santi, A. Chini, M. Dammann, P. Brückner, M. Mikulla, G. Meneghesso, E. Zanoni, IEEE Trans. Electron Dev. **60**, 3166 (2013)
14. P.B. Klein, S.C. Binari, K. Ikossi, A.E. Wickenden, D.D. Koleske, R.L. Henry, Appl. Phys. Lett. **79**, 3527 (2001)
15. K. Tanaka, H. Umeda, H. Ishida, M. Ishida, T. Ueda, Jpn. J. Appl. Phys. **55**, 054101 (2016)
16. M. Meneghini, A. Stocco, R. Silvestri, G. Meneghesso, E. Zanoni, Appl. Phys. Lett. **100**, 233508 (2012)
17. M. Meneghini, C. de Santi, T. Ueda, T. Tanaka, D. Ueda, E. Zanoni, G. Meneghesso, IEEE Electron Dev. Lett. **33**, 375 (2012)
18. Y. Puzyrev, S. Mukherjee, J. Chen, T. Roy, M. Silvestri, R.D. Schrimpf, D.M. Fleetwood, J. Singh, J.M. Hickley, A. Paccagnella, S.T. Pantelides, IEEE Trans. Electron Dev. **61**, 1316 (2014)
19. M. Meneghini, I. Rossetto, F. Hurkx, J. Sonsky, J.A. Croon, G. Meneghesso, E. Zanoni, IEEE Trans. Electron Dev. **62**, 2549 (2015)
20. T. Brazzini, M.A. Casbon, H. Sun, M.J. Uren, J. Lees, P.J. Tasker, H. Jung, H. Blanck, M. Kuball, Appl. Phys. Lett. **106**, 213502 (2015)
21. S.R. Bahl, J. Joh, L. Fu, A. Sasikumar, T. Chatterjee, S. Pendharkar, in *IEEE International Electron Devices Meeting (IEDM)*, San Francisco, Dec 2016, 20.5
22. J. Joh, N. Tipirneni, S. Pendharkar, S. Krishnan, in *IEEE International Reliability Physics Symposium (IRPS)*, Apr 2014, 6C.5
23. S.R. Bahl, D. Ruiz, D.S. Lee, in *IEEE International Reliability Physics Symposium (IRPS)*, Pasadena, Apr 2016, 4A.3
24. M. Meneghini, A. Stocco, R. Silvestri, N. Ronchi, G. Meneghesso, E. Zanoni, in *(IRPS)*, Anaheim, Apr 2012, 2C.2
25. G. Meneghesso, M. Meneghini, A. Stocco, D. Bisi, C. de Santi, I. Rossetto, A. Zanandrea, F. Rampazzo, E. Zanoni, Microelectron. Eng. **109**, 257 (2013)

26. I. Hwang, J. Kim, S. Chong, H.-Sik Choi, S.-Kyu Hwang, J. Oh, J.K. Shin, U.-In Chung, IEEE Electron Dev. Lett. **34**, 1494 (2013)
27. JEDEC standard, JC-70 Wide Bandgap Power Electronic Conversion Semiconductors.
28. [On line:] https://e2e.ti.com/blogs_/b/powerhouse/archive/2015/05/02/understanding-mosfet-data-sheets-part-2-safe-operating-area-soa-graph
29. J. Evarts, P. Jacqmaer, R. Gelagaev, J. Das, M. Germain, J. Van den Keybus, J. Driesen, in *Proceedings of Power Conversion and Intelligent Motion (PCIM) Europe*, Nuremberg, May 2010, 309
30. B. Lu, T. Palacios, D. Risbud, S. Bahl, D.I. Anderson, in *Compound Semiconductor Integrated Circuit Symposium (CSICS)*, Waikoloa, 2011, 1
31. S. Kaneko, M. Kuroda, M. Yanagihara, A. Ikoshi, H. Okita, T. Morita, K. Tanaka, M. Hikita, Y. Uemoto, S. Takahashi, T. Ueda, in *International Symposium on Power Semiconductor Devices and ICs (ISPSD)*, Kowloon, May 2015, 41
32. Z. Liu, B. Li, F.C. Lee, Q. Li, IEEE Trans. Ind. Electron. **64**, 9114 (2017)
33. K. Tanaka, T. Morita, H. Umeda, S. Kaneko, M. Kuroda, A. Ikoshi, H. Yamagiwa, H. Okita, M. Hikita, M. Yanagihara, Y. Uemoto, S. Takahashi, H. Ueno, H. Ishida, M. Ishida, T. Ueda, Appl. Phys. Lett. **107**, 163502 (2015)
34. H. Okita, M. Hikita, A. Nishio, T. Sato, K. Matsunaga, H. Matsuo, M. Mannoh, Y. Uemoto, in *International Symposium on Power Semiconductor Devices and ICs (ISPSD)*, Prague, May 2016, 23
35. Y. Uemoto, M. Hikita, H. Ueno, H. Matsuo, H. Ishida, M. Yanagihara, T. Ueda. T. Tanaka, D. Ueda, IEEE Trans. Electron Dev. **54**, 3393 (2007)
36. K. Tanaka, T. Morita, M. Ishida, T. Hatsuda, T. Ueda, K. Yokoyama, A. Ikoshi, M. Hikita, M. Toki, M. Yanagihara, Y. Uemoto, in *IEEE International Reliability Physics Symposium (IRPS)*, Monterey, Apr 2017, 4B-2
37. R. Degraeve, B. Kaczer, G. Groeseneken, Microelectron. Reliab. **39**, 1445 (1999)
38. S. Dalcanale, M. Meneghini, A. Tajalli, I. Rossetto, M. Ruzzarin, E. Zanoni, G. Meneghesso, in *IEEE International Reliability Physics Symposium (IRPS)*, Monterey, Apr 2017, 4B.1
39. D.C. Sheridan, D.Y. Lee, A. Ritenour, V. Bondarenko, J. Yang, C. Coleman, in *Proceedings of Power Conversion and Intelligent Motion (PCIM)*, 2014
40. Y.-F. Wu, J. Guerrero, J. McKay, K. Smith, in *IEEE Workshop on Wide Bandgap Power Devices and Applications (WiPDA)*, Knoxville, 30, 2014
41. A. Ikoshi, K. Tanaka, M. Toki, H. Yamagiwa, K. Suzuki, D. Arisawa, M. Hikita, M. Yanagihara, Y. Uemoto, T. Ueda, in *Applied Power Electronics Conference (APEC)*, San Antonio, Mar. 2018, IS16-02
42. A. Ikoshi, M. Toki, H. Yamagiwa, D. Arisawa, M. Hikita, K. Suzuki, M. Yanagihara, Y. Uemoto, K. Tanaka, T. Ueda, in *IEEE International Reliability Physics Symposium (IRPS)*, San Francisco, Mar. 2018, 4E.2
43. D. Veereddy, T. McDonald, J. Ambrus, A. Diy, S. Cardwell, B. Pandya, R. Garg, M. Imam, in *IEEE International Reliability Physics Symposium (IRPS)*, Monterey, Apr 2017, 162
44. T. Nagahisa, H. Ichijoh, T. Suzuki, A. Yudin, A.O. Adan, M. Kubo, Jpn. J. Appl. Phys. **55**, 04EG01 (2016)
45. X. Huang, D.Y. Lee, V. Bondarenko, A. Baker, D.C. Sheridan, A.Q. Huang, B.J. Baliga, in *International Symposium on Power Semiconductor Devices & IC's (ISPSD)*, Waikoloa, June 2014, 273

Chapter 6
Impact of Parasitics on GaN-Based Power Conversion

Johan T. Strydom

6.1 GaN Device-Level Parasitics

GaN power devices today are high-electron-mobility transistors (HEMTs). These are field-effect transistors (FETs) and, as such, behave in a similar manner to silicon MOSFETs. Due to the similarities between GaN FETs and silicon MOSFETs, the nature of their device-level parasitics is not new, and much of the device improvements have been made to address the same parasitics, and are based on the cumulative knowledge developed for MOSFETs over the last thirty-plus years. What is different, though, is the relative speed of these GaN devices, and this in turn brings additional challenges. For calibration, consider the relative gate charge profiles of some 100 V GaN FETs and silicon MOSFETs shown in Fig. 6.1. It shows that GaN devices have at least five times lower gate charge than silicon MOSFETs.

To best understand the device-level parasitics, we can start with MOSFET figures of merit (FOMs). The idea of a MOSFET FOM was created to enable comparison of competing device technologies in a manner that would be independent of die size [1–9]. These are expressed as a product between the device on-resistance ($R_{DS(on)}$) and a total charge of a specific parasitic capacitance and are intended to be proportional to the in situ device losses, where the charge component in question is considered the dominant parasitic loss component, during operation. The first of these FOMs considered only total gate charge Q_G [3], while additional FOMs specific to hard-switching [5, 6] targeted the gate to drain charge Q_{GD}. FOMs for soft-switching applications have also been proposed [4, 7] in an attempt to better approximate expected system performance in this application.

J. T. Strydom (✉)
Texas Instruments, Santa Clara, CA, USA
e-mail: j-strydom@ti.com; jstrydom@ieee.org

G. Meneghesso et al. (eds.), *Gallium Nitride-enabled High Frequency and High Efficiency Power Conversion*, Integrated Circuits and Systems,
https://doi.org/10.1007/978-3-319-77994-2_6

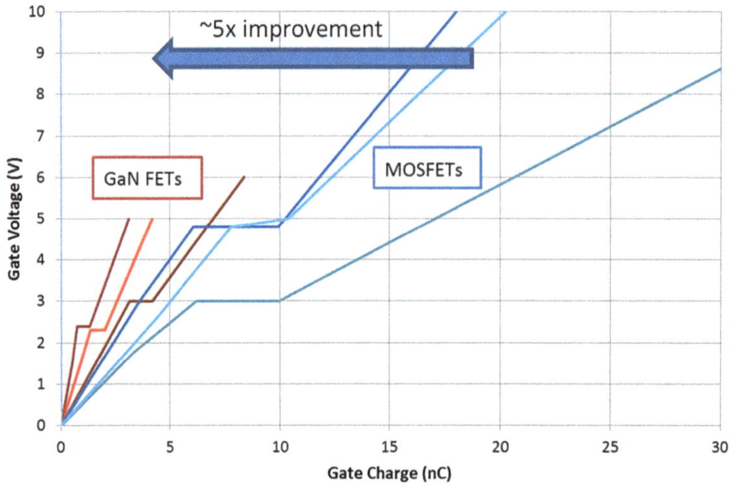

Fig. 6.1 Gate charge profile for 100 V GaN FETs and silicon MOSFETs normalized at 10mΩ

To better understand the importance of internal device parasitics, it is useful to look at the evolution of the MOSFET over time. It is possible to still get an estimation of the original silicon MOSFET FOMs by considering current-generation radiation-tolerant MOSFETs – these MOSFETs have been virtually unchanged since the early 1980s. Comparing these with current state-of-the-art silicon MOSFETs [8] in Fig. 6.2a, a 20x reduction in the original $R_{DS(on)} \cdot Q_G$ FOM can be seen, while the hard-switching related $R_{DS(on)} \cdot Q_{GD}$ FOM shows an even more impressive 40x improvement (Fig. 6.2b) over the same time. Since a reduction in Q_{GD} is more important relative to Q_G for hard-switching applications (the most common use), this parasitic reduction has been disproportionality targeted. For comparison in Fig. 6.2, the current and first generations of GaN FETs are also shown, and it is clear that a similar disproportional improvement in Q_{GD} is being made with conduction FOM having improved 2x–2.5x and the switching FOM having improved by 3x–4x.

6.1.1 Miller Charge Ratio

The system advantage of such an aggressive reduction in Q_{GD} is clear when considering its impact in the traditionally dominant overlap losses during hard-switching [10, 11]. However, an equally important reason for Q_{GD} reduction is to improve the Miller charge ratio (CR) [12] of the MOSFET to minimize unwanted dv/dt-induced turn-on. In short, a device can be made immune to dv/dt-induced turn-on if the Miller CR (defined in Eq. 6.1) of the device at its DC working voltage (V_{DS}) is less than unity.

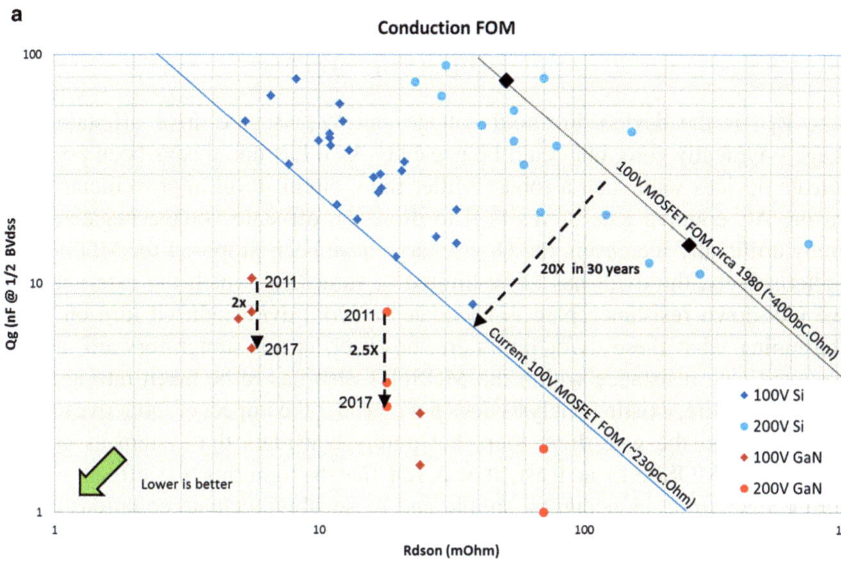

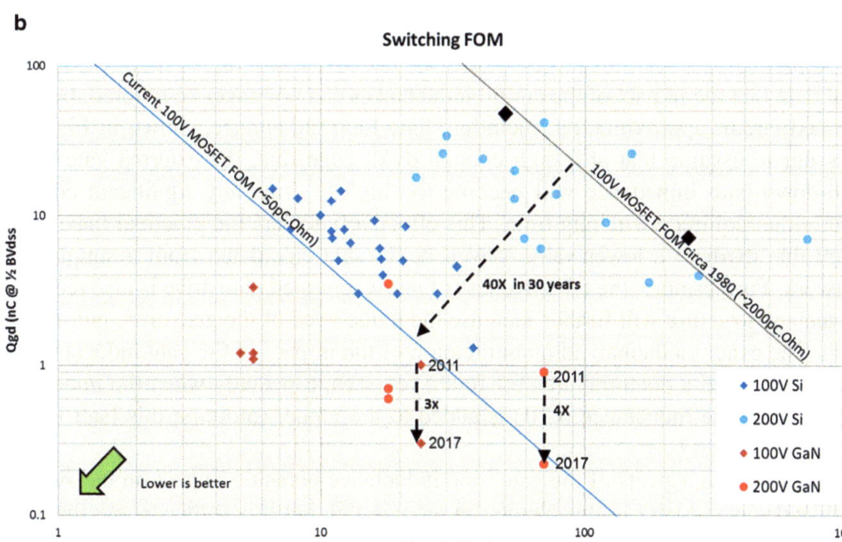

Fig. 6.2 Improvements over time for 100 V and 200 V MOSFETS and GaN FETs for different FOMs. (**a**) Conduction FOM. (**b**) Hard-switching FOM

$$\text{Miller CR} = \frac{Q_{\text{GD}}\,(V_{\text{DS}} - V_{\text{TH}})}{Q_{\text{GS1}}\,(V_{\text{TH}} - V_{\text{GL}})} < 1 \qquad (6.1)$$

where V_{TH} is the device threshold voltage and V_{GL} is gate drive off-state gate voltage – typically zero, but could be negative. As it has not always been possible to create devices with such a robust Miller ratio, circuit-level improvements such as using AC coupled gate drives [13] to drive the off-state voltage negative and thereby artificially increasing the Q_{GS1} charge have been proposed for MOSFETs. Or alternatively, the dv/dt has to be limited to values for which the external gate drive pull-down resistance path is adequate to avoid dv/dt-induced turn-on [14]. Determining what these dv/dt limits are, however, is not straightforward, as the distributed gate resistance within the MOSFET also has to be taken into account [15]. Furthermore, existing analysis does not consider the impact of inductive circuit elements within the gate drive loop, but purely considers the system as an RC network. For MOSFETs at least, it was felt that by ignoring the effects of the MOSFET electrode inductances, simple and reasonably accurate equations could be derived. This simplification was also due to the fact that inclusion of the parasitic inductances made the problem so complicated that a simple equation was not possible [14]. But the "reasonableness" of these simplifications comes into question as the device switching speed increases, as in the case with GaN.

Applying these learnings to GaN devices that have much higher dv/dt capability, it is clear that the impact of the circuit inductances can no longer be ignored. For the lumped circuit approximation, including gate loop inductance, shown in Fig. 6.3, it is fair to assume that at a high enough dv/dt edge rate, the external gate drive pull-down path impedance will become too high to draw any significant current within the required switching time. This effectively lowers the practical maximum allowable dv/dt to a lower value – well below that calculated from a simple RC network. Furthermore, the effectiveness of an AC coupled gate drive is also reduced as such as structure will further increase the inductance of the gate drive pull-down path. The exact mathematical quantification of the impact of the gate inductance is still lacking and a potential area for future research, especially when the impact of common source inductance (CSI) is added to the model (more on this later in the chapter).

The different aspects of the gate loop inductance parasitic impact on the system will be discussed later in this chapter in more detail, but it is suffice to say that at a device level, any effort to reduce the Miller CR of a discrete device will be beneficial to the overall system performance and allow an increase in the maximum allowable dv/dt switching speed.

6.1.2 Internal Gate Resistance

The advantage of reduced internal gate resistance (R_{G}) on system performance has been known for MOSFETs for a long time [16], and the addition of a metal

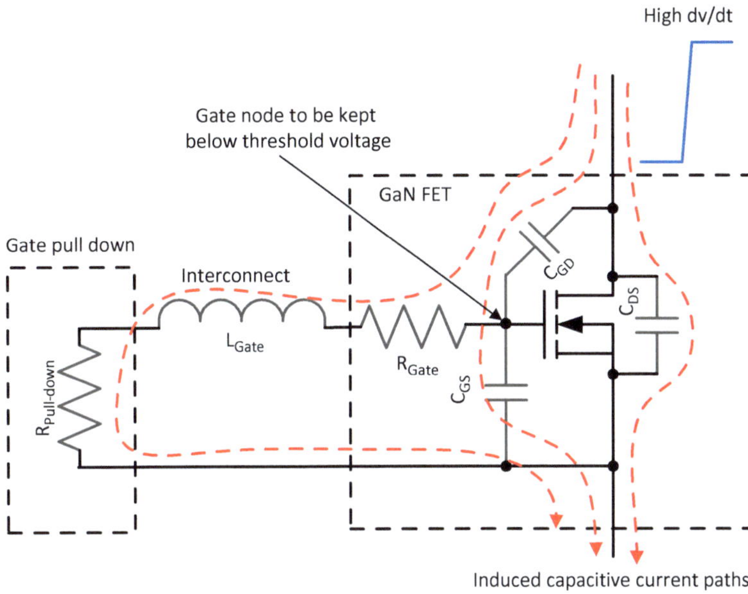

Fig. 6.3 Lumped circuit diagram showing effect of Miller CR during a high dv/dt event for GaN FETs

gate bussing structure is now standard on MOSFETs for reducing internal gate resistance. The advantages and trade-offs for minimizing R_G are well documented for MOSFETs [15]. With GaN, as mentioned in the previous section, the gate pull-down loop effectiveness against dv/dt-induced turn-on is limited by the external gate loop inductance. As such, the trade-offs for minimizing R_G for GaN would shift toward reducing C_{GD} at the expense of increased R_G and/or increased C_G and C_{OSS} (implemented through reduced gate bussing or increased gate shielding or a combination of both). However, since the gate capacitance achievable with GaN is much lower than silicon ($R_{DS(on)}.Q_G$ FOM comparison), it is still possible to create low gate resistance without significantly impacting gate capacitance [17].

6.1.3 Output Capacitance (C_{OSS}) and Diode Reverse Recovery (Q_{RR})

The remaining device-level charge-related parasitics are diode reverse recovery charge (Q_{RR}) and output capacitance charge (Q_{OSS}). When considering diode reverse recovery measurement, these two charges are difficult to separate as they are both included in the Q_{RR} measurement. It is difficult to distinguish which portion of the measured charge is Q_{RR} and which portion is Q_{OSS} because there is no physical indication when the Q_{RR} recombination process completes [9]. It is

Output charge	Q_{OSS}	V_{DD}= 30V, V_{GS}= 0V	32	43	54	nC
Reverse Diode						
Diode continuous forward current	I_S	T_C= 25°C	-	-	100	A
Diode pulse current	$I_{S,pulse}$		-	-	400	
Diode forward voltage	V_{SD}	V_{GS}= 0V, I_F= 50A		0.9	1.2	V
Reverse recovery time	t_{rr}	V_R= 30V, I_F= 50A, di_F/dt = 100A/μs	14	35	56	ns
Reverse recovery charge	Q_{rr}		14	29	58	nC

Fig. 6.4 Example extract of a datasheet showing output charge, Q_{OSS}, larger than Q_{RR} (which would include Q_{OSS} in actual measurement)

possible, however, to measure the Q_{OSS} charge separately and subtract this quantity from the total measured charge to estimate the actual diode reverse recovery charge component. There is today still some ambiguity on how these parameters are reported in manufacturers' datasheets, and when both are given in the MOSFET datasheet, the question should be asked as to whether the datasheet value of Q_{RR} is the actual total measured charge or whether the Q_{OSS} charge has already been subtracted to improve perceived device performance. An example of this, a recreated extract of a datasheet for a 60 V silicon MOSFET, is shown in Fig. 6.4 where the reverse recovery charge Q_{RR} is less than the output charge Q_{OSS}. This is only possible if Q_{OSS} has already been subtracted for the value shown for Q_{RR}.

It has been widely publicized that GaN HEMT devices have zero diode reverse recovery [18–21] as these devices have no internal P-N junctions to recover. These devices do, however, behave in a diode-like manner in 3rd quadrant operation, as the channel will conduct in the reverse direction once the drain voltage drops sufficiently below the gate potential [22]. When testing these GaN devices for diode reverse recovery, a non-zero charge will be measured, but this charge will be equal to the Q_{OSS} of the device [23]. A comparison between the conceptual silicon MOSFET and GaN FET diode recovery waveforms is shown in Fig. 6.5. Note the change in the peak reverse current point location. This hard-switching loss component is significant, and the removal of it also allows increased di/dt operation to further reduce switching losses, without fear of increasing Q_{RR}. With the removal of reverse recovery with GaN devices and with this a significant reduction in overlap switching losses, the need for a better understanding of the output capacitance-related losses is required. Due to the nonlinearity of the C_{OSS} with voltage, there is a difference between the energy stored in the output capacitance, given as E_{OSS}, and the charge stored in the output capacitance, Q_{OSS}, and their related equivalent linear capacitances $C_{OSS(er)}$ and $C_{OSS(tr)}$. These quantities have a direct impact on the switching losses [23, 24].

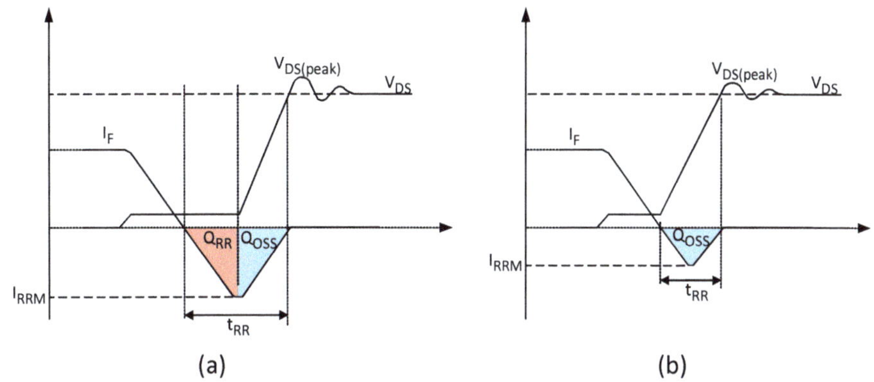

Fig. 6.5 Conceptual diode reverse recovery charge test waveforms for FET devices (**a**) with and (**b**) without Q_{RR}. Total shaded areas are measurement Q_{RR}

As with the other FOM metrics, GaN devices also show an advantage in the soft switching applications [7]. However, it is important to note that since GaN devices are lateral, their output capacitances do not continue to reduce with increasing drain bias as is the case with vertical silicon devices due to the much higher metal bussing capacitance of the lateral device. As such, the relative output capacitance energy storage (E_{OSS}) advantage of GaN FETs diminishes with device voltage rating, even as their overall performance advantage increases. Consider, as example, the normalized E_{OSS} and Q_{OSS} plotted versus voltage for both 100 V GaN and silicon devices shown in Fig. 6.6. At 50 V, the Q_{OSS} of GaN is about half that of silicon, but this does not translate into a similar E_{OSS} advantage. In fact, by 90 V, the E_{OSS} advantage of GaN has disappeared. The 600 V devices, compared in Fig. 6.7, show a similar result. For super-junction MOSFETs, the large output capacitance at low voltage dominates Q_{OSS} and gives GaN FETs a distinct (>5x) advantage. As with the 100 V case, however, this does not translate into an E_{OSS} advantage. The latest generation super-junction device [25] even has lower E_{OSS} than GaN above 450 V.

6.1.4 Specific On-Resistance (R_{SP})

The specific on-resistance of a device is a critical performance parameter and is directly related to the device cost. This is not directly a parasitic parameter, but advances in reducing the die size for a given on-resistance will directly impact the device parasitics, as well as other electrothermal considerations – more on this as the end of this chapter when discussing future trends.

What is worth mentioning here is that an improvement in specific on-resistance requires a reduction or change in the device geometry, which in turn results in a reduction in the device parasitics [26]. Thus any efforts made toward the reduction

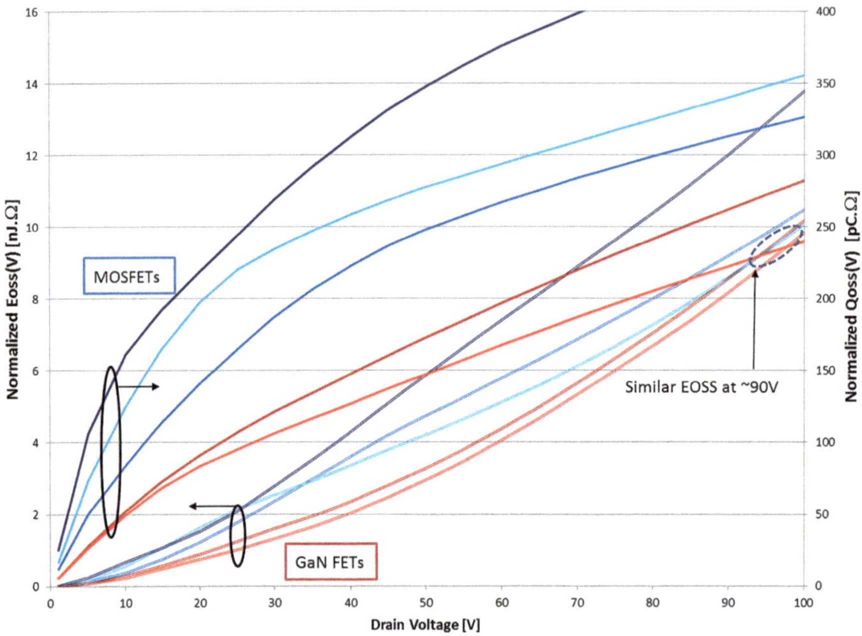

Fig. 6.6 Comparison between 100 V lateral GaN FETs (red) and 100 V vertical MOSFETs (blue) showing normalized E_{OSS}(V) and Q_{OSS}(V), showing the impact of the relatively higher C_{OSS} of GaN at higher voltage

of the device R_{SP} will have the additional benefit of resulting in an improved performance device as well. This is the case not just for silicon MOSFETs [26] but also for super-junction devices [27] and more recently GaN FETs as well. For the 100 V and 200 V figures of merit data presented in Fig. 6.2, most of these improvements have come from the recent die size reduction [28], although not all parasitics scale directly with die size.

Such a "generational" improvement in R_{SP} is typically an indicator that the given device technology is maturing and moving away from a technology demonstrator toward a robust, reliable product. Another example that this is happening is the announcement of the first AEC-Q101 automotive qualified GaN FET [29], which comes after both the initial creation and later portfolio expansion of their second-generation GaN devices.

6.2 GaN Package Parasitics

The impact of package- and board-level parasitics for GaN FETs has been an active area of research from the very start [8, 30] of GaN power device development and continues to build on the cumulative knowledge base built up with silicon MOSFETs. However, there seems to be a view that GaN devices themselves would

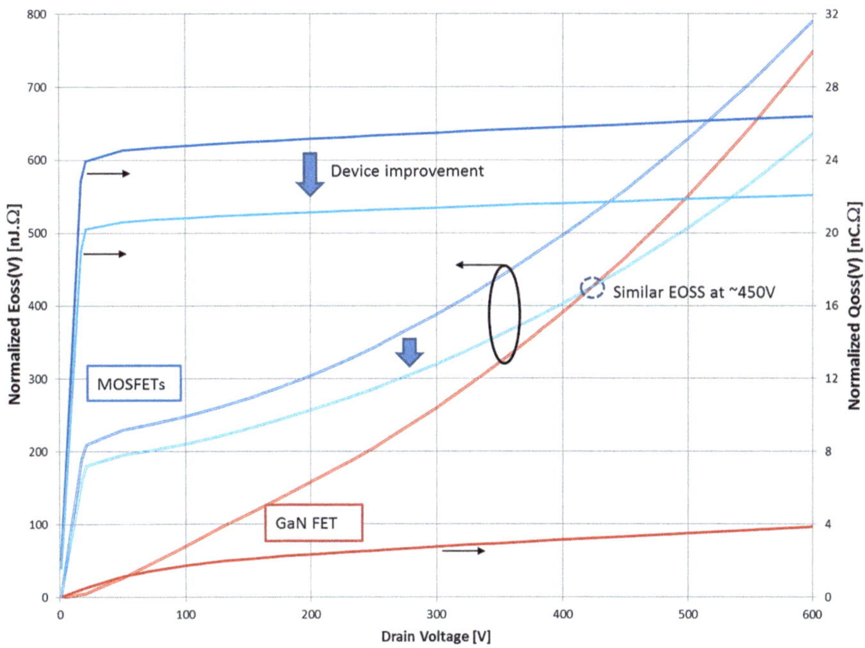

Fig. 6.7 Comparison between a 600 V lateral GaN FET (red) and the latest 600 V vertical super-junction MOSFETs (blue) showing normalized E_{OSS}(V) and Q_{OSS}(V), showing the impact of the relatively higher C_{OSS} of GaN at higher voltage

be incredibly fast, if it weren't for those "pesky packages" and "meddling parasitics" [31]. In contrast, MOSFETs are typically considered to be internally limited and that the next generation of MOSFET with better FOMs is all you need to achieve the as yet unobtainable system performance [32]. But to be honest, this revolution in thinking of the power device as being externally limited did not start with GaN, but has been growing in silicon over the last decade or so, mainly at low voltage where the impact [33] of these parasitics is greatest. But with GaN, due to significant improvement in device-level parasitics, these package- and board-level parasitics have become even more critical.

6.2.1 *MOSFET Package Evolution*

It wasn't all that long ago that power silicon MOSFETs were only available in standard wire-bonded, leaded packages, such as TO-220 and SO-8 [34, 35]. But, over time these packages became inadequate. The main concerns with such a low-cost, low-performance packaging approach [36, 37] can basically be divided into three general areas: package resistance(s), package inductance(s), and thermal resistance(s) – both down to board and up to top of case.

In the evolution of different MOSFET device packages, the aim would always be to improve one or more of these aspects of the package to enhance the device performance. Conceptual package cross-sectional diagrams showing the evolution of some of the key package developments are shown in Fig. 6.8 and will be referenced in the next sections.

Package Resistance

For package resistance, the main contributing components are source-side and drain-side resistance. As the on-resistance of the MOSFET decreased rapidly over time through improvements in both specific on-resistance and increased die size capability, the standard SO-8 packaging the packaging resistance could be as high as one third of the overall device resistance [34]. Most of which was dominated by the source-side wire-bond AC resistance [38]. One method to eliminate this was through the development of copper clip or Copperstrap™ [39] interconnect technology. A better alternative is to remove the source connection from the package entirely, as with DirectFET® [40] or a flip-chip BGA approach for lateral devices such as low-voltage silicon MOSFETs [38] or GaN [8].

Package Inductances

Here the packaging inductance can be broken into three separate terminal inductance components as shown in Fig. 6.9a. The package inductance on the source side on the device becomes common to both the gate drive current path and power loop current path. This "common" source inductance (CSI) has a significant impact on the device switching performance during di/dt [37, 41] as the change in drain current induces a voltage that opposes that of the applied gate drive voltage, thus slowing down both turn-on or turn-off. The copper clip reduces CSI, while packages like DirectFET® and wafer-level chip-scale packages (WLCSP) such as BGA structures can virtually eliminate it [8].

An alternative method for reducing CSI within a package where the source-side package inductance can't be eliminated is through the addition of a separate pin and path for the gate current to return, as shown in Fig. 6.9b. Such a pin is referred to as a "gate return" or "Kelvin" source [42] and is becoming more commonplace, even in traditionally "slow" switching devices, such as a Kelvin emitter for IGBTs [43, 44]. For GaN FETs, the reduction of CSI package inductance has been critical, and the addition of a Kelvin source was available on the first discrete parts [45, 46].

What hasn't been covered within this package discussion so far is the cascode device. In this special case, the high-voltage depletion-mode GaN device (a normally on device) is connected in cascade with a low-voltage silicon MOSFET to create a normally off high-voltage equivalent switch [48] as shown in Fig. 6.10. To optimize such a device package, it is necessary to determine the impact, and how to minimize the influence, of the internal interconnect inductances [49, 50]. Although

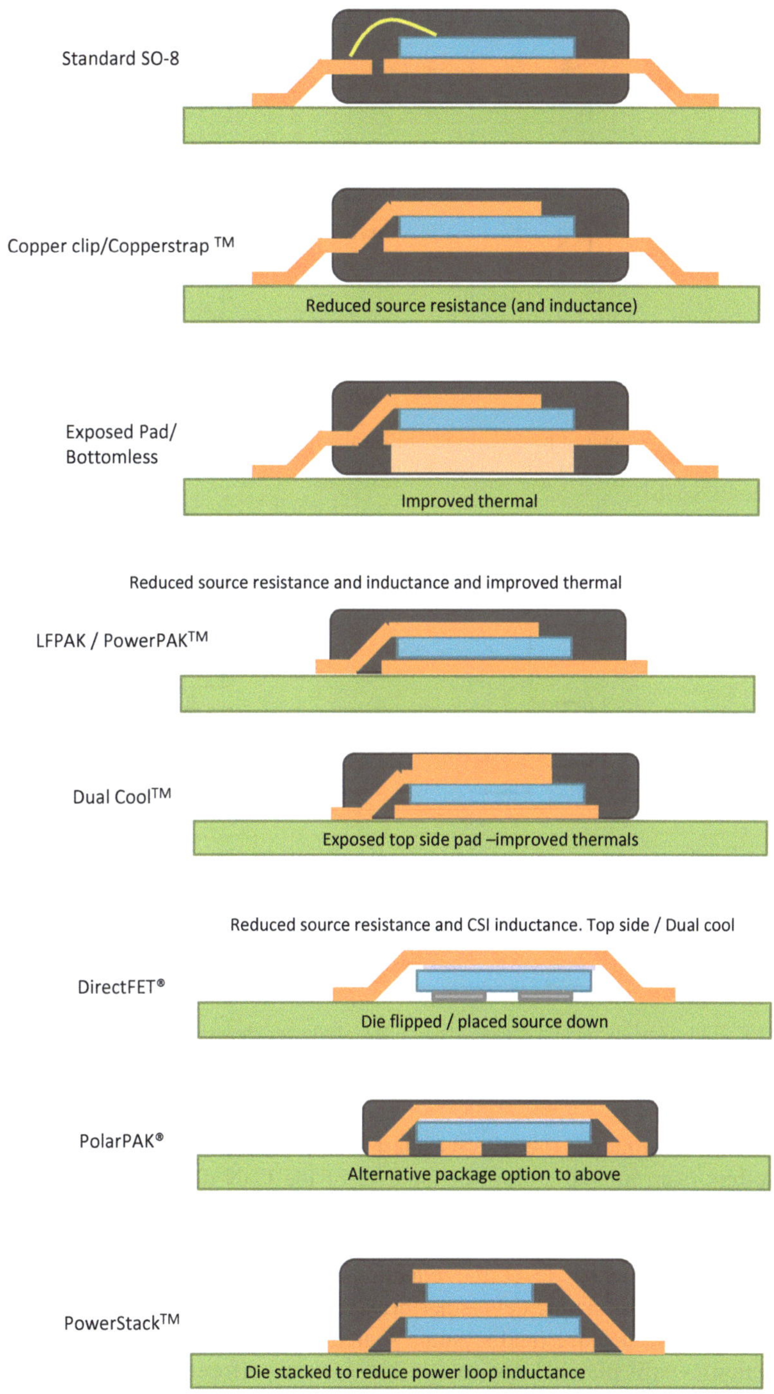

Fig. 6.8 Vertical MOSFET package evolution to address package electrical resistance, inductance, and thermal resistance

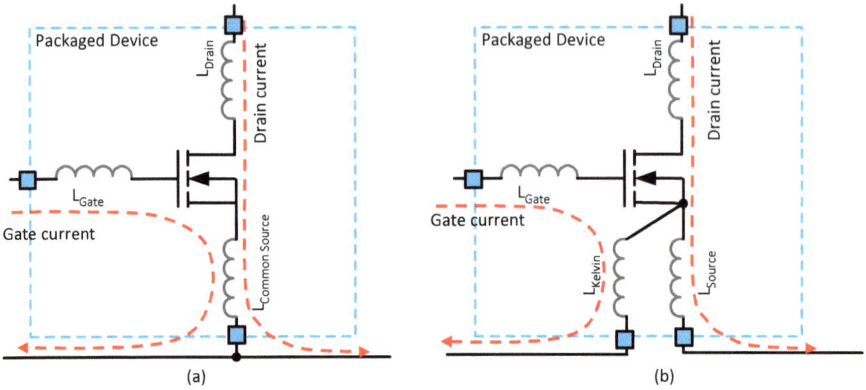

Fig. 6.9 Equivalent circuit of packaged MOSFET showing separate package inductance components (**a**) and alternative extra pin package with Kelvin gate return to eliminate common source inductance (CSI) (**b**)

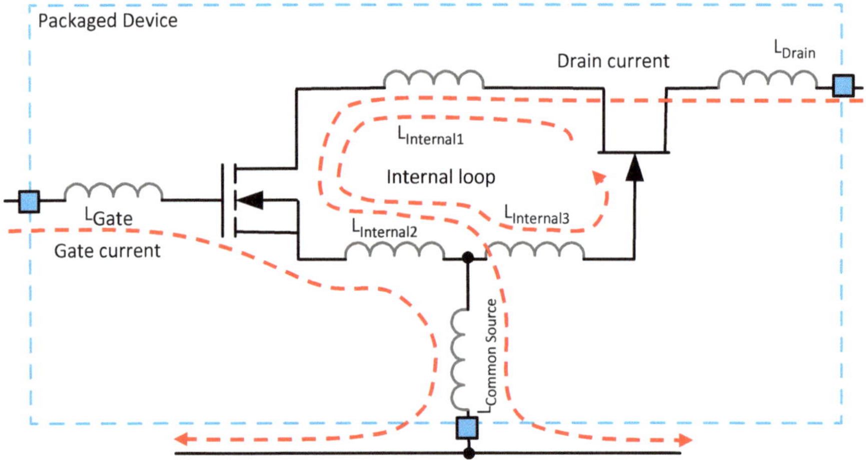

Fig. 6.10 Equivalent circuit of a packaged cascode GaN device showing parasitic inductances and the different high-frequency loops between transistors

it is possible to minimize the relevant internal loop inductances, the cascode device still suffers from limited control of the gate of the GaN device itself, prompting one depletion-mode supplier to use direct gate control [17] and only using the low-voltage silicon device for safety.

In short, on a package level, the reduction of the CSI is the key inductive component to avoid limiting the device performance. The other parasitic inductances – gate and drain inductance (and separate source inductance in the case of a Kelvin connection) – still play a large role in the system performance. However, these components tend to get combined with similar series board-level elements and will be discussed as part of the board-level parasitics in the next section.

Package Thermal Resistance

With the SO-8 package, the die and lead-frame encased in mold compound, heat removal was limited. The aim of subsequent packages was to first create a direct thermal path between the die and printed circuit board (PCB), such as the bottomless SO-8 and PowerPAK devices. To further improve the thermal performance, the thermal path to the top of the device was improved to allow topside cooling as well – such as the DirectFET® and the Dual Cool™ [47] approach. With the WLCSP GaN devices, these thermal issues were present from the start as the GaN device die was much smaller than their equivalent silicon counterparts [8]. In spite of the reduced die size, the WLCSP GaN devices, by virtue have having no additional packaging, show thermal performance equal or better than other silicon package solutions [51]. For higher-voltage GaN FETs, where a package is needed to meet creepage and clearance requirements, the use of either topside [52] or bottom-side [46] cooling die-in-laminate packages has been adopted. These separate packaging approaches allow customer flexibility in choosing their preferred heat removal path. "Dual cool" isn't really an option as GaN devices are lateral, and there will always be a disproportionate thermal advantage "down" through the die substrate versus "up" through the numerous "glass" isolation layers.

6.2.2 *Board-Level Parasitics that Directly Impact GaN Device Performance*

Power Loop Inductance(S)

Going beyond the package-level parasitics, the next most important parasitic from a device performance standpoint is total power loop inductance [53], shown in Fig. 6.11. As can be seen, this is not a single parasitic component, but is rather buildup of board-level parasitic interconnect inductances, package-level drain and source inductances, and the high-frequency (HF) bus decoupling capacitors (normally ceramic capacitors) and their own related parasitics. Thus when considering the package-level parasitics, the reduction of non-common source- and drain-side inductance would also have shown a performance benefit, but this benefit would have been indirectly through the reduction of the overall power loop inductance. For the PowerStack™ structure shown in Fig. 6.8, both vertical silicon MOSFETs within the half-bridge are integrated into a single package [54]. Although this gives a significant reduction in power loop inductance (and common source inductance), and therefore switching loss, it does increase the thermal resistance of the smaller high-side device that is now placed on top of the larger low-side MOSFET. As presented though, the improvement in electrical performance is stated to far outweigh the increase in thermal impedance [54], but this may not be the case for all such stacked die trade-offs.

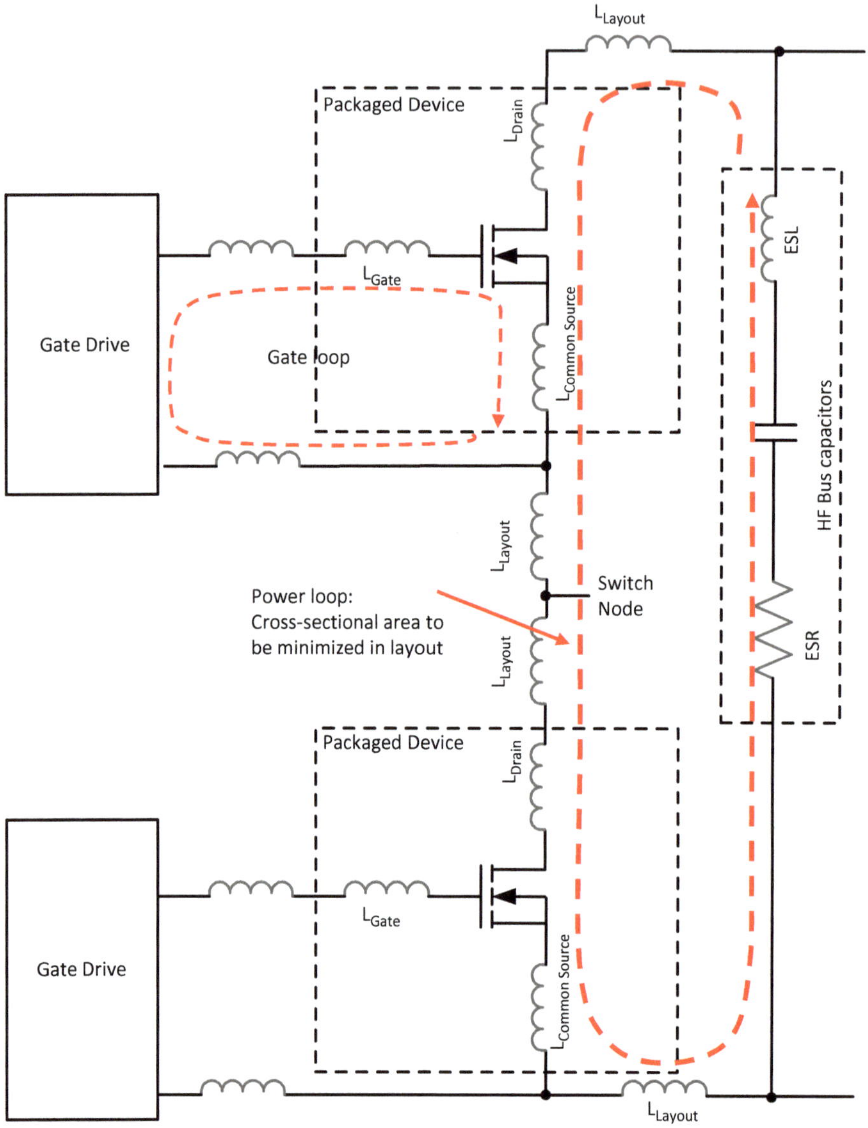

Fig. 6.11 Equivalent circuit of a half-bridge circuit showing the individual inductive components that create the overall power loop inductance (not shown are the associated resistances with each of these inductive components)

For GaN FET, due to the reduced die size relative to silicon, it is necessary to do address board-level parasitic without compromising thermal performance. At low voltage, the packaging inductance is more critical to device performance, and the advantages of WLCSPs have been shown through a combination of reduced packaging inductance and interdigitating drain and source terminals [55]. The alternating source and drain pins reduced inductance through flux cancellation [22] by having equal amplitude and opposite direction currents flowing in close proximity to each other through the pins as shown in Fig. 6.12a. For high-voltage devices, the practicality of these kinds of packaging improvements is limited due to the required voltage spacing between high-voltage pins. However, a board-level layout improvement [53] that also reduces power loop inductance through a similar flux cancellation scheme is still possible, as shown in Fig. 6.12b. At low voltage, both the package- and board-level implementations can be combined to further reduce power loop inductance. Furthermore, the monolithic integration of both devices into a single WLCSP further reduces power loop inductance [56], as shown in Fig. 6.12c. Additionally, monolithic integration also improves thermal performance through hot-spot reduction and allows the creation of high aspect ratio devices, with low package inductance, that would not be possible were it not monolithically integrated [56].

As power handling capability is to be further increased, the relative impact of the power loop inductance is also increased as both the rate of change of current (di/dt) and energy stored within the parasitic inductance are increased as we attempt to avoid a proportional decreased switching speed. One such approach, shown in Fig. 6.12d, is to split the power loop into two symmetrical loops [57], which effectively halves the relevant inductance [58].

However, once the power-level increases to the point where multiple power devices themselves need to be paralleled, alternative approaches are required. Once there are two devices to parallel, it becomes impossible to connect a single, common gate drive to them without adding, and increasing, common source inductance [59], as shown in Fig. 6.13. As the common source inductance degrades switching performance, especially for fast-switching GaN devices, it becomes more advantageous to resort to a multiphase solution instead to achieve higher power [60]. It is, however, theoretically possible to parallel devices without the significant penalty of increased common source and power loop inductance. In practice, this can be reasonably approximated through the paralleling, not of devices, but of device power loops [61] as shown in Fig. 6.14. The key insight to this approach is to create symmetry between multiple power loops that are connected in parallel. There will still be the unavoidable interconnection impedances between devices in such a layout, but if both the device matching and the layout symmetry are sufficient, the dynamic voltages at these nodes become almost equal. Thus they become virtually connected and any impedance between them carries negligible high-frequency current. Another way to interpret this is to consider these interconnection impedances, which would otherwise form part of both the power loop and common source inductance, are pushed outside of the high-frequency (switching edge rate) commutation path and are only carrying lower-frequency currents related to the power transfer and switching repetition frequency. This approach has even been combined with monolithic device integration, in which case the option of mirrored symmetry is lost, but identical power loops can still be created [62]. This means that symmetry of the interconnection impedances between power loops could not

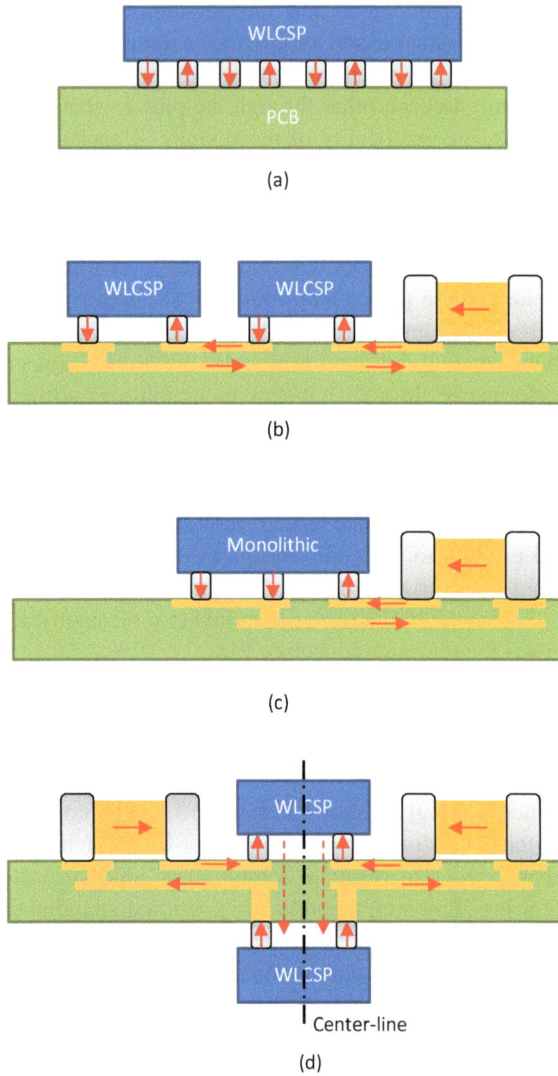

Fig. 6.12 Conceptual device and board layout showing the equal and opposing currents during switching that allows magnetic field cancellation in reduced package and power loop inductances. (**a**) WLCSP with interdigitated drain and source pins reduces package inductance. (**b**) "Optimum" power loop with board-level magnetic field cancellation [53]. (**c**) Further reduction through monolithic integration of both GaN power devices [56]. (**d**) Reduction through two symmetrical/mirror image loops [57]

be achieved, but in spite of this, a significant performance improvement was still achieved; it shows that creating matched high-frequency power loops is critical and, in this case at least, some flexibility in the matching of the interconnect impedance could be tolerated.

At higher voltages, where the relative impact of these inductances is reduced, there has been some success in paralleling devices, rather than power loops [63, 64]. In one case, the analysis presented still aims to reduce the discussed inductance components and considers their layout priority for minimization as "extremely high" [63], while in the other, paralleling more than two devices was considered "very

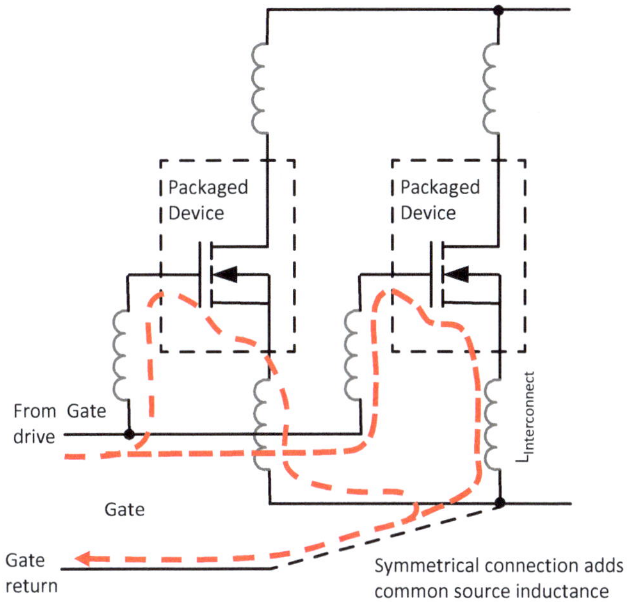

Fig. 6.13 Unavoidable creation of common source inductance when paralleling two or more devices driven from the same gate driver

hard" to make the gate circuit symmetrical [64] and proposed, for such a case, an interleaved multiphase approach instead.

Gate Loop Inductance(s) and Resistance

The gate loop inductance is important for a variety of reasons: Firstly, the gate inductance can directly impact switching performance, slowing down the effective gate driver speed. Secondly, the gate inductance, together with the device gate capacitance, creates a resonant tank that can create an electrical overvoltage at the device gate (mainly an issue for pGaN gate-based GaN devices [22]). Lastly, the gate loop inductance can result in misoperation of the device, such as the case mentioned during the discussion on Miller charge ratio in the beginning of this chapter.

The impact of gate loop inductance, common source inductance, as well as gate loop resistance is all interconnected, and their interaction can be quite complex [65]. Some of the more important aspects will be discussed here with reference to the gate loop structure shown in Fig. 6.15.

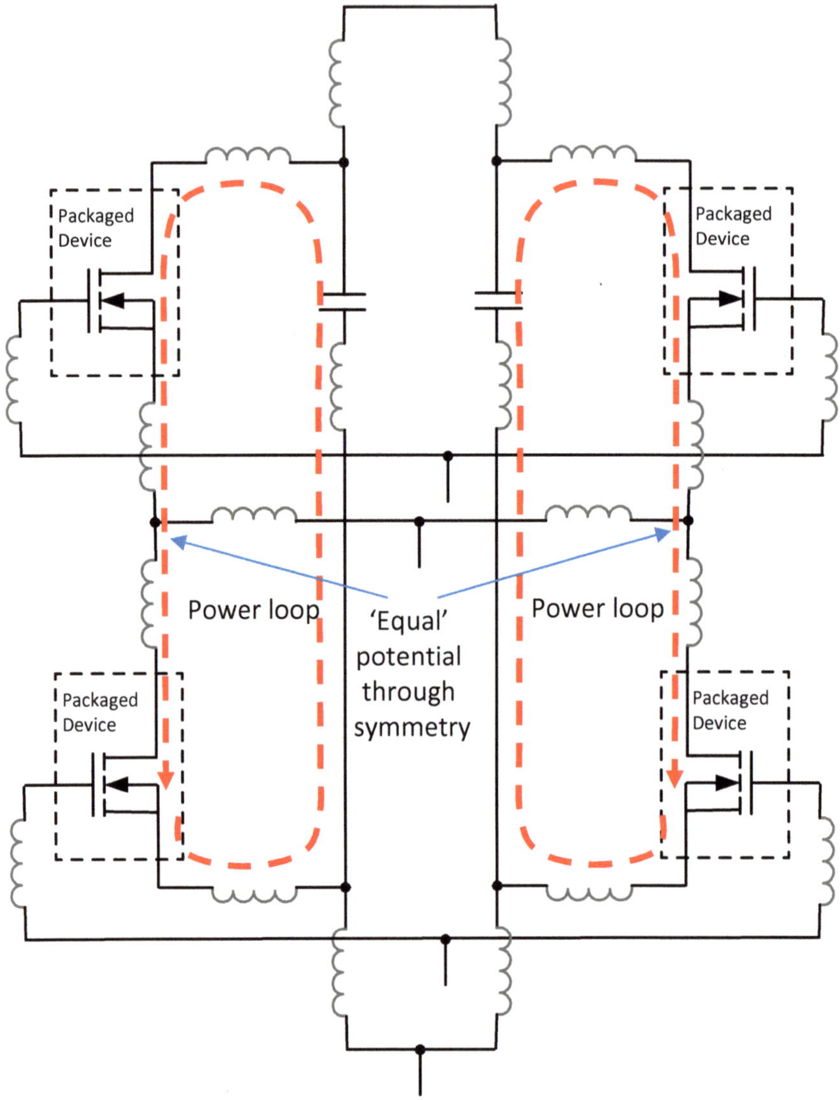

Fig. 6.14 Representation of theoretical approach of paralleling device power loops, rather than devices, to avoid switching performance degradation due to high-frequency interconnection impedance

From Fig. 6.15, it can be seen that the gate loop inductance values for turn-on and turnoff can be significantly different, with the turn-on loop requiring an additional section that passes through the supply's high-frequency decoupling capacitor. A comprehensive discussion on these loops and their respective resistive damping requirements is given in [19] and shows that for both loops, the gate resistance

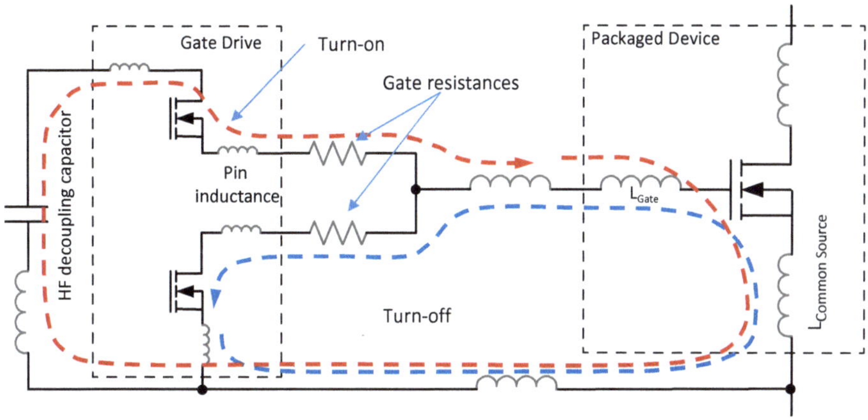

Fig. 6.15 Schematic showing gate loop with relevant lumped circuit components

is added to achieve near-critical damping at the resonant frequency. Much of this damping can be realized through the frequency-dependent resistance (skin and proximity effects) of the packaging- and board-level interconnects, but depending on the actual inductance and device capacitance values, some additional resistance is typically required. Since the turn-off loop has less inductance and can also tolerate having less damping [19], the additional resistance required is less than that for the turn-on loop, and such an arrangement is best realized by separating out the gate drive output path into two separate pull-up and pull-down pins [66]. This also allows the minimization of the turn-off gate loop inductance to maximize the effectiveness of the gate drive pull-down in attempting to suppress Miller charge-related dv/dt induced turn-on, as discussed earlier in this chapter.

It is worth noting that although gate loop inductances are to be minimized, as with common source inductance, the minimization of the gate loop inductance without first minimizing common source inductance can lead to system instability and even failure of the device. This can be best explained by considering the circuit shown in Fig. 6.16. In this case, the large CSI generates a step change voltage within the gate driver loop at device turn-on. This negative voltage initially appears across the small gate loop inductance and starts reducing (and then reversing) the gate drive current. Once the gate current has reversed, the gate starts discharging and the device's source current reduces. At this point, the voltage across the CSI changes polarity and creates a step change in the opposite direction. Depending on the damping coefficient of this loop, this oscillation can either continue/increase or die down. An increasing oscillation will lead to device failure, while an underdamped oscillation can still increase switching loss and cause unwanted switching behavior. By increasing the gate loop inductance in this case, the gate drive current will decrease slowly, slowing the positive di/dt and thus reducing the step voltage before gate current reversal is achieved. In other words, the gate loop

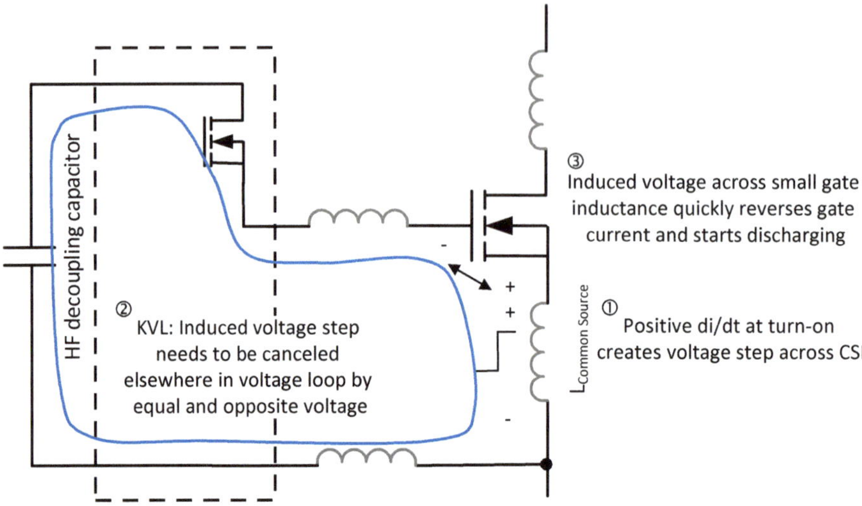

Fig. 6.16 Schematic showing impact of small gate loop inductance interaction with large common source inductance during device turn-on

inductance and gate capacitance form a low-pass filter and limit the effect of the CSI voltage step on the gate voltage. The same mechanism also exists at device turnoff.

It has been shown [67] that for a synchronous rectifier device, when the body diode is being commutated, however, the initially induced voltage step is actually beneficial to the device as it induces an additional negative voltage across the device gate – improving Miller ratio. However, as with the case above, this voltage step can cause an underdamped oscillation in the gate loop that in turn will cause the gate to ring positive. If this positive voltage ring exceeds the device threshold, an unwanted turn-on of the device will be induced [19]. Since skin and proximity effects increase with frequency, it may be possible to achieve sufficient gate loop damping through geometric layout alone; however, elimination of CSI is still a preferred option.

6.3 External Components and Parasitics that Impact Performance and Operation

Up till now, the focus in this section has been on board-level parasitics that directly determine switching behavior as they carry either gate drive or device drain current. Expanding our scope beyond these, however, there are additional system-level parasitics that impact either system performance (reduce efficiency) or system operation (cause misoperation or failure).

6.3.1 Switch Node Related Parasitic Capacitance

When considering a more complete system, as shown in Fig. 6.17, there are a number of parasitic capacitances that could impact the switch node's effective capacitance. The most important, from an efficiency perspective at high voltage, is the equivalent parallel capacitance of the magnetic component connected to the switch node [68]. This capacitance manifests itself as additional C_{OSS} capacitance in parallel with the off-state device. Not only does this increase Q_{OSS} related losses, it also increase the dv/dt related power loop current, resulting in higher peak commutation current and higher voltage overshoot [68].

Other capacitances that can have a significant impact include the bootstrap [69] or isolated supply capacitance [70, 71] as well as any capacitance associated with the logic signal isolation or level shifter [69]. These capacitances however generate additional impact when not directly connected to the power device switching node, but rather through some intervening layout or parasitic inductance. This added parasitic inductance thus induces a pulsating voltage in response to the pulsating capacitive current that in turn is generated by the switching-node dv/dt as shown in Fig. 6.18. Depending on the capacitive current path across these parasitic inductances, these induced voltages can themselves affect the system operation. The extent of these can become rather complex and are best managed through minimizing the inductance of the interconnections between the switching node and these said parasitic capacitances.

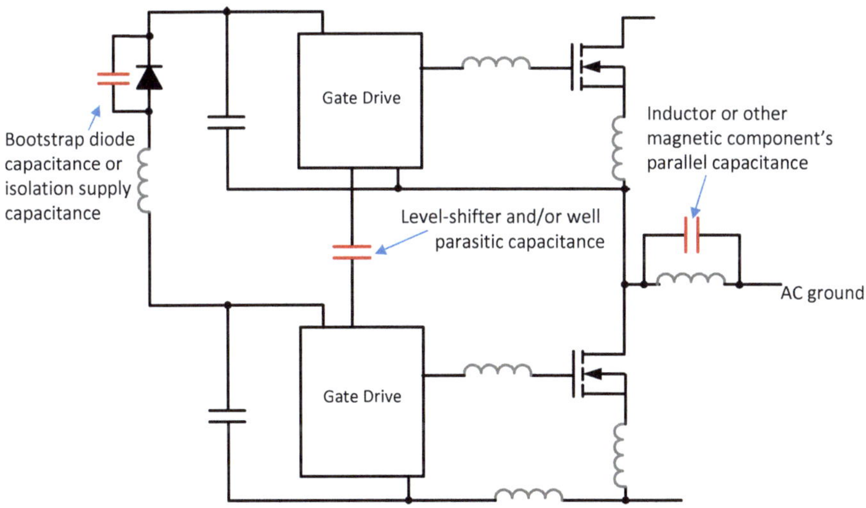

Fig. 6.17 Partial system-level schematic showing potential parasitic capacitances that influence the effective switch node capacitance

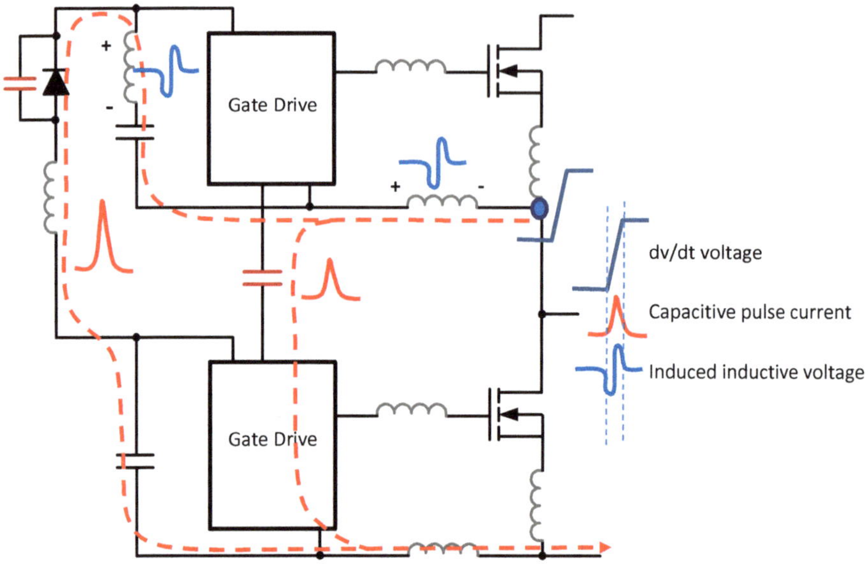

Fig. 6.18 Example of induced voltage generation due to capacitive currents during high dv/dts

6.3.2 *Bootstrap Supply Operation and Parasitics*

When using a bootstrap diode with GaN devices, additional constraints are placed on the system. Firstly, due to the fast switching speed on GaN, the diode reverse recovery of a silicon bootstrap diode will quickly become significant [69]. Use of a silicon Schottky diode is possible at low voltage or requires a SiC Schottky diode for higher voltages. At the time of writing, unfortunately, there are no small-signal SiC diodes available for this purpose, and thus using a larger SiC power diode will incur significant capacitive currents and losses as discussed above. However, this may still be preferable over using a high voltage silicon PiN diode with its related Q_{RR} losses.

When combining a bootstrap supply with using e-mode GaN devices, which have limited gate drive voltage headroom, there are multiple issues [72] and solutions [66, 73] for maintaining high-side voltage close to regulation. Alternatively, additional regulation stages can be added [74] to avoid these issues but do add cost and complexity to the solution.

6.4 GaN Integration and the Impact of Improvements in FOM on Parasitics

The idea of monolithically integrating devices to minimize parasitic inductances, whether these are package or board level, seems a logical next step in GaN development. Since GaN HEMTs are lateral devices, they lend themselves to the potential higher-level circuit integration similar to lateral silicon devices. Research into monolithically integrated GaN devices can be traced back as least as far as 2008 [75] for GaN-on-silicon devices, with higher levels of integration, like the (partially) integrated gate drivers and half-bridge power devices [76] already shown in 2014. Other commercial examples include the monolithic half-bridge integration already mentioned [56], monolithic half-bridge with bootstrap diode [77], a driver-and-power-device integration [78], and, more recently, a complete half bridge with two gate drivers, level shifter and bootstrap into a multi-chip package [74] (although integration is still spread over multiple die). These different levels of integration are shown in Fig. 6.19.

To try and understand the required level of integration, and how this will change over time, one approach has been to consider the impact of these GaN device improvements as they progress toward their material limit [79], which is still more than 1000x away in the case of specific on-resistance [19]. To understand these implications, consider a representative device at the current state of the art as shown in Fig. 6.20. The device has a specific on-resistance, R_{ON}, and a gate width W, device length L, and a specific sheet resistance. This results in a benchmark device

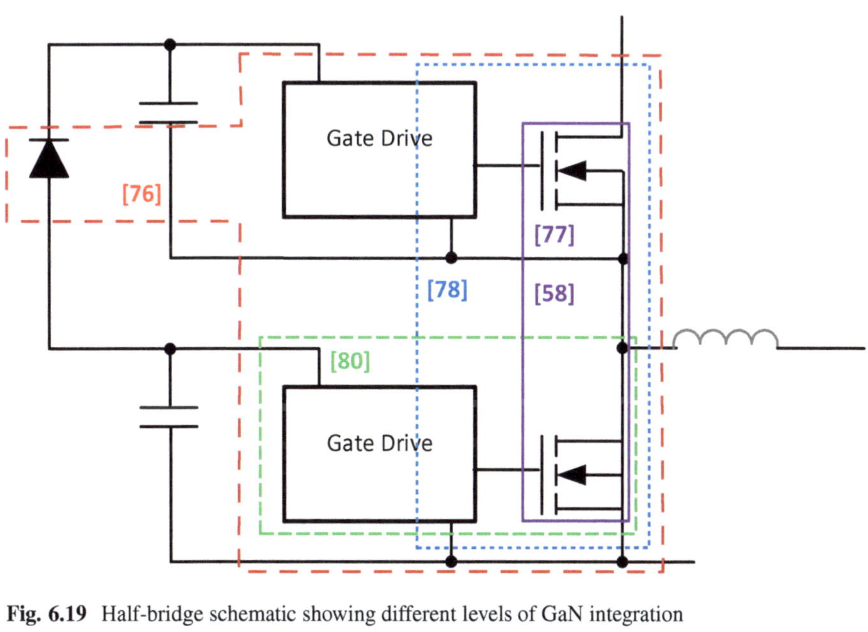

Fig. 6.19 Half-bridge schematic showing different levels of GaN integration

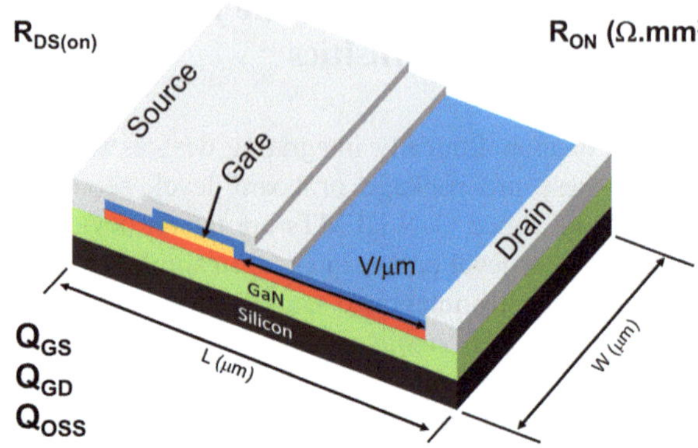

Fig. 6.20 Representative drawing of a lateral device at the current state of the art

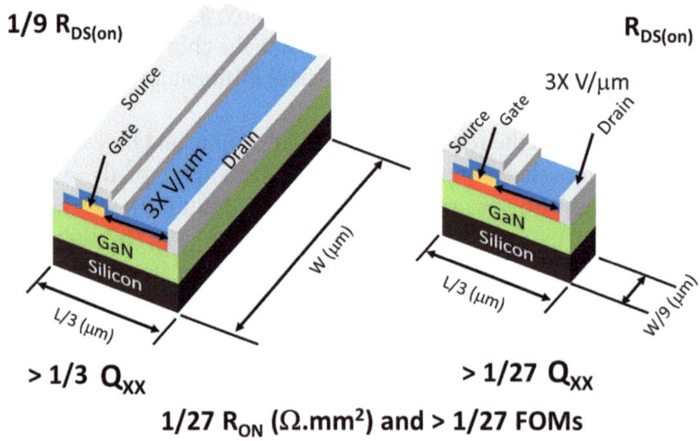

Fig. 6.21 Implications of devices technology improvement toward the material limit

with an on-resistance $R_{DS(on)}$ and device equivalent charge values Q_{GS}, Q_{GD}, and Q_{OSS}. Now assume that through advances in device technology, the device length is reduced by a factor three,as shown in Fig. 6.21. For the same gate width, the device becomes 1/3 the size with 1/3 the device length. To achieve this device length reduction, the sheet resistance also has to reduce by 1/3. Furthermore, as the device is 1/3 the size, all charge-related quantities are also reduced by some extent but approach 1/3 of the original device. To use this improved technology to recreate the benchmark device resistance, the resultant device would be 1/27 the size of charge quantities somewhere between 1/9 and 1/27 (a smaller device will result in lower capacitance, but reduced geometry spacing will also increase capacitance – the interplay of these will depend on the exact geometry and technology improvements). Through such a simplified approach, a rough estimate of the relative improvements in device can be estimated. Suffice to say that a three times improvement in device length (related to peak electric field capability) will result in a cubic ($27 = 3^3$) improvement in specific on-resistance and a more-than-quadratic (somewhere between 9 and 27 times) improvement in device capacitances and also the device figures of merit. In real life, the impact of metal bussing, contact resistance, and other non-idealities will mean the device will be bigger than the 1/27 estimated here, but the trend is clear.

Considering the optimum device on-resistance for a given application, any improvement (reduction) in device FOMs will result in the optimum on-resistance also reducing, compared to the benchmark technology [80], as some of the frequency dependent loss improvement is traded for reduced conduction loss. In this example scenario above, the equivalent optimum device may have a 3–5 times lower on-resistance and still have 3–5 times lower equivalent device charge values.

Based on these assumptions, there are a number of implications as GaN technology matures:

- Nominal on-resistance for any given design will decrease over time – making the impact of non-active area resistance (metal bussing, packaging, solder bumps etc.) more significant.
- Device parasitic capacitance will likewise decrease, making the impact of package and external capacitive parasitics more significant.
- Improvement in device FOMs assumes an equivalent reduction in frequency-dependent losses, including hard-switching (overlap) losses. This requires an equivalent increase in switching speed and thus a related increase in both di/dt and dv/dt during the switching interval, making the impact of common source inductance, power loop inductance, and gate loop inductance proportionally more significant. Any improvement in FOM would require an equivalent improvement in inductive parasitics to maintain their same relative impact on switching behavior. Thus in future, as now, there will be a continuing struggle to reduce inductive parasitic components, possibly only realizable through higher levels of integration.
- The higher required dv/dt also implies the need for higher common-mode transient immunity (CMTI) capability for the related gate drivers, level shifters, and isolated supplies. Some isolated gate drivers are already moving in this direction [81].
- The die size will decrease much faster that the equivalent reduction in conduction and switching losses – even accounting for a shift in the optimum device to a lower on-resistance. This implies that although losses will be lower, the loss energy density will increase and further improvements in thermal heat spreading and heat removal will be required. This will require more three-dimensional packaging and system layout designs to keep electrical and thermal paths orthogonal to each other to minimize their cross-coupling impact.
- Higher di/dt and dv/dt also imply increased frequency spectrum generation for electromagnetic interference. With the required reduction in parasitic inductances, the ringing frequencies will also move higher up into the radiated frequency spectrum. Filtering and containment will become more shielding focused, with components colocated inside the shielding or integrated with the GaN devices requiring higher electromagnetic interference immunity from the internal devices and circuits.

- The increased switching speed does not necessarily imply an increase in switching frequency, as this is more dependent on the development of improvement magnetic materials [82].

These all have the potential of becoming significant areas of research, if the potential of GaN is to be realized.

6.5 Summary

In this chapter the various levels of parasitic elements related to GaN devices were discussed. On the most basic level, the non-ideal characteristics of these devices, related to their device capacitances, are considered as parasitic. Beyond the GaN semiconductor device itself, the package adds additional resistive and inductive parasitics, while the package layout and pin-out affect the board-level layout which in turn adds additional parasitics to the system. These above parasitics all have direct impact on the device performance and system efficiency and their impacts are mostly understood.

When attempting to parallel multiple devices, additional parasitic elements are created, and there are limited options when trying to minimize their impact. Their interactions are far less understood and offer a potential area of research. This is also true for when considering secondary effects of these parasitics. For the primary effects, it is simple to consider the induced currents in capacitive parasitic elements and induced voltages across inductive ones, but understanding the impact the induced current has as its path covers other inductive components, thus inducing voltages which in turn could create additional currents, becomes far less clear. As the development of GaN progresses toward the material limit, these lesser effects will in turn become more significant as the parasitics related to the primary effects are reduced or even eliminated through integration.

References

1. R.W. Keyes, Figure of merit for semiconductors for high-speed switches. Proc. IEEE **60**, 225 (1972)
2. B.J. Baliga, Semiconductors for high-voltage, Vertical Channel FET's. J. Appl. Phys. **53**, 1759–1764 (1982)
3. B.J. Baliga, Power semiconductor device figure-of-merit for high frequency applications. IEEE Electron Device Lett **10**, 455–457 (1989)
4. I.J. Kim, S. Matsumoto, T. Sakai, T. Yachi, New power device figure-of-merit for high frequency applications, in *Proceedings of International Symposium on Power Semiconductor Devices ICs*, (Yokohama, 1995), pp. 309–314
5. A.Q. Huang, New unipolar switching power device figures of merit. IEEE Electron Device Lett **25**, 298–301 (2004)

6. Y. Ying, Device selection criteria based on loss modeling and fig 6.of merit, Thesis of Master of Science in Electrical Engineering of Virginia Tech, 2008
7. D. Reusch, J.T. Strydom, Evaluation of gallium nitride transistors in high frequency resonant and soft-switching DC-DC converters. 2014 IEEE applied power electronics conference and exposition - APEC 2014, Fort Worth, pp. 464–470, 2014
8. J.T. Strydom, eGaN-Silicon Power Shoot-Out: Comparing Figure of Merit (FOM), Power Electronics Magazine, September 2010, http://powerelectronics.com/discrete-power-semis/egantm-silicon-power-shoot-out-part-1-comparing-figure-merit-fom
9. A.Q. Huang, Hard Commutation of Power MOSFET, Application Note AN 2014–03, Infineon Technologies, March 2014
10. Vishay, Power MOSFET basics: Understanding gate charge and using it to assess switching performance, Appl. Note. AN608, Dec 2004
11. On-Semiconductor, MOSFET gate-charge origin and its applications, Appl. Note AND9083/D, April 2012
12. T. Wu, Cdv/dt induced turn-on in synchronous buck regulations, Tech. Rep., International Rectifier, 2007
13. Q. Zhao, G. Stojcic, Characterization of Cdv/dt induced power loss in synchronous Buck DC–DC converters. IEEE Trans. Power Electron. **22**(4), 1508–1513 (2007)
14. Cross Conduction in Modern Power MOSFETs - Application Center, http://www.maplesoft.com/applications/view.aspx?SID=5698
15. A. Elbanhawy, Parasitic gate resistance and switching performance, 2006 CES/IEEE 5th International Power Electronics and Motion Control Conference, Shanghai, pp. 1–4, 2006
16. Electronics Weekly, Advanced Power Technologies, News release, March 2001, http://www.electronicsweekly.com/news/archived/resources-archived/apt-takes-power-mosfets-into-next-generation-through-metal-2001-03/
17. Efficient Power Conversion Corporation, EPC2001C – enhancement mode power transistor, datasheet
18. Texas Instruments, GaN TECHNOLOGY PREVIEW, LMG3410 600-V 12-A Single Channel GaN Power Stage, revised datasheet, June 2016
19. A. Lidow, J.T. Strydom, M. De Rooij, D. Reusch, *GaN Transistors for Efficient Power Conversion*, 2nd edn. (Wiley, New York, 2014)
20. E. Faraci, Enabling industrial and automotive Multimegahertz buck converters with GaN, Power Electronics Magazine, September 2015, http://powerelectronics.com/gan-transistors/enabling-industrial-and-automotive-multimegahertz-buck-converters-gan
21. GanSystems, Technology description, http://www.gansystems.com/technology.php
22. A. Lidow, J.T. Strydom, M. De Rooij, Y. Ma, *GaN Transistors for Efficient Power Conversion*, 1st edn. (Power Conversion Publications, El Segundo, 2011)
23. P. Di Maso, L. Lu, Gan E-HEMTs enable innovation in power switching applications, Industry Presentations, IS13. 2017 IEEE applied power electronics conference and exposition – APEC, 2017
24. J. Glaser, D. Reusch, Deadtime losses in eGaN® Fets and Silicon MOSFETs – how freedom from reverse recovery can cut your losses. Industry Presentations, IS13, 2017 IEEE applied power electronics conference and exposition – APEC, 2017
25. F. Stueckler, Leadless MOSFET power package achieves near-GaN switching losses, Power Electronics Magazine, February 2017, http://www.powerelectronics.com/discrete-power-semis/leadless-mosfet-power-package-achieves-near-gan-switching-losses
26. R.K. Williams, M.N. Darwish, R. Blanchard, R. Siemieniec, P. Rutter, Y. Kawaguchi, The trench power MOSFET: Part I—History, technology, and prospects. IEEE Trans. Electron Devices **64**(3), 674–691 (2017)
27. J. Hancock, F. Stueckler, E.A. Vecino, CoolMOS™ C7: Mastering the Art of Quickness, Application Note AN 2013–04, Infineon Technologies, April 2013
28. A. Lidow, Generation 5 eGaN® technology: A Quantum leap into a new universe of performance!, Application Note AN022, Efficient Power Conversion Corporation, 2017

29. N. Flaherty, eeNews, Power Management, News release, 28 March 2017, http://www.eenewspower.com/news/transphorm-launches-first-automotive-qualified-gan-fets
30. GanSystems, GaN*PX*® packaging technology description, http://www.gansystems.com/ganpx_packaging_new.php
31. S. Sandler, How to measure the world's fastest power switch, Electronic Design Magazine, November 13, 2014, http://www.edn.com/design/test-and-measurement/4437360/2/How-to-measure-the-world-s-fastest-power-switch
32. Infineon Technologies, Infineon Started World's First Volume Production of Next Generation Automotive Power MOSFETs on 300-Millimeter Thin-Wafer Technology market news, http://www.infineon.com/cms/en/about-infineon/press/market-news/2015/INFATV201505–059.html
33. Y. Xiao et al., Analytical modeling and experimental evaluation of interconnect parasitic inductance on MOSFET switching characteristics. IEEE APEC 2004
34. ST Electronics, New packaging concepts for low voltage power MOSFETs lead to performance improvement in advanced DC-DC Converters, Application Note 2066, 2004, http://www.st.com/content/ccc/resource/technical/document/application_note/f7/c6/db/ef/bc/9b/49/d2/CD00045978.pdf/files/CD00045978.pdf/jcr:content/translations/en.CD00045978.pdf
35. J. Lee, Package parasitics influence efficiency, Power Electronics Technology Magazine, November 2005, http://www.powerelectronics.com/passive-components/package-parasitics-influence-efficiency
36. M. Pavier, A. Sawle, A. Woodworth, R. Monteiro, J. Chiu, C. Blake, High frequency DC:DC power conversion: the influence of package parasitics, Applied Power Electronics Conference and Exposition, 2003. APEC '03. Eighteenth Annual IEEE, Miami Beach, pp. 699–704 vol.2, 2003
37. A. Elbanhawy, Are traditional MOSFET packages suitable for the new generation of DC-DC converters, The 4th International Power Electronics and Motion Control Conference, 2004. IPEMC 2004., Xi'an, pp. 316–319 Vol.1, 2004
38. A. Elbanhawy, W. Newberry, Influence of Skin Effect on MOSFET Losses, IECON 2006 - 32nd Annual Conference on IEEE Industrial Electronics, Paris, pp. 2313–2317, 2006
39. C. Smith, New MOSFET technology cuts thermal resistance, improves efficiency, PCIM Magazine, pp. 38–43, March 1999
40. A. Sawle, M. Standing, T. Sammon, A. Woodworth, DirectFET: A proprietary new source mounted power package for board mounted power, p-473, PCIM Europe Proceedings, Nurnberg, July 19th – 21st, 2001
41. M. Pavier, A. Woodworth, A. Sawle, R. Monteiro, C. Blake, J. Chiu, Understanding the effect of power MOSFET package parasitic on VRM circuit efficiency at frequencies above 1 MHz, in *Proc. PCIM Eur*, (May 2003), pp. 279–284
42. F. Stueckler, E.A. Vecino, CoolMOSTM C7 650V switch in a kelvin source configuration, Application Note AN 2013–05, Infineon Technologies, May 2013
43. V. Scarpa, F. Brucchi, Kelvin emitter configuration further improves switching performance of TRENCHSTOP™ 5 IGBTs, Bodo's Power, pp. 34–35, Dec 2014,
44. V. Crisafulli, A new package with kelvin source connection for increasing power density in power electronics design. Power Electronics and Applications (EPE'15 ECCE-Europe), 2015, 17th European Conference on. IEEE, pp. 1–8, 2015
45. Efficient Power Conversion, EPC1001 – Enhancement-mode power transistor, data sheet, March 2011
46. GaN Systems, GS66508P, Bottom-side cooled 650 V E-mode GaN transistor, Preliminary datasheet, March 2017
47. Fairchild Semiconductor (Now On-Semiconductor), Dual CoolTM Package PowerTrench® MOSFETs, Product Brief, PB4000–009, 2015
48. F. Recht, Z. Huang, Y. Wu, Characteristics of Transphorm GaN Power Switches, Application Note AN-0002, Transphorm Inc., undated

49. F.C. Lee, W. Zhang, X. Huang, Z. Liu, W. Du, Q. Li, A new package of high-voltage cascode gallium nitride device for high-frequency applications. 2015 IEEE international workshop on integrated power packaging (IWIPP), Chicago, pp. 9–15, 2015
50. Z. Liu, X. Huang, F.C. Lee, Q. Li, Package parasitic inductance extraction and simulation model development for the high-voltage Cascode GaN HEMT. IEEE Trans. Power Electron. **29**(4), 1977–1985 (2014)
51. D. Reusch, J.T. Strydom, A. Lidow, Thermal evaluation of chip-scale packaged gallium nitride transistors. 2016 IEEE applied power electronics conference and exposition (APEC), Long Beach, pp. 587–594, 2016
52. GaN Systems, GS66516T, Bottom-side cooled 650 V E-mode GaN transistor," Preliminary datasheet, March 2017
53. D. Reusch, J.T. Strydom, Understanding the effect of PCB layout on circuit performance in a high frequency gallium nitride based point of load converter, 2013 Twenty-eighth annual IEEE applied power electronics conference and exposition (APEC), Long Beach, pp. 649–655, 2013
54. M. Romig, O. Lopez, 3D packaging advancements drive performance, power and density in power devices, Texas Instruments, White Paper, SLIT126, 2011
55. J.T. Strydom, The eGaN FET-silicon power shoot-out: 2: Drivers, layout, Power Electronics Magazine, Jan 2011, http://www.powerelectronics.com/discrete-power-semis/egan-fet-silicon-power-shoot-out-2-drivers-layout
56. D. Reusch, J.T. Strydom, A. Lidow, Monolithic integration of GaN transistors for higher efficiency and power density in DC-DC Converters. International exhibition and conference for Power electronics, intelligent motion, renewable energy and energy management (PCIM Europe), pp. 986–993, 2015
57. K. Wang, L. Wang, X. Yang, X. Zeng, W. Chen, H. Li, A multi-loop method for minimization of parasitic inductance in GaN-based high-frequency DC-DC converter. IEEE Trans. Power Electron. **PP**(99), 1–1, 2017
58. G.J. Krausse, DE-Series fast power MOSFET, an introduction, Directed Energy, Inc., Tech Note 9300–0002 Rev 3, 2002
59. J.B. Forsythe, Paralleling of power MOSFETs for high power output, International Rectifier Application Note
60. D, Baba, Benefits of a multiphase buck converter, Texas Instruments, Analog Applications Journal, 1Q 2012, http://www.ti.com/lit/an/slyt449/slyt449.pdf
61. D. Reusch, J.T. Strydom, Effectively paralleling gallium nitride transistors for high current and high frequency applications, 2015 IEEE applied power electronics conference and exposition (APEC), Charlotte, pp. 745–751, 2015
62. D. Reusch, J.T. Strydom, J. Glaser, Improving high frequency DC-DC converter performance with monolithic half bridge GaN ICs. 2015 IEEE energy conversion congress and exposition (ECCE), Montreal, pp. 381–387, 2015
63. J. Lu, H. Bai, A. Brown, M. McAmmond, D. Chen, J. Styles, Design consideration of gate driver circuits and PCB parasitic parameters of paralleled E-mode GaN HEMTs in zero-voltage-switching applications. 2016 IEEE applied power electronics conference and exposition (APEC), Long Beach, pp. 529–535, 2016
64. Z. Wang, Y. Wu, J. Honea, L. Zhou, Paralleling GaN HEMTs for diode-free bridge power converters. Applied power electronics conference and exposition (APEC), IEEE, pp. 752–758, 2015
65. A. Elbanhawy, Fairchild semiconductor (now on-semiconductor), limiting cross-conduction current in synchronous buck converter designs, Application Note, AN-7019, Sept. 2005
66. Texas Instruments, LM5113, 100 V 1.2-A / 5-A, Half-bridge gate driver for enhancement mode GaN FETs, datasheet, June 2011
67. B. Yang, J. Zhang, Effect and utilization of common source inductance in synchronous rectification. 2005 IEEE applied power electronics conference and exposition, (APEC), Vol. 3, pp. 1407–1411, 2005

68. N. Galanos, J. Popovic, J.A. Ferreira, M. Gerber, Influence of the magnetic's parasitic capacitance in the switching of high-voltage cascode GaN HEMT. 9th international conference on integrated power electronics systems, CIPS 2016, Nuremberg, pp. 1–6, 2016
69. J.T. Strydom, D. Reusch, Design and evaluation of a 10 MHz gallium nitride based 42 V DC-DC converter, Applied power electronics conference and exposition (APEC), pp. 1510–1516, 2014
70. W. Zhang, X. Huang, F.C. Lee, Q. Li, Gate drive design considerations for high voltage cascode GaN HEMT. Applied power electronics conference and exposition (APEC), 2014 Twenty-Ninth Annual IEEE. IEEE, pp. 1484–1489, 2014
71. V-S. Nguyen, P. LeFranc, J-C. Crebier, Gate driver architectures for high speed power devices in series connection. PCIM Europe 2017; international exhibition and conference for power electronics, intelligent motion, renewable energy and energy management, pp. 1–8, 2017
72. P.M. Roschatt, S. Pickering, R.A. McMahon, Bootstrap voltage and dead time behavior in GaN DC–DC Buck converter with a negative gate voltage. IEEE Trans. Power Electron. **31**(10), 7161–7170 (2016)
73. D. Reusch, M. de Rooij, Evaluation of gate drive over voltage management methods for enhancement mode gallium nitride transistors, in *Applied Power Electronics Conference and Exposition (APEC)*, (IEEE, 2017), pp. 2459–2466
74. L. Xue, J. Zhang, Active clamp flyback using GaN power IC for power adapter applications, Applied power electronics conference and exposition (APEC), pp. 2441–2448, 2017
75. W. Chen, K-Y. Wong, K.J. Chen, Monolithic integration of lateral field-effect rectifier with normally-off HEMT for GaN-on-Si switchmode power supply converters, in IEDM Tech. Dig., pp. 141–144, Dec 2008
76. S. Ujita, Y. Kinoshita, H. Umeda, T. Morita, S. Tamura, M. Ishida, T. Ueda, A compact GaN-based DC-DC converter IC with high-speed gate drivers enabling high efficiencies. IEEE 26th international symposium on power semiconductor devices & IC's (ISPSD), pp. 51–54, 2014
77. Efficient Power Conversion Corporation, EPC2107 – Enhancement-Mode GaN Power Transistor Half-Bridge with Integrated Synchronous Bootstrap, Revised datasheet, July 2017
78. D. Kinzer, Unlocking the power of GaN, applied power electronics conference and exposition (APEC), 2016, Industry Session IS16
79. D. Pattanayak, D. Kinzer, A. Lidow, D. Anderson, J.T. Strydom, Power electronics: Beyond the silicon limit. Annual symposium, Santa Clara Valley-San Francisco Chapter of Electron Devices Society, Sept 6, 2013
80. J.T. Strydom, eGaN® FET- Silicon Power Shoot-Out Volume 11: Optimizing FET On-Resistance, Power Electronics Magazine, Oct. 2012, http://www.powerelectronics.com/discrete-power-semis/egan-fet-silicon-power-shoot-out-volume-11-optimizing-fet-resistance
81. Silicon Labs, Si827x , 4 Amp ISOdriver with High Transient (dV/dt) Immunity, Datasheet, Preliminary Rev. 0.5
82. A.J. Hanson, J.A. Belk, S. Lim, C.R. Sullivan, D.J. Perreault, Measurements and performance factor comparisons of magnetic materials at high frequency. IEEE Trans. Power Electron. **31**(11), 7909–7925, 2016

Chapter 7
GaN in AC/DC Power Converters

Fred Wang and Bo Liu

With low power loss and high switching capability, GaN device technology can benefit power electronics converters in a number of ways. GaN devices can substitute Si devices in a converter for improved efficiency and power density. In addition, GaN devices can simplify the converter topology, leading to lower cost and higher reliability. GaN devices can also introduce system-level benefits as a result of their fast speed and therefore better dynamics. They can also enable new applications. This chapter presents how GaN technology can benefit AC/DC converters, covering both the single-phase and three-phase cases. Associated challenges in using GaN devices and potential solutions are discussed.

7.1 GaN-Based Single-Phase AC/DC Converter

7.1.1 Direct Device Substitution and Topology Simplification

In general, GaN devices have low on-resistance, low output capacitance, and low input capacitance and gate charge. Enhancement-mode (E-mode) GaN devices have no reverse recovery issue associated with antiparallel diodes in Si cases; and cascode GaN devices have only a low-voltage Si MOSFET in series and thus slight reverse recovery effect. Therefore, reverse recovery loss and associated parasitic

F. Wang (✉) · B. Liu
Center for Ultra-wide-area Resilient Electric Energy Transmission Networks (CURENT), The University of Tennessee, Knoxville, TN, USA
e-mail: fred.wang@utk.edu

G. Meneghesso et al. (eds.), *Gallium Nitride-enabled High Frequency and High Efficiency Power Conversion*, Integrated Circuits and Systems,
https://doi.org/10.1007/978-3-319-77994-2_7

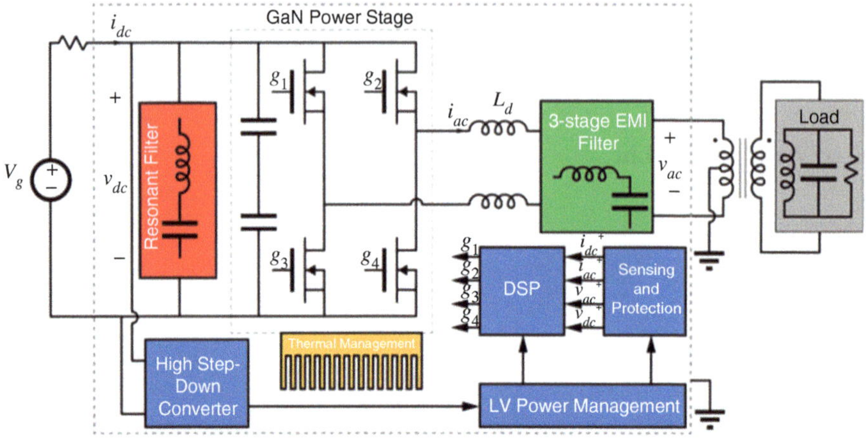

Fig. 7.1 Schematic of GaN-based 2 kW single-phase inverter [2]

ringing during device turn-off can be largely avoided in GaN converters. These properties help GaN-based power converters achieve low conduction and switching loss compared to Si counterparts [1].

Direct replacement of Si devices by GaN devices in a single-phase AC/DC converter can reduce power loss and improve power density with reduced cooling. This concept can be illustrated with a converter developed for the Google Little Box Challenge, as shown in Figs. 7.1 and 7.2. A 2 kW single-phase photovoltaic (PV) inverter using 650 V GaN devices on a conventional hard-switching full-bridge topology, with a modestly increased switching frequency at 100 kHz, achieved the peak efficiency of 97.6% and power density of 102 W/in^3 [2].

The excellent properties of GaN devices can also revitalize some simple and mature topologies that in many cases have fallen out of favor in Si MOSFET-based converters due to their performance limitations. One example is totem-pole bridgeless PFC, as shown in Fig. 7.3. This topology is the simplest among various boost-type power factor correction (PFC) rectifier variants, with only two rectifier diodes or two MOSFETs, two active switches, and one AC boost inductor. However, the topology is seldom used in hard-switching continuous-current mode (CCM) due to the high turn-on switching loss and reverse recovery spikes of Si MOSFETs and ringing of the antiparallel body diode [3]. Instead, it is mainly used in soft-switching operation mode, so that the aforementioned issues can be mitigated through zero-voltage switching (ZVS). However, high-current ripples in this mode still result in significant conduction loss and turn-off loss, limiting the application in low frequency and low power level.

Replacing Si devices with GaN devices, the significantly reduced junction capacitance and zero reverse recovery charge in E-mode GaN or slight reverse recovery charge in cascode GaN, together with the high switching speed, efficiently reduce the switching loss and voltage/current stress, making CCM totem-pole PFC

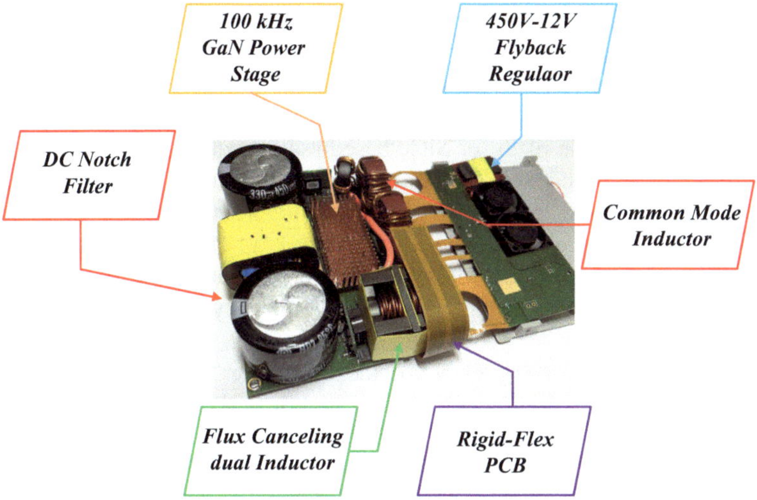

Fig. 7.2 Benchtop GaN-based 2 kW single-phase inverter [2]

Fig. 7.3 Schematic of totem-pole PFC

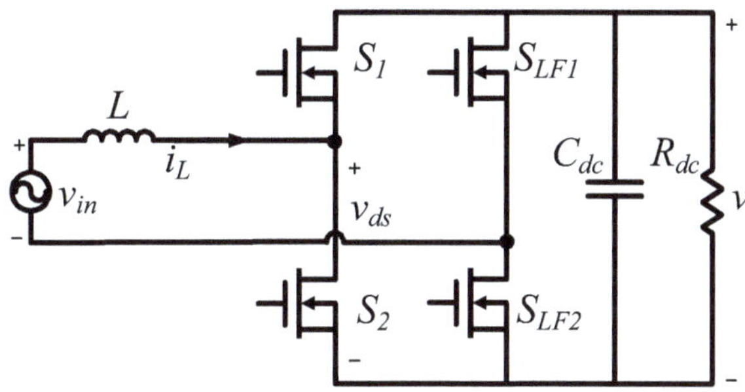

a practical solution. As reported in [4] and [5], GaN-based PFC can achieve 99% power efficiency at 50 kHz, 1 kW, or at 100 kHz, 2.4 kW, in hard-switching CCM operation, whereas the corresponding Si-based PFC cannot be efficiently operated in CCM at such frequencies and requires more complicated critical mode (CRM) or hybrid CCM/discontinuous mode (DCM) to achieve 94.6% efficiency at 200 W [6] or 98.3% efficiency at 3 kW [7].

With low switching loss, GaN-based converters can be switched at higher frequency for reduced passives and higher power density while still maintaining a decent efficiency. The benefit can be combined with soft-switching, given that the turn-off loss of the GaN is small and a significant portion of turn-on loss from junction capacitive energy can be eliminated with ZVS operation. Meanwhile, the drawback of high conduction loss in ZVS schemes will be insignificant owing to low on-resistance of GaN devices. These benefits can be illustrated in the CRM totem-pole PFC case. By pushing switching frequency to 1∼3 MHz range with CRM operation, 98.8% peak efficiency and >200 W/in^3 power density are achieved on a 1.2 kW totem-pole PFC [3].

7.1.2 Enabling New Applications and Topologies

Wireless power transfer (WPT) is used here as an example to illustrate the applications that can be enabled by GaN-based single-phase converters. Traditionally, class E converters made of Si- or SiGe-based radio-frequency (RF) MOSFET are adopted in WPT applications. Due to the limited device rating and high current stress of resonance tank, these WPT systems are often rated at low power and used in RF fields. GaN devices have enabled high-power WPT converters. Reference [8] reported a 6.78 MHz or 13.56 MHz grid-tied WPT charger up to 10 kW. In [9], a 4 kW, 13.56 MHz inverter stage with 96.5% efficiency is achieved using 650 V E-mode GaN HEMT.

Different topologies using GaN for the transmitter side are proposed as in [10, 11], focusing on improving the overall efficiency with fewer conversion stages, as illustrated in Fig. 7.4. Figure 7.4a topology consists of a two-stage GaN-based 6.78 MHz transmitter with CRM totem-pole PFC rectifier and ZVS-tank-assisted full-bridge inverter. The benchtop test demonstrates 98.6% and 93% power efficiencies for the rectifier and the inverter, respectively, at 100 W full load, as shown in Fig. 7.8, achieving 91.7% overall efficiency. Fig. 7.4b topology further combines the two-stage by eliminating one phase leg and operating the rectifier in DCM totem-pole PFC mode. The tested overall efficiency at 100 W full load is 92.1%, slightly higher than the two-stage with fewer components and could be further optimized [11]. These results clearly demonstrate the potential of GaN converters in WPT system to improve its typical low 50% $\sim$ 70% system efficiency [12].

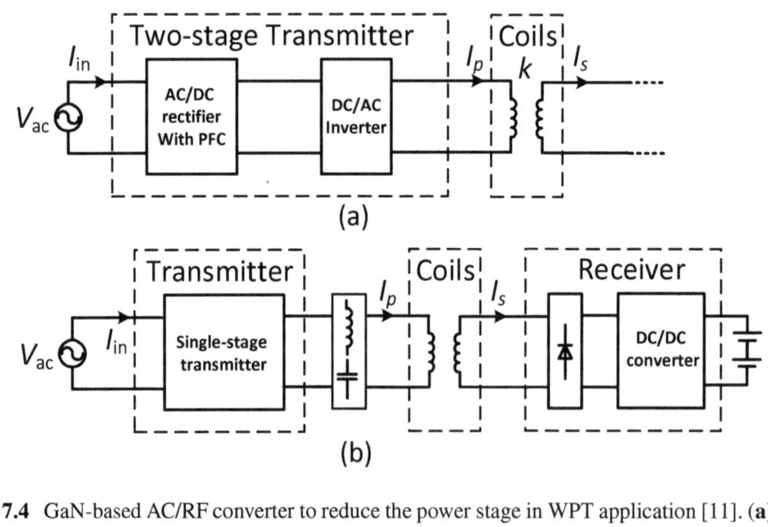

Fig. 7.4 GaN-based AC/RF converter to reduce the power stage in WPT application [11]. (**a**) two-stage transmitter, (**b**) single-stage transmitter

Owing to low switching loss and high speed of GaN, some conventional control techniques used in grid-tied AC/DC application can also be transformed into high-frequency applications. A recent work in [13] that introduces the selective harmonic elimination (SHE) pulse width modulation to GaN-based WPT systems shows a good example. A technique which was used to selectively eliminate low-frequency harmonics in grid-tied inverters now provides a single RF inverter the capability of dual-output frequencies simultaneously, covering both the wideband dual-mode (100 kHz and 6.78 MHz) and narrowband dual-mode (87–300 kHz), showing attractiveness in low-cost multi-receiver WPT applications. This extended multifrequency programmed pulse width modulation (MFPWM) is also applied in the electrosurgical power supply where a 50 kHz ultrasonic output is used for dissection and sealing and a 500 kHz RF output is used to cut/coagulate tissue [14].

Operated at high switching frequencies, GaN-based converters have better dynamic performance, wide harmonic suppression capability, and faster tracking response owing to the high control bandwidth. In addition, the elimination of antiparallel diode and the fixed on-resistance in both forward and reverse direction ensure AC/DC efficiency independent of the power factor, greatly simplifying the loss distribution and thermal design and idealizing the output impedance behavior. These, in turn, enable other smart functionalities of GaN-based AC/DC, such as smart resistor concept [15], where a GaN converter is controlled as an active resistor to compensate and mask the nonlinearity and negativity of AC or DC load with decent transient response and efficiency, ensuring stability of the remaining system.

7.1.3 Challenges and Potential Solutions

Using GaN device technology to push for high switching frequency single-phase AC/DC converter also introduces many design challenges. As an example, several intrinsic issues exist in CRM boost-type PFC, and they become significantly magnified when running at MHz switching. New challenges associated with digital control of such high-frequency converters also arise. These key design aspects for high-frequency single-phase PFC are extensively studied in recent years and are summarized here.

ZVS Extension

For a CRM totem-pole PFC in Fig. 7.3, when the main switch S_2 is turned off at the zero crossing, ideally, a resonance between the AC inductor and device junction capacitors will discharge S_2 junction capacitance and realize ZVS of S_2, as shown in Fig. 7.5. However, the ZVS condition is only valid when input peak voltage V_{in} is lower than $0.5V_o$. Above that, ZVS of the main switch is lost. The existence of non-ZVS region will lead to unacceptable high switching loss especially when totem-pole PFC is pushed to high switching frequencies.

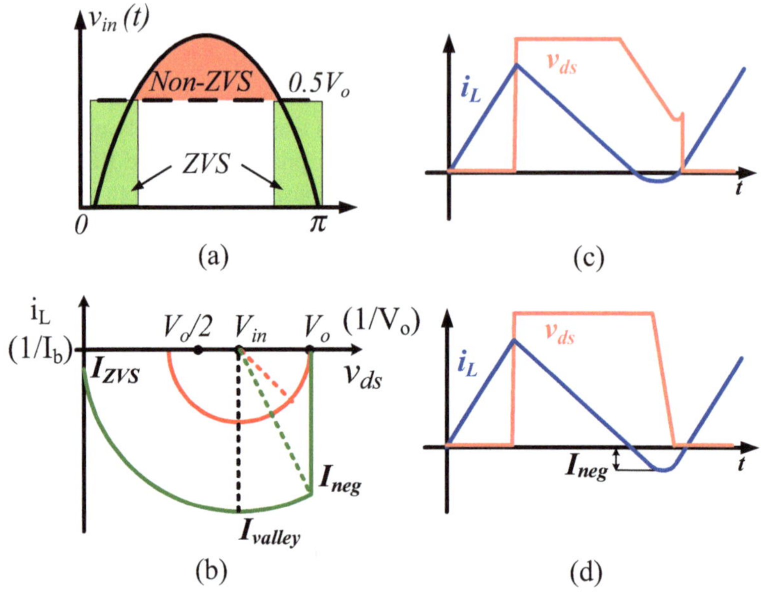

Fig. 7.5 Illustration of ZVS issue and TCM scheme in totem-pole PFC. (**a**) Non-ZVS region; (**b**) from CRM to TCM to extend ZVS; (**c**) voltage and current in non-ZVS case; (d) voltage and current in ZVS extension case

To achieve the ZVS operation, S_1 has to be conducted for a longer time so that enough negative current I_{neg} flows through the inductor and stores enough initial energy to fully discharge C_{oss} of S_2 during the LC resonance after SR switch is turned off. This modified scheme is often named as triangular-current-mode (TCM) or quasi-square-wave (QSW) mode [6, 16–20]. In a practical implementation, a margin for turn-on instant has to be kept before the inductor current crosses zero in case of any delay or timing mismatch along the modulation and gate path. This requires extra negative $I_{neg,}$ leading to higher circulating current and conduction loss. Therefore, the margin should be small.

Zero-Crossing Distortion and Digitally Controlled Variable On-Time Modulation

With high switching frequency, the current harmonics and power quality issue may also become pronounced. In totem-pole PFC, the distortion mainly comes from the violation of constant on-time control assumption. Ideally, if the negative current I_{neg} before the turn-on of S_2 is zero, and I_{ZVS} after the conduction of S_2 "body diode" is negligible, the peak current of the inductor is nearly double of the average current, considering that the resonant current in a partial cycle of the

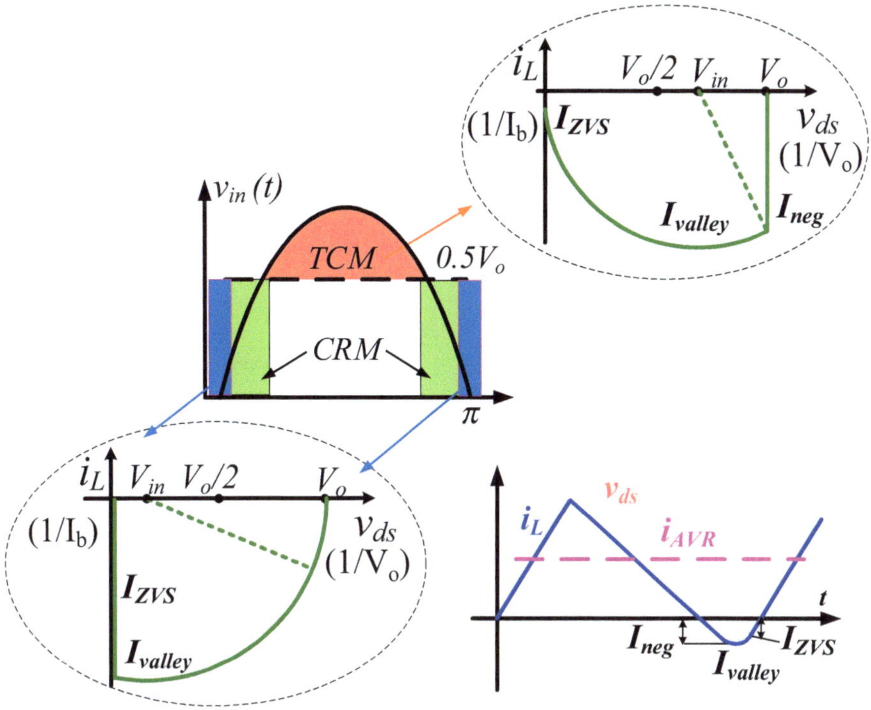

Fig. 7.6 Illustration of the current distortion mechanism in CRM-TCM-based totem-pole PFC

very high-frequency resonance is negligible. Consequently, the average of CRM switching current always follows the sinusoidal shape of the input voltage over a whole line cycle, as in (7.1).

$$V_{\mathrm{in}} \approx L\frac{I_{\mathrm{peak}} - 0}{T_{\mathrm{on}}} \quad \Rightarrow \quad I_{\mathrm{peak}} \approx \frac{V_{\mathrm{in}} T_{\mathrm{on}}}{L} \tag{7.1}$$

However, practically, two types of distortions exist over one line cycle. In the natural ZVS region when $V_{\mathrm{in}} < 0.5V_{\mathrm{o}}$, as shown in Fig. 7.6, there is always a certain amount of negative I_{ZVS} around the ZVS switching instant. This is because the radius of the resonance circle $r = V_{\mathrm{o}}\text{-}V_{\mathrm{in}}$ is larger than V_{in}. Especially around zero crossing region, high I_{ZVS} could be comparable to the positive peak current, thus canceling the effective input average current and leading to current distortion. Only as V_{in} increases, the component becomes smaller and eventually becomes null at the boundary of CRM and TCM.

In the $V_{\mathrm{in}} > 0.5V_{\mathrm{o}}$ region, although TCM helps ZVS realization, the introduced negative current I_{neg} becomes the second negative component that violates the ideal assumption of constant on-time control, whereas the I_{ZVS} term may be minimized by well controlling the I_{neg}.

These two negative components increase significantly at higher switching frequencies. As analyzed in (7.2) and (7.3), their magnitudes are inversely proportional to the square root of the AC inductance. At higher frequencies, the input AC inductance has to be decreased to ensure CRM operation, thus leading to higher distortion of the fundamental current.

$$I_{neg} \approx j_{neg} I_{base} = j_{neg} \frac{V_{base}}{\sqrt{L / C_{oss\_eq}}} \tag{7.2}$$

$$I_{ZVS} \approx j_{ZVS} I_{base} = j_{ZVS} \frac{V_{base}}{\sqrt{L / C_{oss\_eq}}} \tag{7.3}$$

j_{neg} or j_{ZVS} in (7.2) and (7.3) is the normalized current referred to the base current and is only a function of input and output voltage, solvable from the normalized trajectory plot and irrespective of the switching frequency.

To compensate the distortion in MHz PFC application, extra turn-on time should be exerted to the main switch to offset the impact of the two above negative current components primarily in the zero-crossing region and TCM region. This leads to the variable on-time control in a line cycle [6, 18–20]. Accurate calculation of the desired on-time is feasible but involves massive computation and high-performance digital controller [6]. Other methods such as look-up table could be applied to simplify computation efforts and reduce control cost [18, 19].

Digitally Controlled Interleaving and Ripple Cancellation

An obvious issue of single-phase CRM totem-PFC is the high switching current ripple due to the nature of CRM operation. High DM noise on the AC side must be filtered by the DM filter to comply with electromagnetic interference (EMI) standards, leading to bulky filter size and high passive loss. In addition, with controlled on-time instead of the fixed switching period, the switching frequency will vary as input voltage changes or load current changes, leading to wideband noise. Therefore, a single-stage DM filter may have to be designed based on the noise at the lowest frequency, again increasing filter efforts.

Interleaving of two-phase CRM totem-pole PFC thus becomes beneficial as shown in Fig. 7.7. By shifting the two phases 180° apart, the fundamental switching ripple will be canceled, reducing the filter size by 50% theoretically.

Two types of interleaving phase controls are often applied for variable frequency PFC, i.e., open-loop interleaving and closed-loop interleaving [19]. The challenge associated with GaN-based MHz totem-pole PFC is that with limited computation speed and resources, the interleaving phase control may result in noticeable phase error at high frequency. The minimum-phase variation could be one switching

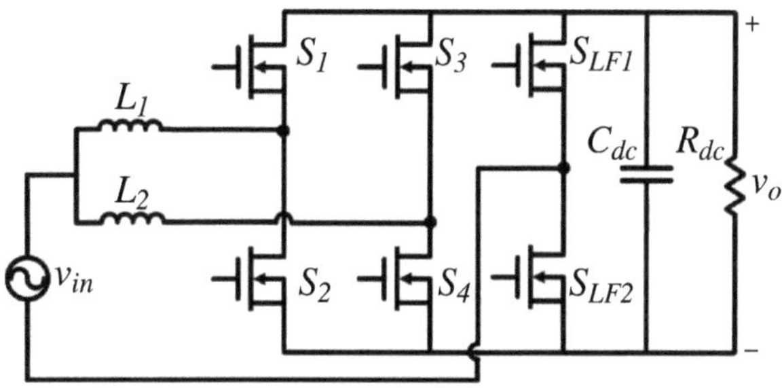

Fig. 7.7 Schematic of two-phase interleaved totem-pole PFC

cycle or more by the relatively slow closed-loop controller or one clock cycle by the discrete-step DSP clock of the open-loop controller. The phase error, in turn, deteriorates the ripple cancellation performance and could lead to a significant amount of residual noise at the fundamental switching frequency, enlarging the size of EMI filter. In general, closed-loop interleaving control is more robust and preferred in low-frequency design, since it detects the phase timing of the two phase legs and actively tunes the on-time of the slave phase leg and therefore guarantees its soft-switching and CRM transition [19]. To utilize this control in MHz high-frequency converters, the control delay has to be reduced either by adopting advanced high-speed microcontrollers with core frequency above hundreds of MHz or using FPGA-type controller ICs that support parallel computation. Otherwise, open-loop interleaving control is much faster and is preferred in this application. However, since open-loop interleaving does not ensure the safe CRM transition of the slave phase leg, loss of CRM may occur and induce input current oscillation on the slave phase leg. According to [18], the turn-on instant synchronization of the slave phase leg should be used since it has the self-stabilization ability in case of oscillations induced by delay disturbance and CRM loss, while the turn-off instant synchronization will introduce subharmonic oscillation and should be forbidden.

7.2 GaN-Based Three-Phase AC/DC Converter

7.2.1 Benefits

Due to the limited current rating of the existing GaN devices, there have not been many applications for GaN-based three-phase AC/DC converters. Still, clear benefits can be observed with the relatively limited examples. Similar to single-phase applications, applying GaN devices in three-phase AC/DC can help achieve higher efficiency compared to its Si counterpart. In the three-phase PV inverter application, a 10 kW, 400 V DC rated two-level voltage source inverter (VSI) is designed with 650 V top-cooled surface mount GaN and operated moderately at 50 kHz switching frequency and demonstrates 98.8% peak efficiency excluding EMI filters [21, 22]. With the superior performance, GaN is also developed for the motor drive application recently [23]. The study compared the performance of GaN- and Si IGBT-based 3 HP 230 V AC induction motor at 100 kHz and 15 kHz, respectively. The results show slightly higher efficiency in GaN-based inverter affected by the inclusion of output filter to handle *dv/dt* stress of GaN, but significant efficiency improvement of the whole motor system due to the reduction of heating on the motor as a result of lower switching harmonics in the GaN converter. A 1.5 kW 300 V DC rated VSI with sine-output filter and 650 V GaN devices is demonstrated in [24], showing 97% efficiency including the filter loss at 100 kHz switching frequency. The increased switching frequency results in a smaller filter which helps reduce harmonics and loss in the motor.

One new application reported in [25, 39] is battery charger intended for aircraft applications, where GaN-based AC/DC converter is used as active front-end to regulate the intermediate DC bus voltage. GaN technology can help shrink the size and weight of the converter in this application by pushing the switching frequency to hundreds of kHz, leading to smaller filters required in aircraft applications. Switching at 450 kHz using GaN devices, for example [25], instead of 70 kHz with SiC JFET [26] or 68 kHz with Si devices [27] for a three-phase Vienna-type rectifier, weight and volume of passive components are significantly reduced, still maintaining high power efficiency.

7.2.2 *Challenges and Potential Solutions*

Although the fast switching GaN devices introduce many benefits, they also pose severe challenges in applications to hard-switching three-phase AC/DC converters. A comprehensive understanding of the characteristics and associated effects and a reformulation of design procedures developed for Si-based AC/DC converters are necessary. Some of the key issues and potential solutions are presented here.

Effect of Junction Capacitance and High Switching Frequency on Voltage Distortion

The fast switching intensifies the impact of parasitics on switching commutation, negatively affecting power quality of the power converter. For example, when a GaN-based AC/DC or DC/AC converter is operated at high switching frequency, nonideal commutation can cause severe voltage distortion [25, 27, 29] (Fig. 7.8).

Higher switching frequency exacerbates the voltage distortion as a result of shorter switching period. This effect is worse in switch-diode configured converters such as Vienna-type rectifiers, and it can also be accompanied by dead-time effect in phase-leg configured voltage source converters or overlap time effect in current source converters (Fig. 7.9).

Since the average of the ideal PWM voltage over a switching cycle represents the average fundamental voltage, the distortion voltage due to the slow charging

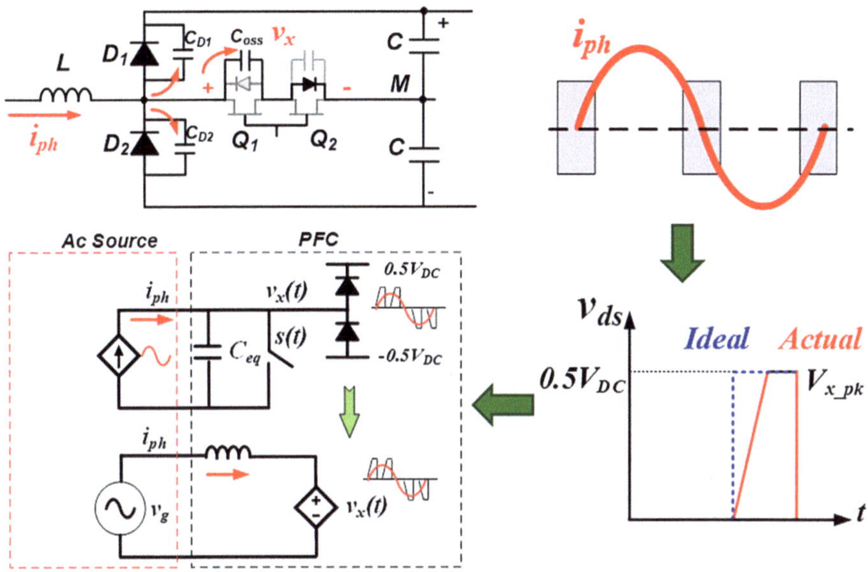

Fig. 7.8 Illustration of voltage and current distortion in a three-phase Vienna-type rectifier due to the PWM loss caused by charging of device junction capacitance at turn-off

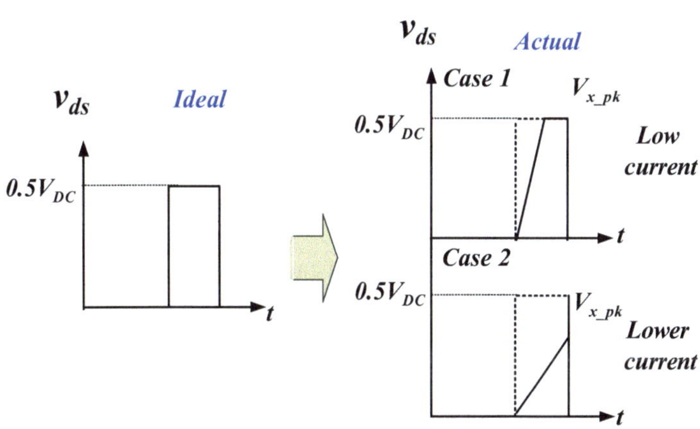

Fig. 7.9 Ideal and actual PWM voltage shape during turn-off transient

process introduces volt-second loss. From this perspective, a feedforward turn-off compensation scheme can be applied [25], to reshape the actual PWM voltage such that the compensated PWM voltage has the same volt-second as the ideal case. A simple approach, as shown in Fig. 7.10, is to equalize the two-shaded volt-second areas since the other areas are commonly shared. The compensated duty cycle for both SPWM and SPWM+3rd schemes is unified in (7.4), with the only difference in a phase angle-related factor α [25]. In (7.4), d_{th} denotes the boundary duty cycle,

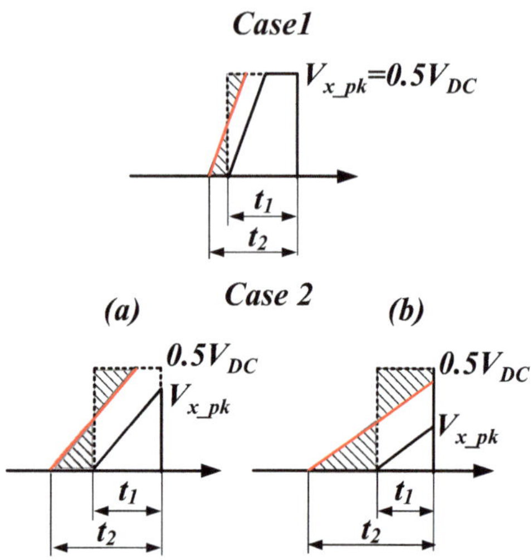

Fig. 7.10 Two voltage distortion cases associated with different current level and the proposed turn-off compensation scheme considering the voltage shape before and after compensation. Dashed line: ideal PWM voltage. Solid black line: actual uncompensated turn-off voltage. Red line: compensated voltage based on the shaded volt-second area

and R_{target} represents equivalent input impedance of PFC, and C_{eq} is the charge-based equivalent output capacitance over the voltage range [0, $V_{\mathrm{x\_pk}}$] of T-shape connected devices per phase leg.

$$\Delta d = \begin{cases} \frac{R_{\mathrm{target}} C_{\mathrm{eq}}}{2T_s} \frac{\alpha(\theta)}{d}, & (\ d \geq d_{\mathrm{th}}) \\ \sqrt{\frac{2R_{\mathrm{target}} C_{\mathrm{eq}}}{T_s} \alpha\,(\theta)} - d, & (d < d_{\mathrm{th}}) \end{cases} \tag{7.4}$$

The presented approach not only compensates the distortion but also provides an accurate analytic model for output PWM voltage distortion as a function of DC link voltage, AC operation point, device junction capacitances, and converter switching frequency. Taking switching frequency as an example, theoretical impact of switching frequency on voltage distortion in a Vienna-type rectifier is illustrated in Fig. 7.11. For the given device and operation condition, only when fs is above 100 kHz, this distortion gets pronounced (Figs. 7.12 and 7.13).

The whole control diagram of the three-level rectifier including distortion compensation is illustrated in Fig. 7.14.

Figure 7.15 shows the test results. Without an accurate voltage error compensation scheme, an obvious distortion was observed in the input currents, mainly consisting of 6 $k \pm 1$ (k is a positive integer) order harmonics due to the six zero crossings of the three-phase currents within a line cycle. After applying the proposed scheme, those harmonics were significantly suppressed, and the current THD was reduced from 10.3% to 3.0%.

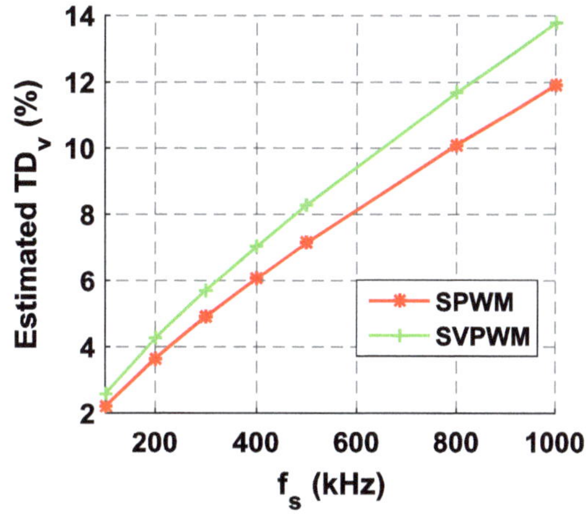

Fig. 7.11 Modeled distortion factor vs. switching frequency

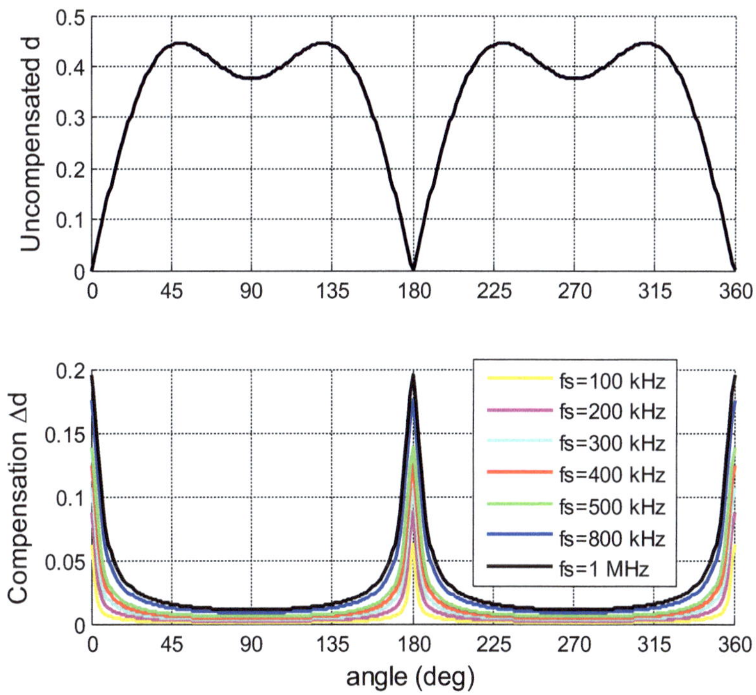

Fig. 7.12 Duty cycle compensation vs. f_s for SVM

Current Sampling

Another challenge in applying GaN to high-frequency high-density AC/DC converters is sampling. Because of the high switching speed of GaN devices, high *dv/dt* and *di/dt* related transient noises and ripples are augmented in a hard-

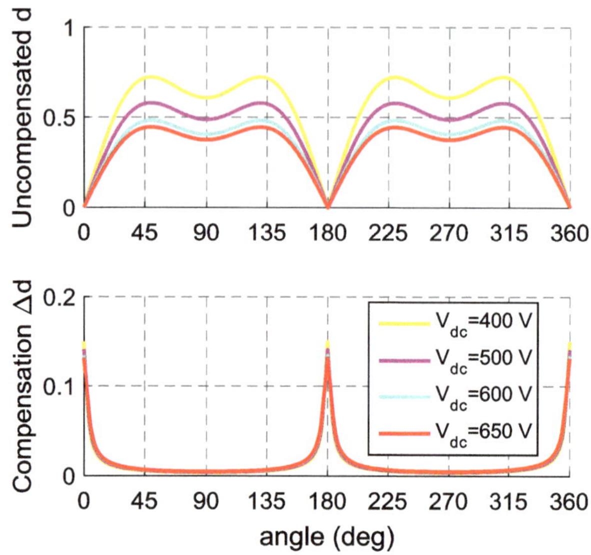

Fig. 7.13 Duty cycle compensation considering nonlinear C_{oss} for SVM

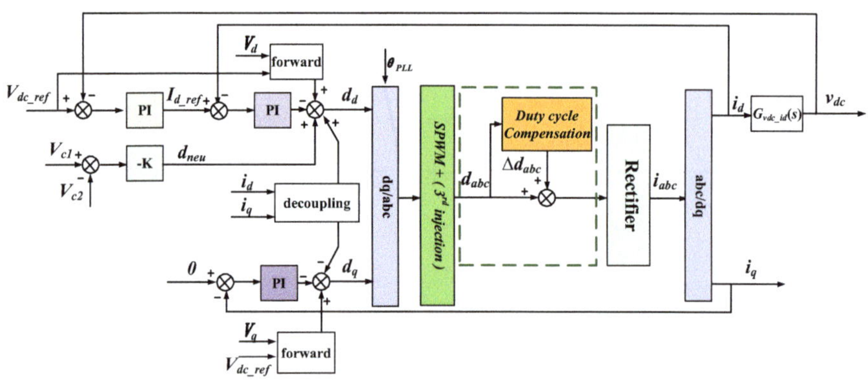

Fig. 7.14 Control diagram of three-phase PFC with the presented modulation compensation scheme

switching three-phase AC/DC converter. In addition, as analog-to-digital converter (ADC) conversion time becomes relatively long for high-frequency converter, signal measurement quality is more sensitive to sampling timing.

Synchronizing ADC with PWM carrier is a typical way for power electronic converters to avoid noise spikes and associated ringing of the measured signals around switching instants. Current sampling instant, in particular, should be carefully selected since an improper sampling instant will lead to inaccurate current feedback information and jeopardize the current loop control. A traditional realization is to align ADC sampling in the middle of a turn-on interval through triangle-shape-based modulation carrier. By center-aligning the timing, in principle, the average current value would be accurately acquired.

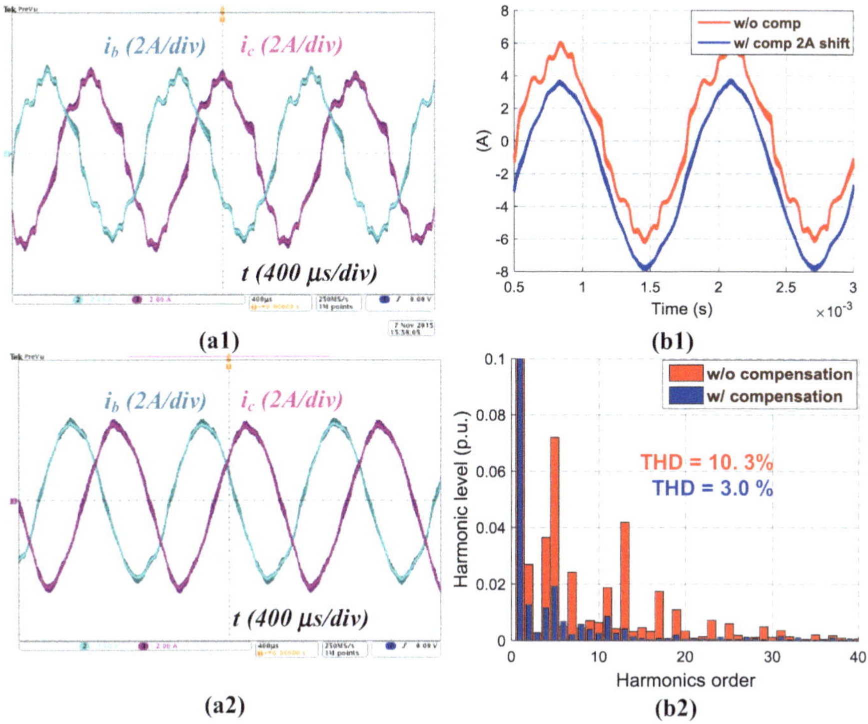

Fig. 7.15 Experimental comparison of current quality with and without the presented compensation scheme. (**a1**-**a2**). Experimental waveforms of i_b and i_c currents without and with compensation. (**b1**-**b2**). Comparison of i_b current without and with compensation. (blue curve in (**b1**) is shifted down by 2 A)

However, the actual duty cycle in AC/DC converters is time-varying in the range from 0 to 1, resulting in varying time window for sampling conversion. Alternating the sampling instant between the rising edge and falling edge of the switching ripple when duty cycle crosses 0.5 is proposed for a single-phase PFC to make the sampling interval longer and leave ample time for ADC sampling-and-hold and digital conversion [30]. Unfortunately, this single-phase approach cannot be conveniently applied for three-phase converters as will be analyzed below. Even worse, since the switching period in GaN-based high-frequency converters is very short, almost comparable to the sampling period of a commercial ADC, the selection of longer sampling interval becomes essential and more challenging.

Issues in Existing Sampling Methods

Since the triangle carrier-based modulation is widely adopted in DSP coding and acts as the synchronization source of sampling instants, the well-known 2-stack

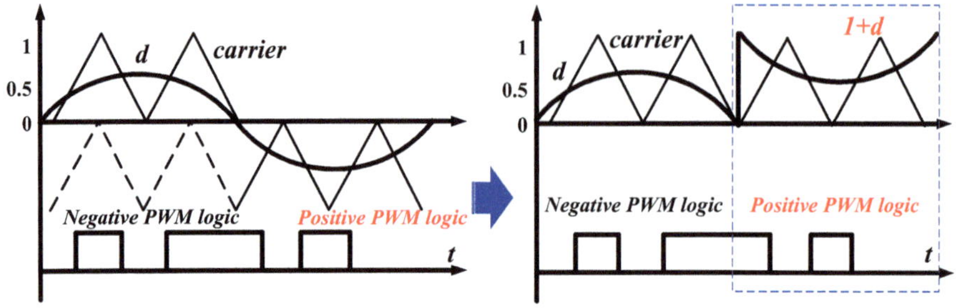

Fig. 7.16 Modulation implementation in Vienna-type rectifier

carrier-based modulation in DSP for the three-level Vienna-type converter is implemented. To avoid the negative region of the carrier, an equivalent modification can be made as shown in Fig. 7.16. Different from two-level converters, the comparator logic in Vienna-type converter should be toggled per half line cycle to match the phase current direction. Here, the instantaneous duty cycle d is expressed in (7.5),

$$d = M \sin(\theta) \tag{7.5}$$

where M is the modulation index defined as the ratio of peak phase voltage V_{N} over the half DC-link voltage (7.6)

$$M = \frac{V_{\mathrm{N}}}{V_{\mathrm{DC}}/2} \tag{7.6}$$

Although sampling at the middle point of either turn-on or turn-off interval can represent the averaged value, it is expected to sample at the long interval considering the relatively time-consuming ADC process in high switching converters as shown in Fig. 7.17. From this concept, a sampling scheme for one phase of the Vienna-type rectifier is proposed, as illustrated in Fig. 7.18. If $M < 0.5$, in positive half line cycle, sampling conversion should be triggered when the carrier counter CTR reaches its period value PRD (CTR = PRD), while in the negative half cycle, sampling point should be placed at CTR = 0. When $M > 0.5$, this scheme can still be applied except for some modifications. As shown in Fig. 7.18b, in the positive half line cycle, if the instantaneous duty cycle d is higher than 0.5, sampling conversion should be reversed, i.e., triggered at CTR = 0; in the negative line cycle, if d is below −0.5, sampling conversion should be at CTR = PRD. Thus, the sampling instant is always placed at the long interval in one switching cycle.

However, when applying this scheme to three phases, three sampling instants will conflict. Taking one region with phase angle <30° as an example, as illustrated in the shadow region of Fig. 7.19, phase A and B should be sampled at CTR = PRD, while phase C should be at CTR = 0. Thus, three conversion instants are not the

Fig. 7.17 Desired current sensing instant

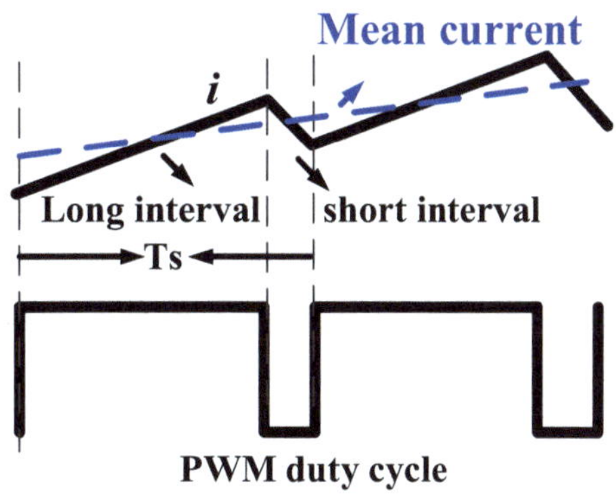

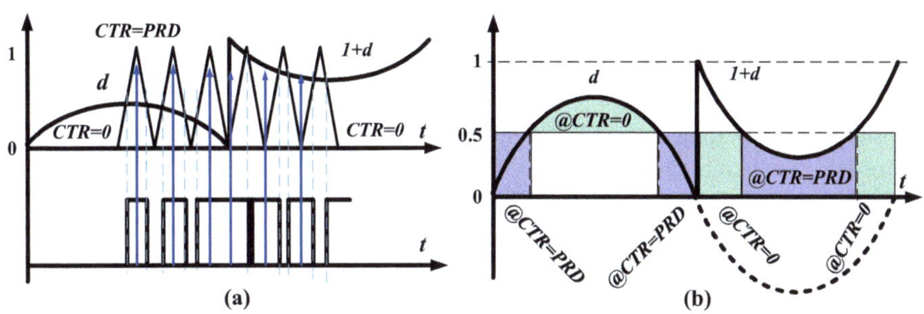

Fig. 7.18 Illustration of a basic hybrid sampling scheme for one phase of Vienna-type rectifiers: (a). $M < 0.5$; (b). $M > 0.5$

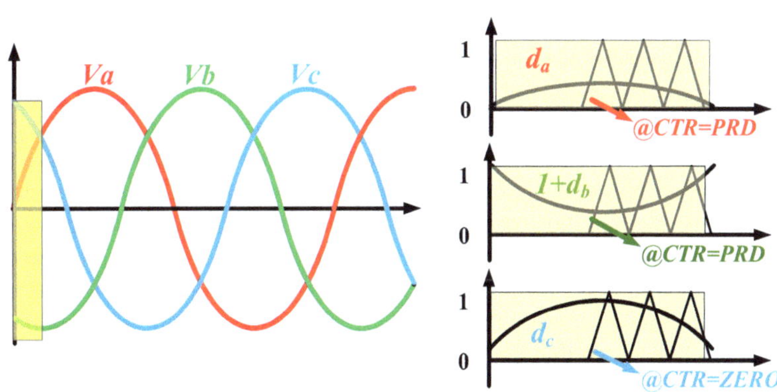

Fig. 7.19 Illustration of the sampling instant conflicts in three-phase converters with $M < 0.5$

same, requiring different sampling instant monitoring, extra triggering sources and independent ADC interrupt control logic, which complicates the controller design regarding both hardware and software.

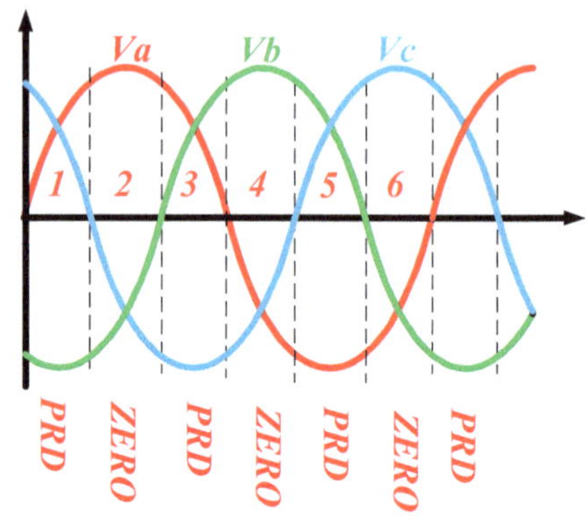

Fig. 7.20 Sector-based sampling scheme for three-phase converters with carrier counter value CTR = zero or PRD

Sector-Based Method

To sample in the long interval and still accommodate all three phases, a simple sector-based scheme can be employed [31]. As shown in Fig. 7.20, a whole line cycle is divided into six sectors. For each sector, there is only one common instant to align ADC samplings of three-phase currents, either at CTR = ZERO or CTR = PRD, depending on its sector number. For example, when the angle is within sector 1, sampling for three phases should be placed at CTR = 0; for section 2, the sampling instant rotates to CTR = PRD. Although this common sampling instant alternates between two nearby sectors, the sector partition can be easily realized using phase-locked-loop (PLL) angle. And there is no need to switch the sampling instants as duty cycle d or modulation index M changes. In addition, three-phase currents share the same ADC triggering signal and thus can be sampled by DSP's internal ADC and processed only by one interrupt.

Though simple, the sector-based scheme cannot guarantee that all three phases are sampled at their long interval simultaneously. This can be clarified through the following probability analysis considering different modulation indexes and line angles for three-phases, illustrated in Fig. 7.21. A carrier/modulation ratio $N = 13$ is asserted, and only half line cycle is plotted due to the symmetrical nature of AC waveforms. For simplicity, the analysis uses only phase A as an example, and phases B and C will follow the same pattern as phase A due to symmetry. Also notice that in Fig. 7.21, when any of the three-phase duty cycle d_{abc} falls in the negative region, it will be shifted to $1+ d_{abc}$.

In Fig. 7.21a, when M is below 0.5 and line angle θ is in sector 1 and 3, the long interval for phase A in a switching cycle occurs around CTR = PRD. Per the sector-based scheme, the sampling instant is also placed at CTR = PRD. However, when θ is in sector 2, the long interval for phase A in a switching cycle should be at CTR = PRD which differs from the schemed triggering logic CTR = 0. Therefore, phase A will be sampled at its short interval in this sector, whereas the other two phases will be sampled at their long intervals.

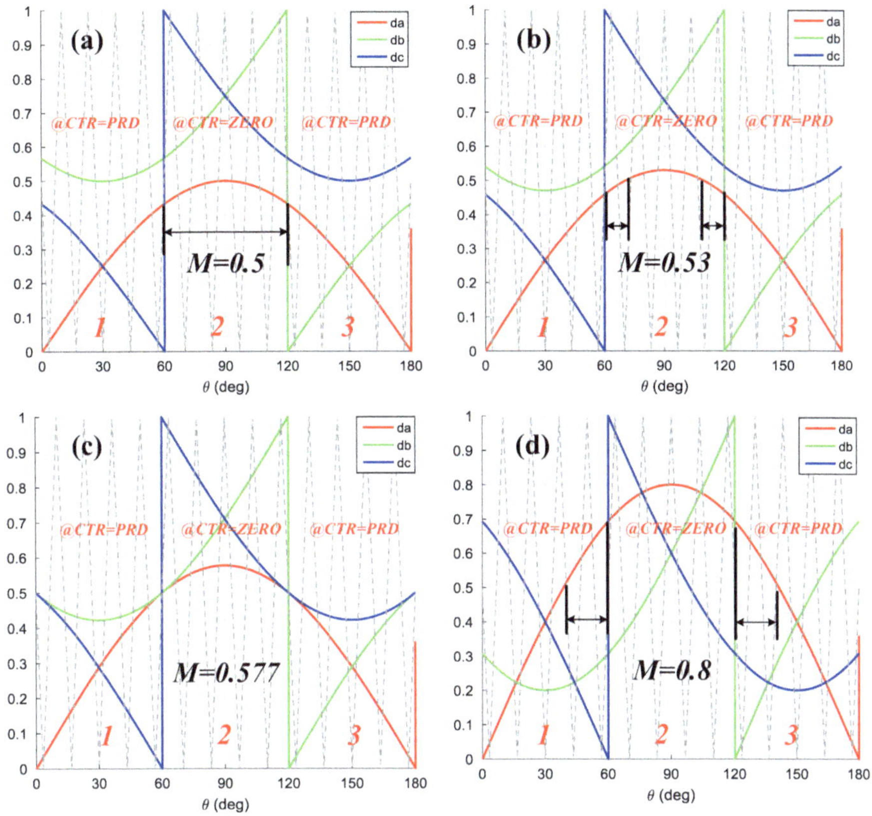

Fig. 7.21 Short intervals of proposed sampling scheme in phase A of a Vienna-type rectifier, with carrier/modulation ratio 13 under different modulation indexes. (**a**) M=0.5; (**b**) M=0.53; (**c**) M=0.577; (**d**) M=0.8

Use λ to depict the probability of sampling instant occurring in the short interval instead of the desired long interval of a switching cycle. It can be defined as the angle range with short interval sampling over the whole line cycle as in (7.7)

$$\lambda = \frac{\theta_{\text{short\_interval\_sampling}}}{360} \tag{7.7}$$

For phase A, if $M < 0.5$, the probability is

$$\lambda_A = \frac{60}{180} = \frac{1}{3} \tag{7.8}$$

From Fig. 7.21a, b, since the peak of d_a locates in sector 2, d_a will cross $d = 0.5$ boundary (above this level, the long interval for phase A occurs at CTR = 0) first in this sector, as M increases. Thus,

Table 7.1 Ratio of sampling instant occurring in short switching interval in a line cycle

$M \leq 0.5$	$0.5 < M \leq 1/\sqrt{3}$	$M > 1/\sqrt{3}$
1/3	$\frac{\left(\sin^{-1}\left(\frac{0.5}{M}\right)-60\right)}{90}$	$\frac{\left(\sin^{-1}(0.866M)-30\right)}{90}$

$$d_{\mathrm{a}} = M \sin(\theta) = 0.5 \tag{7.9}$$

Gradually as M further increases, d_{a} will start going beyond $d = 0.5$ boundary in sector 1 and sector 3, as shown in Fig. 7.21d. The crossing boundary angle is $\theta = 60°$, where d_a crosses 0.5. Thus, the corresponding modulation index M_{th} in this condition can be derived from $M_{\mathrm{th}} \sin(60^{\circ}) = 0.5$, i.e., $M_{\mathrm{th}} = 1/\sqrt{3} = 0.577$.

Therefore, when M is above 0.5 but below 0.577 as the case in Fig. 7.21b, there will be two angle-segments with sampling instant at short intervals in phase A as marked. Due to the central symmetry, the short interval sampling probability factor is

$$\lambda_{\mathrm{A}} = \frac{2 * \left(\theta - 60^{\circ}\right)}{180} \tag{7.10}$$

where

$$\theta = \sin^{-1}\left({}^{0.5}\Big/_{M}\right) \tag{7.11}$$

Similarly, the case with $M > 0.577$ can also be analyzed. The only difference is d_{a} crosses $d = 0.5$ in sector 1 and 3. Thus, there will be two short-interval sampling segments in sector 1 and 3, respectively, as illustrated in Fig. 7.21d.

The ratio of undesired sampling interval over the whole line cycle is finally provided in Table 7.1 and illustrated in Fig. 7.22. It is clear that when modulation index is higher than 0.5, this method can achieve even better performance.

Other important conclusions are as follows: only currents of two phases will be sampled at the long interval each time; for each phase, there is always a portion of line cycle with shorter sampling interval; three phases alternate similarly. For Vienna-type rectifier with carrier-based sinusoidal pulse width modulation (SPWM), the worst cases occur at $M <= 0.5$ and 1, with the maximum chance of one-third line cycle per phase to be sampled at the short interval; and the best case occurs at $M = 0.577$ where all three phases are sampled at the long interval along a whole line cycle.

A comparison with and without the proposed current sampling method is carried out at 115 V_{rms} 800 Hz AC input voltage, 600 V DC output, and 450 kHz switching frequency. Figure 7.23 shows the voltage and current waveforms with the new sampling method, and Fig. 7.24 presents the comparison results of harmonic spectra. The main improvement is at 2nd, 3rd, and 5th order harmonics owing to the accurate capture of the fundamental current component and immunity to noises and errors on the raw sensing signals.

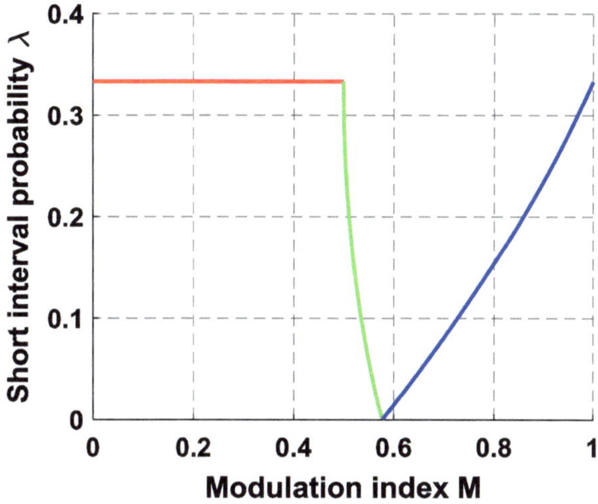

Fig. 7.22 The possibility of sampling instant occurring in short switching interval in a line cycle

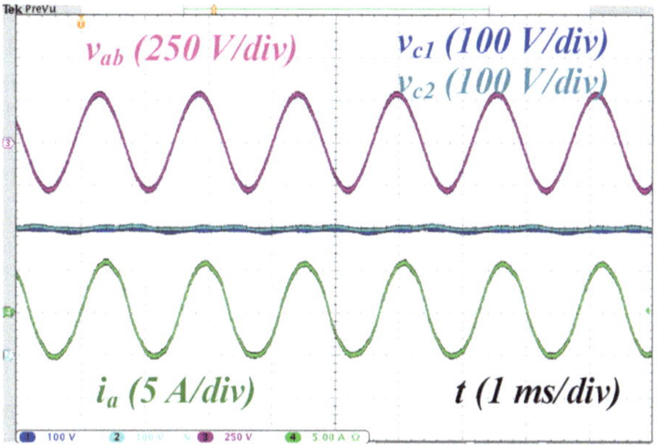

Fig. 7.23 Waveforms with proposed sampling scheme

The sector-based scheme for sampling described for Vienna-type rectifier can be applied to general three-phase converters to avoid the transient noises and ripples with high switching speed devices during the hard-switching operation.

7.3 Cooling Design for GaN-Based Converters

The current GaN devices have low profile surface mount (SMT) packages. This helps reduce the parasitic inductance but adds the difficulty on the cooling system design.

For bottom-cooled devices, heat from the device must be conducted from junction to case and spread out from PCB layers to thermal interface material (TIM) and heatsink. Along the heat transfer path, many factors will affect the thermal resistance. The dominant one is the PCB thermal resistance, which is

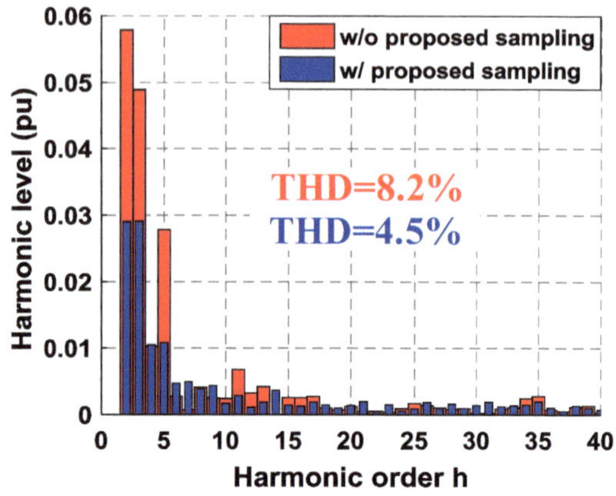

Fig. 7.24 Current harmonic spectrum with and without proposed scheme

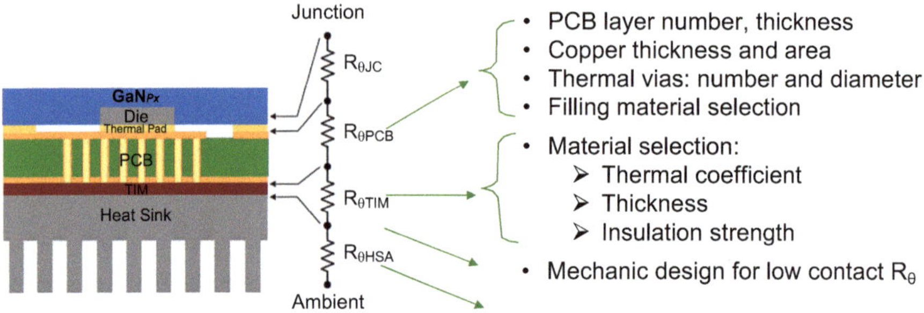

Fig. 7.25 Design concept and main factors for surface mount GaN device cooling

significantly impacted by PCB material and circuit layout, with key contributors listed in Fig. 7.25. To determine this thermal resistance, finite element method (FEM)-based simulation has been used for different types of PCB layouts [32], where the thermal resistance could be between 1.98 °C/W and 11.54 °C/W for FR4 material, depending on the density of vias, the number of layers, and PCB thickness. However, the multi-layer copper beneath the device prevents the vertical layout structure, leading to increased parasitic loop inductance, and induces voltage overshoot.

A three-phase Vienna-type rectifier prototype demonstrates the thermal design of bottom-cooled E-mode GaN as illustrated in Fig. 7.26. To enhance the thermal performance, a combined approach including thinner PCB, multi-layer, filled thermal vias, and a pin-fin heatsink is applied.

Replacing the thermal vias with a copper thermal interface block (TIB) can significantly reduce the thermal resistance between the device package and the heatsink [33]. As given in Fig. 7.27, copper TIBs act as heat spreaders and thermal standoffs to interface between heatsink and device package through PCB cutout. With lower sides around the device contact surface, TIBs also potentially allow the utilization of PCB area beneath devices. Since a TIB is directly contacted to

Fig. 7.26 Cooling design of a three-phase Vienna-type rectifier using bottom-cooled GaN devices

Fig. 7.27 Heatsink connection using thermal interface blocks (TIBs) [32]

the device thermal pad, and bolted to heatsink with insulating pad and screws, the mechanical stress of heatsink connection and TIM compression is applied to TIBs, instead of PCB or device package (Fig. 7.27).

As implemented in Fig. 7.28, the resulting thermal impedance of the copper TIB for each device is less than 0.3 °C/W, instead of 3~5 °C/W of typical thermal via design. If forced cooling is allowed instead of natural convection cooling, a much smaller heatsink would be possible so that TIBs could be integrated into separate heatsinks for each device, eliminating the need for insulated thermal interface material. Such a scheme was implemented in [34], but the TIB was replaced by a PCB copper inlay and a custom-machined heatsink, increasing manufacturing complexity and cost. This solution, however, still does not alleviate the loop parasitic issue. A trade-off, thus, has to be made between thermal and electrical performance.

To avoid this situation, some top-cooled packages for GaN devices gradually emerge in the market. The direct contact with the heatsink on the top surface of

Fig. 7.28 Assembly of gate injection transistor GaN devices, PCB, and TIBs for a full-bridge inverter [33]

the GaN device eliminates the bulky thermal impedance of PCB and enables the vertical layout beneath the device. However, this approach may introduce mechanic stress onto GaN devices, thus requiring accurate torque control and fabrication of the heatsink [35].

In addition to the thermal management itself, when operating GaN devices in the hard-switching converter, the capacitive coupling effect between the heatsink and GaN devices also should not be overlooked [36]. A small coupling capacitance may already show a high percentage of the low output capacitance of GaN devices. Additionally, as illustrated in Fig. 7.29, a common heatsink is often used for the low-profile surface mount GaN devices on one phase leg, or even on all phase legs, to reduce the cost, mechanical effort, and heatsink volume. This, however, forms a Y-connected capacitance network. By applying the Y/Δ transformation, the resultant capacitances are paralleled to each device of the phase leg, respectively. Since the coupling capacitance is formed between the source or drain and heatsink with the electrically insulated thermal interface in between, and the source or drain pad of GaN device is shorted with the device thermal pad to achieve low dynamic $R_{\mathrm{ds,on}}$, three capacitances are not equal as highlighted. Nevertheless, the situation becomes worse in the bottom-cooled device case than that in the top-cooled device case, since a larger area of thermal copper has to be laid on all PCB layers to spread out the heat of bottom-cooled devices, enlarging the coupled capacitance.

At high switching frequency operation, this capacitance negatively affects the switching speed and introduces considerable switching loss to the active switch. One approach to mitigate the capacitive effect is to separate the heatsink for the lower and upper switches in the phase leg [37] so that the coupling path between the switching node and the bus bar can be eliminated, but only if two heatsinks are allowed to have different potentials.

Fig. 7.29 Capacitive coupling between devices and heatsink

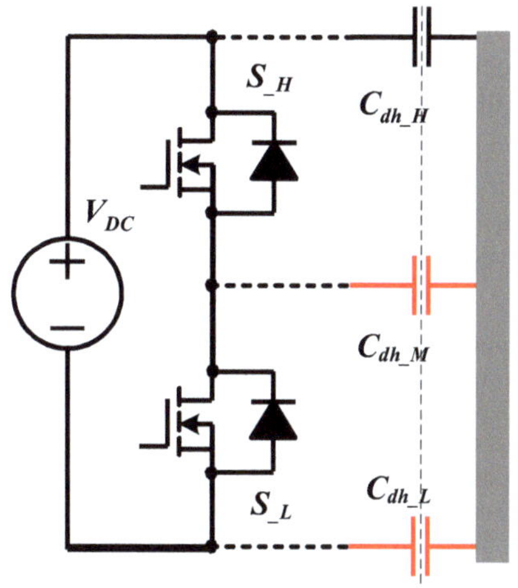

Active cooling approaches can also be used to regulate the converter loss, by switching frequency adjustment, modulation schemes, gate driver *dv/dt* and *di/dt* control [38], or derating operation, at the expense of higher THD or slow control dynamics. In this sense, reliable and nonintrusive temperature sensing technologies become a prerequisite.

7.4 Summary

This chapter overviews the benefits, challenges, and associated potential solutions of applying GaN devices in AC/DC converters. The benefits of GaN device technology can be realized through device substitution, topology simplification, as well as through enabling new system-level functions and new applications. The challenges in using these devices mainly stem from their extremely fast switching speed and small physical size.

For single-phase AC/DC converters, higher efficiency, simpler topology, and higher power density can be achieved thanks to lower switching loss in hard-switching converters and lower conduction loss in soft-switching PFC as well as in other converters. New applications benefiting from GaN technology have emerged, including wireless power transfer systems, medical power supplies, and smart impedance converters. Applying GaN in high-frequency high-density converter also poses new challenges, such as the ZVS range extension, zero-crossing distortion, and digital control. Special design considerations are needed to address these challenges.

For three-phase AC/DC converters, GaN technology has been attempted in PV, motor drive, and battery charger applications. Significant efficiency and/or power density improvement have been demonstrated through direct device substitution and/or increase in switching frequency. New issues for applying GaN in high-frequency three-phase converters include junction capacitance impact and current sampling. Design with consideration for these issues and new solutions is needed. Some potential approaches are presented here.

Finally, the difficulties in cooling GaN-based converters are explained from electrical, thermal, and mechanical design aspects, and different approaches are described and compared.

Acknowledgement This work made use of Engineering Research Center Shared Facilities supported by the Engineering Research Center Program of the National Science Foundation and the Department of Energy under NSF Award Number EEC-1041877 and the CURENT Industry Partnership Program. The authors would like to acknowledge help from Dr. Edward Jones, Mr. Chongwen Zhao, Ms. Ling Jiang, and Dr. Daniel Costinett of CURENT at the University Tennessee, Knoxville (UTK).

Some materials are based on research results of CURENT at UTK from projects sponsored by the Boeing Company and ABB.

References

1. E.A. Jones, F. Wang, D. Costinett, Review of commercial GaN power devices and GaN-based converter design challenges. IEEE Trans. Emerging Sel. Top. Power Electron **4**(3), 707–719 (2016)
2. C. Zhao, B. Trento, L. Jiang, E. Jones, B. Liu, Z. Zhang, F. Wang, L.M. Tolbert, J. Jansen, R. Kress, R. Langley, Design and implementation of a GaN-based, 100-kHz, 102-W/in3 single-phase inverter. IEEE Trans. Emerging Sel. Top. Power Electron **4**(3), 824–840 (2016)
3. Z. Liu, F.C. Lee, Q. Li, Y. Yang, Design of GaN-based MHz totem-pole PFC rectifier. IEEE Trans. Emerging Sel. Top. Power Electron **4**(3), 799–807 (2016)
4. L. Zhou, Y. Wu, U. Mishra, True-bridgeless totem-pole PFC based on GaN HEMTs, in *Procedings of PCIM Europe*, (May 2013), pp. 1017–1022
5. L. Zhou, Y. Wu, J. Honea, Z. Wang, High-efficiency true bridgeless totem pole PFC based on GaN HEMT: Design challenges and cost-effective solution, in *Proceedings of PCIM Europe*, (May 2015), pp. 1–8
6. C. Marxgut, F. Krismer, D. Bortis, J.W. Kolar, Ultraflat interleaved triangular current mode (TCM) single-phase PFC rectifier. IEEE Trans. Power Electron. **29**(2), 873–882 (2014)
7. J. Biela, D. Hassler, J. Minibӧck, J.W. Kolar, Optimal design of a 5 kW/dm3 / 98.3% efficient TCM resonant transition single-phase PFC rectifier, in *Proceedings of IEEE International Power Electronics Conference (IPEC-ECCE Asia)*, (Japan, June 2010)
8. N.K. Trung, K. Akatsu, Design high power and high efficiency inverter operating at 13.56MHz for wireless power transfer systems, in *Proceedings of IEEE Energy Conversion Congress and Exposition*, (Sept 2016)
9. N.K. Trung, K. Akatsu, Design 13.56MHz 10 kW resonant inverter using GaN HEMT for wireless power transfer systems, in *Proceedings of IEEE Energy Conversion Congress and Exposition*, (Oct 2017)
10. L. Jiang, F. Tamjid, D. Costinett, A GaN-based 100 W two-stage wireless power transmitter with inherent current source output, in *IEEE PELS Workshop on Emerging Technologies: Wireless Power (WoW)*, (Knoxville, Oct 2016)
11. L. Jiang, D. Costinett, A. Fathy, S. Yang, A single stage AC/RF converter for wireless power transfer applications, in *Proceedings of IEEE Applied Power Electronics Conference Exposition*, (March 2017)

12. X. Lu, P. Wang, D. Niyato, D.I. Kim, Z. Han, Wireless charging technologies: Fundamentals, standards, and network applications. IEEE Commun. Surv. Tutorials **18**(2), 1413–1452 (2016)
13. C. Zhao, D. Costinett, GaN-based dual-mode wireless power transfer using multifrequency programmed pulse width modulation. IEEE Trans. Ind. Electron. **64**(11), 9165–9176 (2017)
14. C. Zhao, D. Costinett, B. Trento, D. Friedrichs, A single-phase dual frequency inverter based on multi-frequency selective harmonic elimination, in *Proceedings of IEEE Applied Power Electronics Conference Exposition*, (March 2016)
15. K.A. Potty, E. Bauer, H. Li, B. Hu, J. Wang, Smart resistor: Dynamic stabilization of constant power loads in DC microgrids with high bandwidth power converters and energy storage, in *Proceedings of IEEE Applied Power Electronics Conference Exposition*, (March 2017)
16. B. Su, Z. Lu, An interleaved totem-pole boost bridgeless rectifier with reduced reverse-recovery problems for power factor correction. IEEE Trans. Power Electron. **25**(6), 1406–1415 (2010)
17. B. Su, J. Zhang, Z. Lu, Totem-pole boost bridgeless PFC rectifier with simple zero-current detection and full-range ZVS operating at the boundary of DCM/CCM. IEEE Trans. Power Electron. **26**(2), 427–435 (2011)
18. C. Marxgut, J. Biela, J.W. Kolar, Interleaved Triangular Current Mode (TCM) resonant transition, single phase PFC rectifier with high efficiency and high power density, in *Proceedings of Power Electronics Conference (IPEC)*, (2010), pp. 1725–1732
19. Z. Liu, Z. Huang, F.C. Lee, Q. Li, Digital-based interleaving control for GaN-based MHz CRM totem-pole PFC. IEEE Trans. Emerging Sel. Top. Power Electron **4**(3), 808–814 (2016)
20. J.W. Kim, S.M. Choi, K.T. Kim, Variable on-time control of the critical conduction mode boost power factor correction converter to improve zero-crossing distortion, in *Proceedings of IEEE International Conference on Power Electronics and Drive Systems (PEDS)*, (Nov 2005), pp. 1542–1546
21. H. Li, X. Zhang, Z. Zhang, C. Yao, F. Qi, B. Hu, et al., Design of a 10 kW GaN-based high power density three-phase inverter, in *Proceedings of IEEE Energy Conversion Congress and Exposition*, (2016)
22. H. Li, X. Li, Z. Zhang, C. Yao, J. Wang, Design consideration of high power GaN inverter, in *Proceedings of IEEE Workshop Wide Bandgap Power Devices Appl. (WiPDA)*, (2016), pp. 23–29
23. J. Honea, J. Kang, High-speed GaN switches for motor drives. Power Electronics Europe **3**, 38–41 (2012)
24. J. Lautner, B. Piepenbreier, High efficiency three-phase-inverter with 650 V GaN HEMTs, in *Proceedings of PCIM*, (May 2016)
25. B. Liu, R. Ren, E. Jones, F. Wang, D. Costinett, Z. Zhang, A modulation compensation scheme to reduce input current distortion in GaN based high switching frequency three-phase three-level Vienna type rectifiers. IEEE Trans. Power Electron. **33**(1), 283–298 (2018)
26. R. Lai, F. Wang, P. Ning, D. Zhang, D. Jiang, R. Burgos, D. Boroyevich, K.J. Karimi, V.D. Immanuel, A high-power-density converter. IEEE Ind. Electron. Mag. **4**(4), 4–12 (2010)
27. Q. Wang, X. Zhang, R. Burgos, D. Boroyevich, A. White, M. Kheraluwala, Design and optimization of a high performance isolated three phase AC/DC converter, in *Proceedings of IEEE Energy Conversion Congress and Exposition*, (Oct 2016)
28. Z. Zhang, F. Wang, L.M. Tolbert, B.J. Blalock, D.J. Costinett, Evaluation of switching performance of SiC devices in PWM inverter-fed induction motor drives. IEEE Trans. Power Electron. **30**, 5701–5711 (2015)
29. B. Liu, R. Ren, E.A. Jones, F. Wang, D. Costinett, Z. Zhang, A compensation scheme to reduce input current distortion in GaN based 450 kHz three-phase Vienna type PFC, in *Proceedings of IEEE Energy Conversion Congress and Exposition*, (Sept 2016)
30. D.M. Van de Sype, K.D. Gusseme, A.P. Van den Bossche, J.A.A. Melkebeek, A sampling algorithm for digitally controlled boost PFC converters. IEEE Trans. Power Electron. **19**(3), 649–657 (2004)
31. B. Liu, R. Ren, Z. Zhang, F. Wang, D. Costinett, A sampling scheme for three-phase high switching frequency and speed converter, in *Proceedings of IEEE Applied Power Electronics Conference Exposition*, (2018.) Accepted

32. GaN Systems, GN005_PCB Thermal Design Guide for GaN Enhancement Mode Power Transistors, [Online] http://www.gansystems.com/whitepapers.php
33. E.A. Jones, P. Williford, Z. Yang, J. Chen, F. Wang, S. Bala, J. Xu, J. Puukko, Maximizing the voltage and current capability of GaN FETs in a hard-switching inverter, in *Proc. IEEE International Conference on Power Electronics and Drive Systems (PEDS)*, (Dec 2017.) Accepted
34. Cree, CRD-5FF0912P, SiC MOSFET high-frequency evaluation board for 7L D2PAK, 2016, [Online]. Available: http://www.wolfspeed.com/downloads/dl/file/id/930/product/1/sic mosfet_high_frequency_evaluation_board_for_7l_d2pak.pdf
35. GaN Systems, GN001 Application Guide - Design with GaN Enhancement mode HEMT, [Online]: Available: http://www.gansystems.com/whitepapers.php
36. B. Liu, Z. Zhang, E. Jones, F. Wang, Application of GaN in hard-switching converters: Challenges and potential solutions. Power Electron (in Chinese) **51**(9), 3–13 (2017)
37. I. Josifovic, J. Popovic-Gerber, J.A. Ferreira, Improving SiC JFET switching behavior under influence of circuit Parasitics. IEEE Trans. Power Electron. **27**, 3843–3854 (2012)
38. P.K. Prasobhu, V. Raveendran, G. Buticchi, M. Liserre, Active thermal control of a DC/DC GaN-based converter, in *Proceedings of IEEE Applied Power Electronics Conference Exposition*, (March 2017)
39. R. Ren, B. Liu, E.A. Jones, F. Wang, Z. Zhang, D. Costinett, Capacitor-clamped, three-level GaN-based DC–DC converter with dual voltage outputs for battery charger applications. IEEE Trans. Emerging Sel. Top. Power Electron **4**(3), 841–853 (2016)

Chapter 8
GaN in Switched-Mode Power Amplifiers

David J. Perreault, Charles R. Sullivan, and Juan M. Rivas

8.1 Intro, Outline, Terminology

8.1.1 Scope and Overview of the Chapter

Radio-frequency (rf) power is important to a wide range of applications, including radio transmitters, plasma generation, medical imagers (e.g., MRI), power converters, and wireless power transfer (WPT) among myriad other applications. These applications can greatly benefit from the efficiency provided by switched-mode power conversion techniques, though their operating frequencies are high compared to those traditionally utilized in switched-mode power electronics. Fortunately, improved designs along with advances in power semiconductor devices and magnetics are opening the door to much more efficient generation and delivery of power at radio frequencies.

This chapter presents an overview of switched-mode power amplifiers – or radio-frequency inverters – encompassing their design, control, and construction. We focus on the high-frequency (HF, 3–30 MHz) and very-high-frequency (VHF, 30–300 MHz) ranges. In this frequency regime, management and utilization of circuit and device parasitics and careful design of passive components are critical to

D. J. Perreault (✉)
Massachusetts Institute of Technology, Cambridge, MA, USA
e-mail: djperrea@mit.edu

C. R. Sullivan
Thayer School of Engineering, Dartmouth College, Hanover, NH, USA
e-mail: charles.r.sullivan@dartmouth.edu

J. M. Rivas
Stanford University, Stanford, CA, USA
e-mail: jmrivas@stanford.edu

G. Meneghesso et al. (eds.), *Gallium Nitride-enabled High Frequency and High Efficiency Power Conversion*, Integrated Circuits and Systems,
https://doi.org/10.1007/978-3-319-77994-2_8

achieving efficient operation. Lumped-element circuits predominate, but distributed elements (e.g., transmission lines) and rf circuit techniques are valuable in these applications. We explore key aspects of rf power conversion, including power circuit design, design and application of passive components at radio frequencies, selection and efficient drive of power devices at rf, and control methods for modulating power and managing load variations.

8.1.2 Background on the Terms "Power Amplifier" and "Inverter"

In keeping with the typical application needs of this space, we consider the design of power conversion systems that can efficiently generate a sinusoidal rf output from a direct current (dc) input. We accomplish this using switched-mode techniques, that is, by using semiconductor devices acting as switches rather than as current-source elements, in manners that would ideally achieve 100% efficiency. Indeed, it is a fundamental insight that – used properly – an ideal switch is a device that can losslessly transform power between dc and ac waveforms (e.g., [1]).

In the power electronics community, a switched-mode circuit that transforms power from dc to ac waveforms is known as an inverter, a term that historically derives from "inverted rectification" [2, 3]. Here we consider "radio-frequency" inverters in which the active switches transition on and off at most once per ac output cycle owing to the high desired output frequencies. Interestingly, such operation is in keeping with that of the earliest inverter circuits [3]. In the rf community, by contrast, a dc to (radio-frequency) ac circuit is known as a power amplifier. The focus on power amplification arises because at high frequencies, the achievable power gain of a single stage (i.e., generated rf output power divided by power delivered to the transistor gate input) has traditionally been very limited, with multiple cascaded stages necessary to achieve a desired output power (i.e., with the output of one stage feeding the gate input of the next stage). Fortunately, at the frequencies considered here and with recent advances in power semiconductor devices (including GaN devices), power gain has become less of a focus (e.g., compared to efficiency). Nonetheless, it is recognized that resonant gate drives and tapered gate drivers each represent cascading of stages. For purposes of this chapter, then, we will use the terms inverter and switched-mode power amplifier (PA) interchangeably.

8.1.3 Design Considerations

We focus on power amplifier designs that synthesize (approximately) sinusoidal output waveforms. Such power amplifiers may be differentiated by a number of fur-

Table 8.1 ISM bands

Frequency band (MHz)	Center frequency (MHz)
13.553–13.567	13.560
26.957–27.283	27.120
40.66–40.70	40.68
902–928	915[a]
2,400–2,500	2,450
5,725–5,875	5,800
24,000–24,250	24.125

[a]in Region 2

ther factors, such as whether they need provide fixed or variable output frequencies (and what frequency bandwidth they need to operate over). In many applications, such as those using the industrial, scientific, and medical (ISM) band frequencies [4] illustrated in Table 8.1), operation is nearly fixed frequency. An additional differentiating factor is whether a design needs only to provide a fixed drive level or should be able to provide dynamically variable control (so-called linear control) of the output power or voltage. A third key factor is whether a design only need to operate into a known fixed load impedance or must accommodate highly variable loads. We will address each of these aspects, starting from consideration of designs that can synthesize a single-frequency, single-amplitude output into a fixed known load impedance.

8.2 Basic Inverter/Power Amplifier Topologies

8.2.1 Class D, DE "Totem Pole" Topologies

Class D Power Amplifier and Design for Zero-Voltage Switching

The half-bridge inverter circuit is a fundamental building block used in many power electronic systems. One basic approach toward switched-mode rf generation is to use a half-bridge inverter in conjunction with a tuned network (e.g., a resonant tank) to synthesize a sinusoidal output waveform, as illustrated in Fig. 8.1. The switches Q_1 and Q_2 in Fig. 8.1 are shown with reverse diodes, along with capacitors C_1 and C_2 that include (or may entirely comprise) the switch parasitic capacitances. Transistors Q_1 and Q_2 are switched alternately, with identical duty ratios and dead times between their on periods. The resonant tank filters voltage $v_{Q_2}(t)$, yielding a sinusoidal current that delivers power to the load. In the rf community, a multi-switch inverter in which the transistor voltage waveforms approximate square waves and the transistor currents approximate half sine waves is known as a "Class D" power amplifier; the half-bridge inverter of Fig. 8.1 is thus one version of a *Class D* power amplifier.

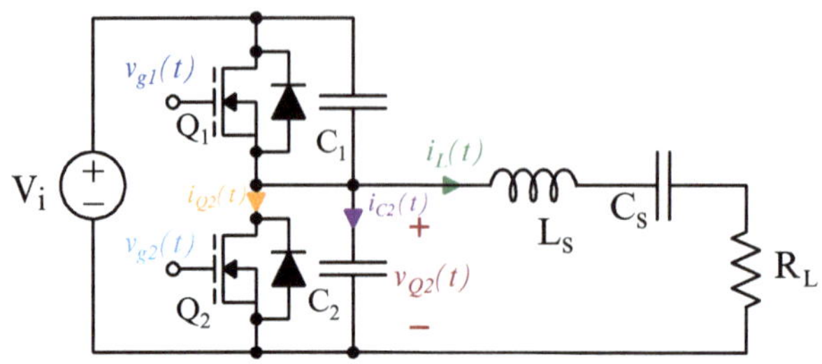

Fig. 8.1 Class D inverter using resonant tank (load) current to provide ZVS soft switching

If the two switches in Fig. 8.1 are operated close to 0.5 duty ratio with minimal dead times between their on periods, $V_{Q_2}(t)$ approximates a square wave at a desired output frequency. However, at high frequencies, the loss associated with discharging the switching device capacitances each cycle becomes prohibitive. To mitigate this loss mechanism, the load network (including resonant tank and load) can be tuned to appear inductive, and the switches can be operated with significant dead time between them. This results in low-loss zero-voltage switching (ZVS switching) at turn-on and turnoff of the devices.

In general, during the turnoff transition, the capacitances C_1 and C_2 across the devices slow device voltage rise during turnoff, which reduces device turnoff loss. Idealizing the switch turnoff as a linear fall in a current over a duration t_f, and considering a device current I_L at turnoff and a net capacitance $C = C_1 + C_2$ across the switch, the turnoff loss can be approximated as $E_{off} = I_L^2 t_f^2/(24C)$. Thus, with sufficient capacitance (relative to device switching speed and load current), voltage rise and loss during switch turnoff can be minimal ("ZVS turnoff"). To take advantage of this reduction in turnoff loss, circuit operation must be arranged that the devices also turn on at (or near) zero voltage, to eliminate turn-on loss due to lossy discharge/charge of the capacitances C_1 and C_2.

As illustrated in Fig. 8.2, if the amplitude of current $i_L(t)$ is sufficiently large and its phase sufficiently lags that of the voltage waveform $v_{Q_2}(t)$, and the dead time between switches is appropriately chosen, then the load network can losslessly charge voltage v_{Q_2} from zero to input voltage V_i, providing an opportunity for Q_1 to turn on at zero voltage once its reverse diode starts to conduct.

This pattern will repeat itself in the other half cycle, yielding the switching pattern illustrated in Fig. 8.3.

Central to realizing zero-voltage switching in the circuit of Fig. 8.1 is a load network that has a resistance and net inductive reactance such that full lossless charging/discharging of capacitors C_1 and C_2 can be completed during the dead time. The smaller the net load inductance (and phase shift ϕ in Fig. 8.2) is, the larger the load current that is necessary to achieve ZVS device turn-on (the ZVS turn-on might be accomplished during reverse diode conduction as in Fig. 8.3). The limiting case, in which the load network impedance and dead time are just such

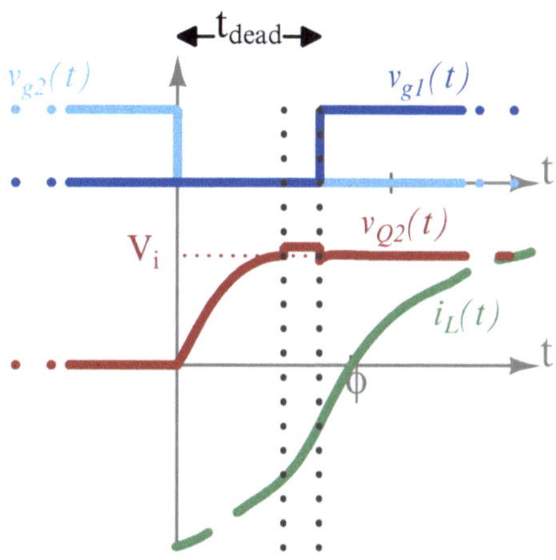

Fig. 8.2 Commutation from Q_2 to Q_1 in the ZVS Class D power amplifier of Fig. 8.1. For ZVS turn-on, the load current $i_L(t)$ must be sufficiently lagging with respect to voltage $v_{Q_2}(t)$, and there must be sufficient dead time to losslessly discharge/charge C_1 and C_2

that the incoming device turns on at zero voltage and zero current (and hence zero dv/dt), is shown in Fig. 8.4.

This is known as "Class DE" switching, owing to its close relation to the Class E power amplifier described in a subsequent section. Class DE operation occurs for a particular combination of duty cycle and load impedance such that the load current reaches zero just as the incoming device turns on at zero voltage; this mode is advantageous in that it reduces the sensitivity of efficiency to the duration and timing of device turn-on. Details of the allowable resistive/inductive load range for ZVS operation are derived in [5]. Assuming that the load network has sufficient quality factor (X_L/R_L) to ensure approximately sinusoidal load current, the range of effective net load impedances that can provide zero-voltage switching is illustrated in Fig. 8.5, along with the locus for the limiting case of Class DE switching. The allowable load impedance range is shown normalized to the impedance of the total device capacitance at the switching frequency $R' = R_L \cdot 2\pi f \cdot (C_1 + C_2)$ and $X' = X_L \cdot 2\pi f \cdot (C_1 + C_2)$. [5] also describes the use of matching networks to map a given load resistance range into the allowable ZVS switching range of the ZVS Class D power amplifier of Fig. 8.1.

While the Class D inverter of Fig. 8.1 is effective and widely used, it has the disadvantage that ZVS operation is highly tied to the load impedance and resonant tank characteristics. Separating the loading required for soft switching from that for power delivery often provides design benefits, especially for flexibility in the operating range. One way to do this is illustrated in Fig. 8.6a. Capacitors C_a and C_b split the bus voltage such that shunt inductor current L_x charges up and down linearly during the on times of the transistors. This provides a quasi-triangular current $i_x(t)$ for ZVS soft switching that is in the correct direction and is at a maximum for the switching transitions, with a peak value that depends linearly on input voltage V_i. Thus, one obtains a known current that can be sized for soft switching, independent of (resistive) load variations in the output network; the fixed

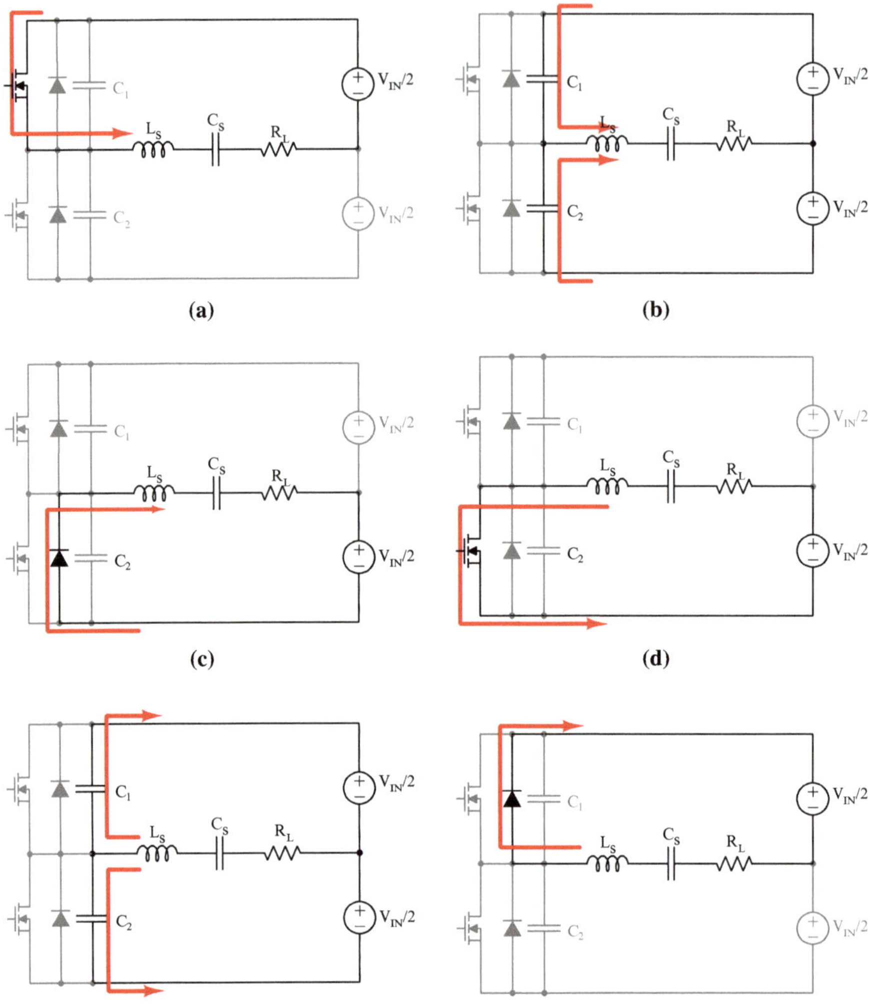

Fig. 8.3 Switching pattern of the ZVS Class D inverter over a complete cycle. (**a**) Mode 1. (**b**) Mode 2. (**c**) Mode 3. (**d**) Mode 4. (**e**) Mode 5. (**f**) Mode 6

nature of the soft-switching current component is a benefit of this design. Tuning the load for resistive loading at the fundamental frequency minimizes any undesirable reactive component of inverter loading. It should be noted that many alternative means of providing shunt currents for ZVS switching like that in Fig. 8.6 are also possible (e.g., by utilizing the magnetizing inductance of a transformer connecting the inverter to the load network to provide current for soft switching).

Fig. 8.4 Zero-voltage switching in the Class D circuit of Fig. 8.1 in the limiting case where the energy in the load network and switching dead time is just sufficiently large to charge to enable ZVS. This condition, in which the incoming device turns on at both zero voltage and zero current (and hence zero dv/dt), is known as "Class DE" switching

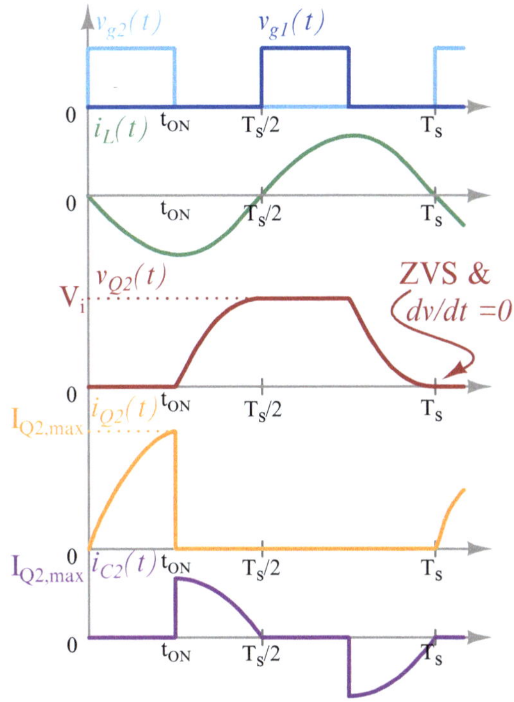

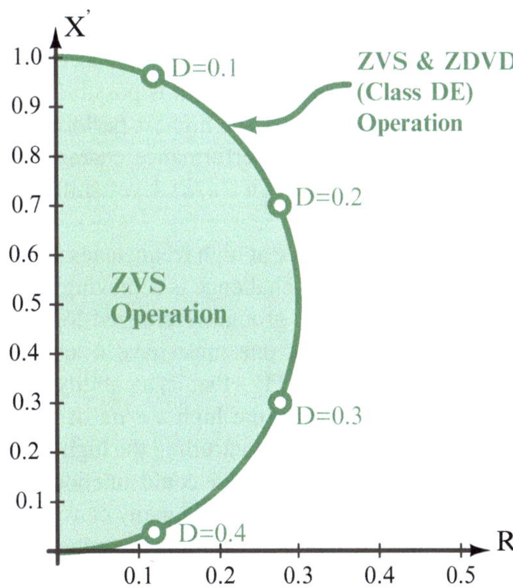

Fig. 8.5 (Adapted from [5]). This figure shows the range of net load network impedances for which zero-voltage switching may be achieved, along with the locus for the limiting case of Class DE switching. The resistive and reactive components of load impedance are normalized to the impedance of the total device capacitance at the switching frequency $R' = R_L \cdot 2\pi f \cdot (C_1 + C_2)$ and $X' = X_L \cdot 2\pi f \cdot (C_1 + C_2)$. It is assumed that the load network quality factor X_L/R_L is sufficiently high to yield approximately sinusoidal current waveforms at the input of the load network

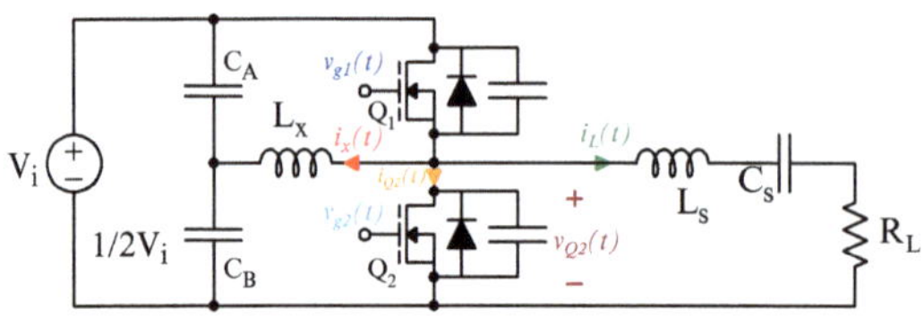

(**a**) ZVS Class D inverter with shunt inductor for soft switching

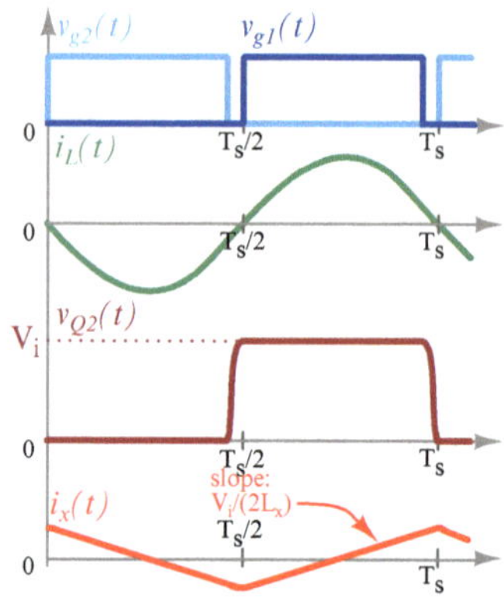

(**b**) Operating waveforms.

Fig. 8.6 (**a**) Class D inverter with shunt inductors. (**b**) Relevant waveforms

High-Frequency Design Considerations and Limitations of the ZVS Class D Inverter

Principle limitations in scaling the Class D inverter to very high frequencies include the practical challenges in level shifting of the gating signal to the high-side switch and of maintaining the needed precision of timing between the two switches. The timing constraints of the switch-drive waveforms in Figs. 8.2 and 8.6b impose requirements on the level shifting and gate driving which become increasingly hard to meet as voltage and frequency increase. While it is possible in some cases to use transformer-based level shifting (e.g., [6]), the highest-performance designs utilize digital isolators and similar circuits. The performance characteristics of a number of digital isolators and gate drives for high dv/dt level shifting are illustrated in Table 8.2.

Limitations in achievable performance at high frequencies arise from the characteristics of these control circuits. A first challenge is achieving reliable level shifting of the gate-drive signal $q_2(t)$ from the ground-referenced logic level to the high-side flying device. To accomplish this, one must have a level shifter with large common-mode transient immunity (CMTI) – that is, an ability to correctly transmit the gate-drive signal while withstanding the high dv/dt of the voltage $v_{Q_2}(t)$ – which serves as the reference potential for controlling the high-side device Q_1. (The use of a complementary devices for the high side could ameliorate this difficulty, but suitable devices are not typically available.) While many conventional level shifters have poor (or even unspecified) CMTI, level shifters are now available with CMTI ratings exceeding 50 V/ns.

In addition to CMTI limits, a further factor limiting frequency is the achievable timing of the two gate-drive waveforms, both in terms of the duration of time it takes to charge and discharge the switch gates and in the variations in delays

Table 8.2 Performance characteristics of digital isolators and gate drives for high dv/dt level shifting

Device	Minimum CMTI (kV/μs)	Propagation delay (ns)	Delay skew (ns)	Min. pulse width (ns)	Pulse-width distortion (ns)	Isolation voltage (V)
Silicon Labs Si8620BT	60	8 (typ.) 13 (max.)	0.4 (typ.) 2.5 (max)	5	0.2 (typ.) 4.5 (max.)	5000
NVE IL711[a]	30	10 (typ.) 15 (max.)	4 (typ.) 6 (max)	10	0.3 (typ.) 3 (max.)	5000
ADI ADUM210	75	6.8 (typ.) 14 (max.)	7 (max.)	6	0.7 (typ.) 3 (max.)	5000
TI ISO721M	25	10 (typ.) 16 (max.)	3 (max.)		0.5 (typ.) 1 (max.)	4000
TI ISO7810	100	10.7 (typ.) 16 (max.)	4.5 (max.)		0.6 (typ.) 4.6 (max.)	2000
AVAGO HCPL-0900	15	10 (typ.) 15 (max.)	4 (typ.) 6 (max.)		2 (typ.) 3 (max.)	2500
TI LMG1210[b]	150	17 (max.)	0.3 (typ.) 1 (max)	4	1 (typ.) 3 (max.)	200

[a] Part has an indeterminate start-up state
[b] Half-bridge driver

that arise from available level shifters and drivers.[1] As illustrated in Table 8.2, the channel-to-channel skew of level shifting and the pulse-width distortion of the shifted signal can be on the order several nanoseconds each, and there is typically a minimum pulse width of several nanoseconds to ensure that any gate-drive transition happens. The switching transition duration imposed by driving the transistor gate capacitances can even be more impactful, especially considering the high gate time constants and total charge of high-voltage devices. (It may be appreciated that wide-bandgap devices, including GaN-on-Si devices and SiC devices, are yielding improvements in terms of gating as compared to the metal-gate Si devices that have often been previously utilized in this space.) Together, these driving times and their variations limit the frequency that can be achieved with the Class D inverter while maintaining switched-mode operation and without incurring loss of ZVS and risk of shootthrough. With presently available level shifters and devices, these factors impose maximum practical switching frequencies for the Class D inverter in the tens of MHz at the tens to hundreds of volts range. To achieve higher efficiencies and performance in the high HF and VHF range, one typically moves to either single-switch inverters or to multi-switch inverters that do not require level shifting, as we describe in subsequent sections.

[1]Absolute delays of gate-drive signals is itself not an issue, unless one seeks to actively adjust switch timing under feedback control.

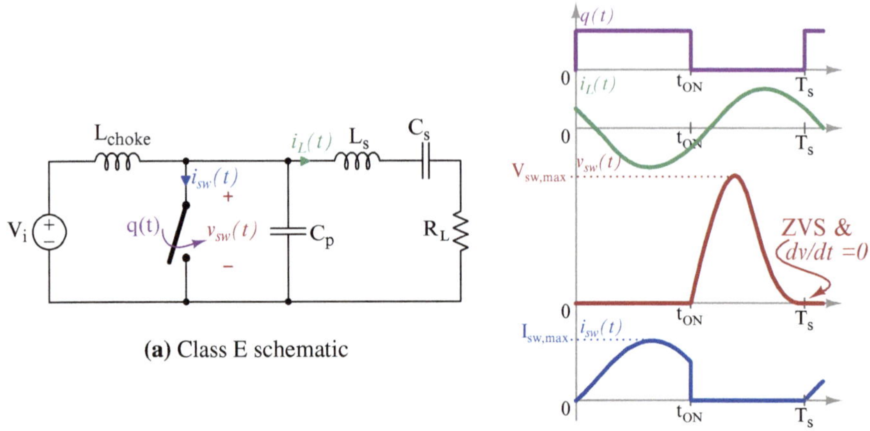

(a) Class E schematic

(b) Class E waveforms

Fig. 8.7 (**a**) Class E schematic. (**b**) Relevant waveforms. Notice that at *turnoff*, the drain voltage achieves zero-voltage-switching and zero dv/dt

8.2.2 Single-Switch Inverters

At switching frequency higher than few tens of MHz, designing a Class D inverter structure becomes increasingly difficult, or even impractical, due to the detrimental effects that board parasitics, timing delays, and CMTI limits in gate-drive circuits have on the circuit behavior. At these frequencies ($\gtrsim$ 10 MHz), topologies that use a single ground-referenced power semiconductor, like the Class E, F, E/F, and Φ_2 inverter topologies, are more suitable. Hence, among the advantages of single-switch resonant topologies, we can list the following: (a) there is no need to synchronize multiple signals, (b) there is no shoot through due to improper dead-time settings, and also importantly, (c) there is no need to float gate-drive signals. These characteristics also facilitate the use of resonant gate-drive techniques, which can provide efficiency benefits at very high frequencies.

Class E rf Inverter

In 1975, Sokal and Sokal [7] introduced the Class E inverter (Fig. 8.7a). The values of the resonant network components are selected to achieve two principal objectives: *a)* chieve ZVS at transistor turn-on $v_q(t)|_{t=t_{on}} = 0$ and *b)* achieve zero dv/dt (ZDVDT) during the turn-on transition $\frac{d}{dt} v_q(t)\|_{t=t_{on}} = 0$.

While the former condition dramatically reduces turn-on switching loss (*ideally ZVS eliminates switching losses)* thus enabling operation at high frequencies, the latter ensures that device losses remain low despite slow, mistimed, or even *jittery* gate-drive signals. Having ZDVDT allows the use of gating signals with relatively long rise/fall times. Under ZDVDT, one can use a gate-driving signal with transition times that can last a significant fraction of the switching cycle without incurring in excessive switching losses. This characteristic makes the Class E design particularly amenable to resonant-drive techniques, which trade gating speed for gating loss. Because of its efficient operation at high switching frequencies and its design simplicity, the Class E inverter is a popular choice in rf designs at frequencies that can reach well into the 10s of GHz.

The traditional Class E amplifier circuit consists of a single ground-referenced switch, connected to the input supply via a large inductor L_{choke}. At the switching frequency, the impedance of L_{choke} is large which results in a small input current ripple. For analysis purposes, the current through L_{choke} can be considered constant. The resonant elements C_p, C_s, and L_s in Fig. 8.7a and the load form a resonant tank circuit tuned to reduce capacitive discharge losses in the switch by achieving ZVS. This is accomplished by selecting values for L_s, C_s, and R_L in Fig. 8.7b that appear inductive at the switching frequency of operation. In a broad range of practical implementations, as in the Class D inverter, one can incorporate the switching device capacitance into the value of C_p. C_p plays an important role in setting the right conditions to achieve ZVS and ZVDT during the turnoff transition of the switching device. Low-distortion applications use a resonant tank with a high-loaded quality factor (Q_L), though a large Q_L tends to degrade efficiency.

Assuming a large Q_L (e.g., $Q_L > 10$) and $t_{on} = \frac{1}{2f_S}$ where f_s is the switching frequency, the values for the resonant components of a Class E rf inverter that achieves ZVS and ZDVDT are [8]:

$$L_s = \frac{Q_L R_L}{2\pi f_s}, \qquad C_s = \frac{1}{2\pi f_s (Q_L - 1.1525) R_L}, \qquad C_p = \frac{1}{34.22 f_s R_L}. \tag{8.1}$$

Typical waveforms of an ideal Class E inverter are depicted in Fig. 8.7b. Notice that $v_{sw}(t)$ achieves ZVS and ZDVDT at $t = T_s$ as the switch turns on. With the components values obtained with Eq. 8.1, the power delivered to the load is $P_o \approx \frac{0.576 V_i^2}{R_L}$.

The Class E switch voltage and current ($v_{sw}(t)$ and $i_{sw}(t)$ in Fig. 8.7a) are relatively high compared to other rf power amplifier topologies, with the voltage ideally peaking at $3.6 \times V_i$ ($V_{sw,max}$ in Fig. 8.7b) when $t_{on} = \frac{1}{2f_S}$ and ZVS and ZDVDT are achieved. In practice, because of the nonlinear characteristics of the switching device capacitance, $V_{sw,max}$ can reach even higher voltages, often $\approx 4 \times V_i$. Under nominal conditions, the peak switch current $I_{sw,max} \approx 1.7 \times \frac{V_i}{R_L}$. Useful design tables and equations for selecting the resonant components of a Class E PA design, as well as tables listing peak switch voltage and current values and other performance metrics for different duty cycles and values of Q_L, are presented in [9]. Reference [10] offers good insight into the practical tuning of a Class E inverter.

It is worth noting that important losses in the classical Class E power amplifier typically include the resonant inductor loss, the gate-drive loss, the device on-state conduction loss, and the device off-state conduction loss through the lossy device capacitance (so-called "R_{oss}" loss, named for the equivalent series resistance of the device output capacitance C_{oss}). When device capacitance comprises a substantial portion of the total capacitance C_p, this latter off-state conduction loss component can be a major contributor to overall loss, as the ESR of the device output capacitance is often substantially greater than the on-state resistance. Likewise, depending upon the device and operating frequency, gate-drive loss can be substantial; this can often be partially mitigated through resonant gate-drive methods [11–13].

As an example, Fig. 8.8 shows LTSpice simulation waveforms of a Class E inverter tuned using the Eq. 8.1, when $f_s = 30$ MHz, $R_L = 1\ \Omega$, and $Q_L = 15$.

As in other resonant switched-mode power amplifiers, the simultaneous ZVS and ZDVDT conditions of a Class E inverter only occur at a single operating point. Relatively small changes in component values will result in switching conditions failing to achieve ZVS and thus increasing the switching losses of the PA. In the Class E inverter of Fig. 8.7a, the load R_L forms part of the resonant tank that sets the ZVS conditions in the circuit. Figure 8.9 shows the simulated drain waveform (V_{sw}) and drain efficiency of the Class E inverter of Fig. 8.8 as R_L varies in the $0.25\ \Omega \leq R_L \leq 5\ \Omega$ range. Notice that for R_L values greater than the nominal $1\ \Omega$, the inverter fails to achieve ZVS which increases switching losses and lowers efficiency.

A high-loaded quality factor (Q_L) also constrains the range of efficient operation of the Class E inverter to a narrow frequency range. Figure 8.10 shows how the drain waveform of the Class E inverter ($v_{sw}(t)$ of Fig. 8.8) distorts and rapidly losses the ZVS as the switching frequency deviates from the nominal designed value (30 MHz).

ZVS Operation of the Class E Inverter with Variable Loads

The components of the Class E inverter have been chosen to maintain ZVS even when the load changes over a significant range [14], although one may give up the benefits of ZDVDT operation. (This is in addition to other methods to limit the impact of load variations, as described in Sect. 8.2.4.) The method outlined in [14] enables soft switching to be maintained across a wide inductive/resistive load range. Conceptually, this transformation is achieved with the addition of resonant components that, through circuit manipulation, are incorporated and combined with existing components in the circuit. In a traditional Class E inverter design, the input inductor L_{choke} is large and plays no role in shaping the voltage across the drain to achieve ZVS through resonant means. If L_{choke} is replaced by a smaller inductor

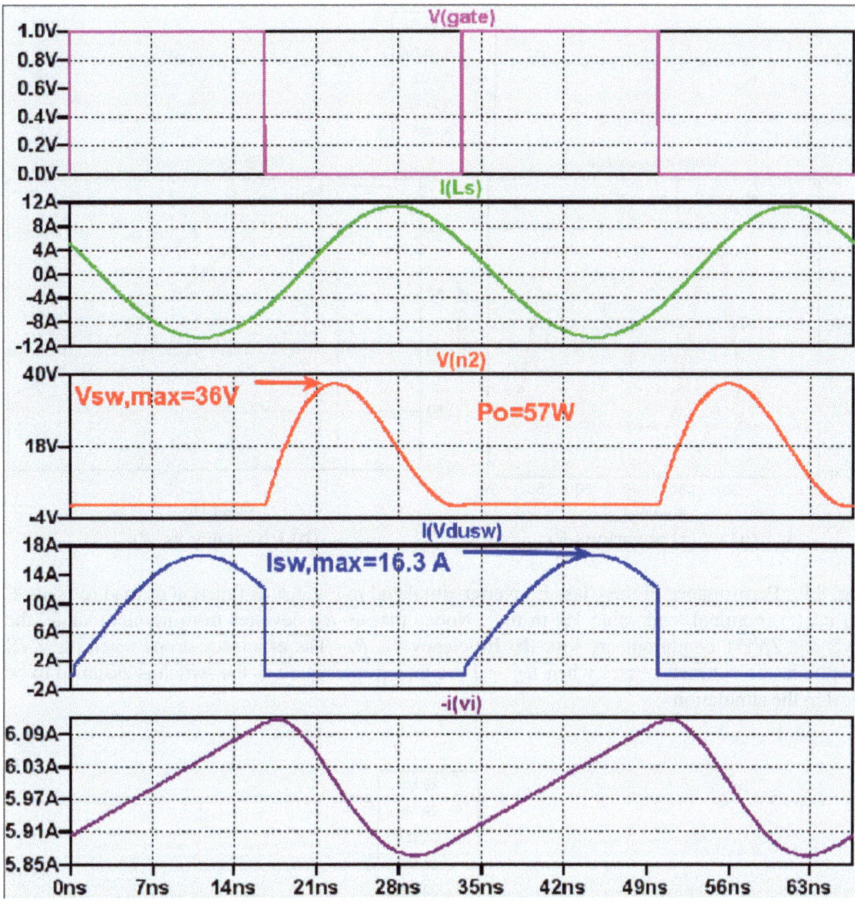

Fig. 8.8 LTSpice simulation waveforms of the Class E circuit of Fig. 8.7a. Here $V_i = 10$ V, $f_s = 30$ MHz, $t_{on} = 16.66$ ns, $Q_L = 15$, $R_L = 1\ \Omega$, $L_{choke} = 10\ \mu$H, $L_s = 79.5$ nH, $C_s = 383$ pF, and $C_p = 974$ pF. Notice that $V_{sw,max} = 3.6 \times V_i$ and that ZVS and ZVDT are achieved. In this simulation, the switch is considered ideal. From top to bottom: $v_{gs}(t)$, $i_L(t)$, $v_{sw}(t)$, $i_{sw}(t)$, $i_i(t)$

(L_F in [14]), the input inductor interacts with other components in the network resonantly to achieve ZVS over a relatively wider R_L range. The reduction of the input inductance results in an increase in the circulating currents in the circuit, which can impact efficiency, but provides the benefit of a wide load range while also allowing faster dynamic response.

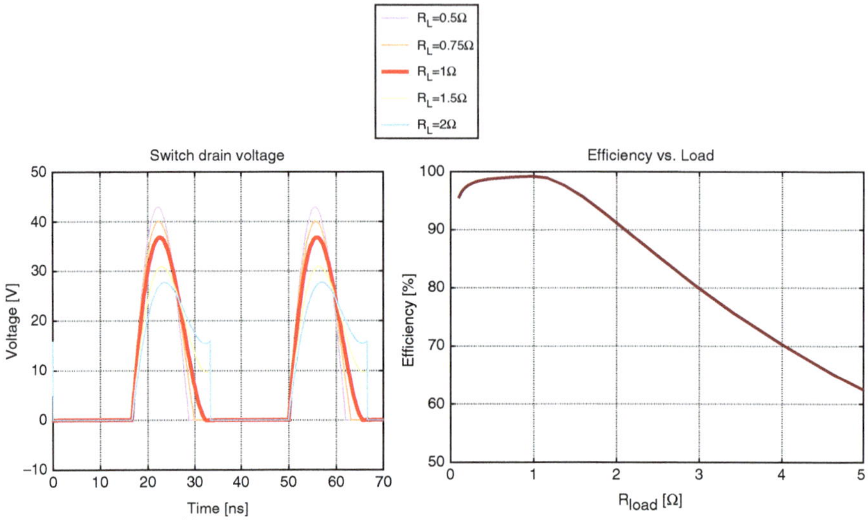

(a) $v_{ds}(t)$ at various R_L

(b) Efficiency vs. R_L

Fig. 8.9 Performance of the Class E inverter simulated in Fig. 8.8 as function of load resistance. (**a**) $v_{sw}(t)$ (nominal load value 1Ω in red). Notice that as R_L deviates from nominal value, the ZVS and ZVDT conditions are lost. (**b**) Efficiency vs. R_L. The efficiency drops when the ZVS conditions are not met. Losses when $R_L < 1$ are underrepresented as the switch is assumed to be ideal in the simulation

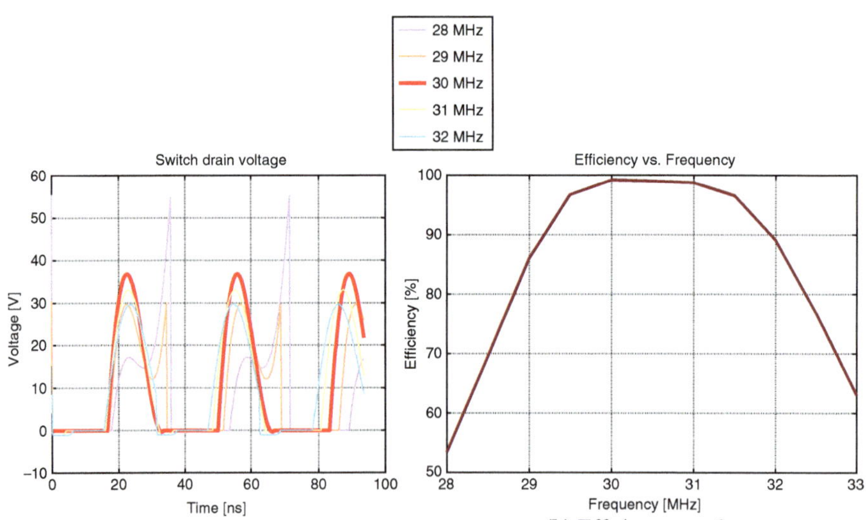

(a) $v_{ds}(t)$ at various f_s

(b) Efficiency vs. f_s

Fig. 8.10 Performance of the Class E inverter simulated in Fig. 8.8 as function of the switching frequency. (**a**) $v_{sw}(t)$ (Nominal $f_s = 30$ MHz in red). Notice that when f_s deviates from the nominal value, the ZVS and ZVDT conditions are lost. (**b**) Efficiency vs. f_s

Higher-Order Tuning

In the Class E inverter, the peak drain voltage can reach or exceed $3.6 \times V_{in}$ [7]. This limits the input voltage that can safely be used with a given semiconductor when considering the device's rating: the maximum input voltage of a Class E rf amplifier implemented using a commercial 650 GaN device is $\approx$ 160 V or less, depending on the voltage derating guidelines specific to the intended application. It is possible to retune the circuit parameters of a Class E circuit and reduce the peak drain voltage by a small amount. This is also done by replacing L_{choke} by a smaller L_f that interacts with C_p and the other resonant components in the circuit network to limit the maximum stress voltage in the semiconductor while maintaining ZVS. This is, in a way, similar to the tuning in [14] to extend the ZVS range over a wider load range. The two tuning methods use the available degrees of freedom of the resonant network to achieve ZVS and either extend the ZVS range or reduce the peak switch voltage at the expense of ZDVDT switching as in the traditional Class E.

With the inclusion of more resonant elements, it is possible to further reduce the peak switch voltage of the rf inverter. The addition of resonant traps and/or tanks tuned at (or close to) the harmonics of the switching frequency can reduce the maximum switch voltage to about $2 \times V_{in}$. This tuning approach is used in the Class F amplifier, where a transistor conducts as a current source for part of the rf cycle (similar to other linear rf amplifiers like the Class AB) but uses multiple tuned filters to shape the harmonic content of the switch voltage to increase drain efficiency. As more and more harmonics are included in the drain voltage shaping process, the drain efficiency of the PA improves, as the switch-voltage waveforms start approaching a square [15].

Similar concepts can be put to use in the design of switched-mode power amplifiers, in which resonant elements play a role in shaping the switch voltage or current to meet a particular performance objective. For example, the Class Φ_2 or E/F_2 rf PA shown in Fig. 8.11a utilizes resonant components (L_{MR}, C_{MR}) tuned to place a *short* at the second harmonic of the switching frequency [12, 16]. L_{MR}, C_{MR} in conjunction with L_F, C_P, and the drain-source capacitance of the switch shape the *"off"* impedance of the drain node to shape the voltage of the switching device to limit the peak voltage to about $2 \times V_{in}$ while maintaining ZVS and ZVDT, like the waveform sketched in Fig. 8.11b.

The tuning procedure outlined in [12, 16] starts with the lumped network of Fig. 8.12a and selects values that place a *short* in Z_{in} at the second harmonic of the switching frequency and an *open* at the fundamental and third harmonic. The impedance Z_{in} of the tuned network is shown in Fig. 8.12b. These are just initial values for the components of the inverter of Fig. 8.12a. In a subsequent tuning step, the values of L_f and C_p and the values of the output resonant network formed by L_s, C_s, and R_L are adjusted to make the drain impedance Z_{ds} inductive at f_s (with $0 < \angle Z_{ds} < +90°$) and capacitive at the third harmonic of f_s (with $-90° < \angle Z_{ds} < 0°$). As an additional constraint, the magnitude of the impedance Z_{ds} at $3f_s$ is set lower in magnitude than the magnitude of Z_{ds} at the fundamental. The inductive impedance at f_s is necessary to achieve ZVS. The peak switch

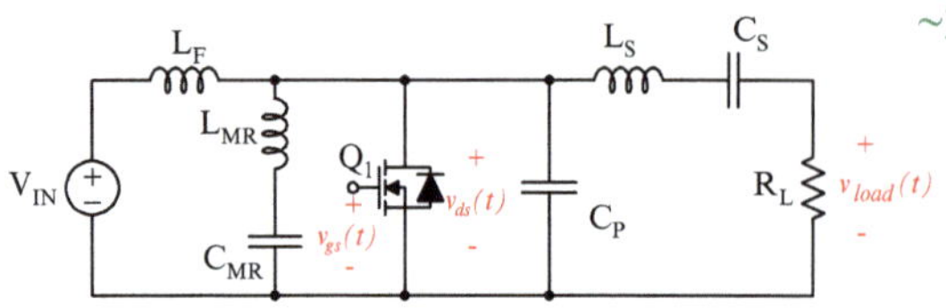

(a) Class Φ_2 schematic

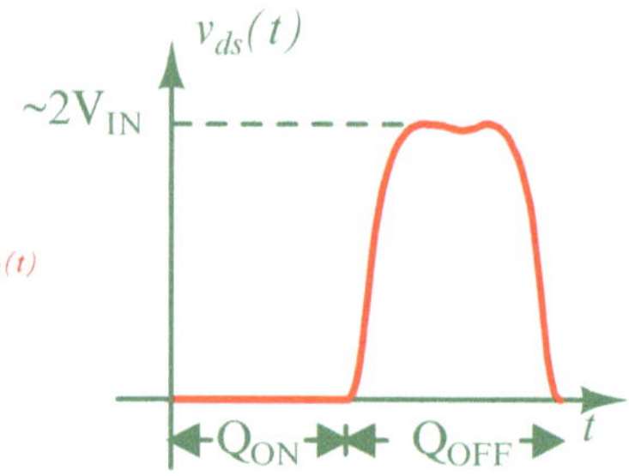

(b) Class Φ_2 Drain waveform

Fig. 8.11 (**a**) Class Φ_2 schematic. (**b**) Relevant waveforms. Notice that at *turnoff*, the drain voltage achieves zero-voltage switching and zero dv/dt

Fig. 8.12 (**a**) Lumped Network. (**b**) Network impedance. Notice the impedance at the fundamental, second, and third harmonic

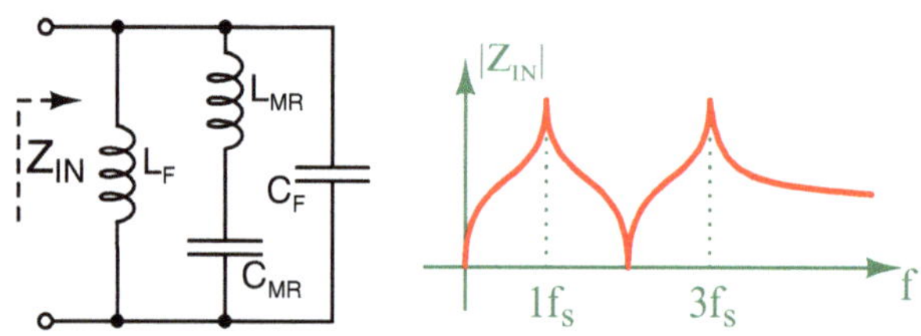

(a) **(b)**

voltage and a near ZDVDT are set by the constraints in Z_{ds} at $3f_s$. The effect the impedance magnitude ratio (between fundamental and third harmonic) has on the drain voltage waveform is shown in Fig. 8.13 which shows simulated drain impedance and drain voltage for three cases that have the same impedance at the fundamental but other inductance values at the third harmonic of f_s[12]. This tuning approach effectively provides a waveform that is dominated by fundamental and third harmonic components of the right amplitudes for an approximately trapezoidal drain voltage. It is also noteworthy that this class of amplifier can also be tuned to provide wide load range capability [14]. It will be appreciated that other harmonic tunings can be employed to provide desired tradeoffs between switch-voltage and current ratings.

Gate Driver for Single-Switch High-Frequency Inverters

An important consideration in the design of a switched-mode rf inverter is the gate drive. Figure 8.14 shows the simplified schematic of a conventional low-side gate drive commonly used in PWM and resonant converters. S_a and S_b are alternately activated to make $v_{sw}(t)$ a signal with fast rising/falling times. A damping resistor R_{ext} is often included to prevent unwanted oscillations stemming from layout and package parasitics. The gate-drive loss of this driving scheme is $P_{g,loss} = f_s V_g Q_g$,

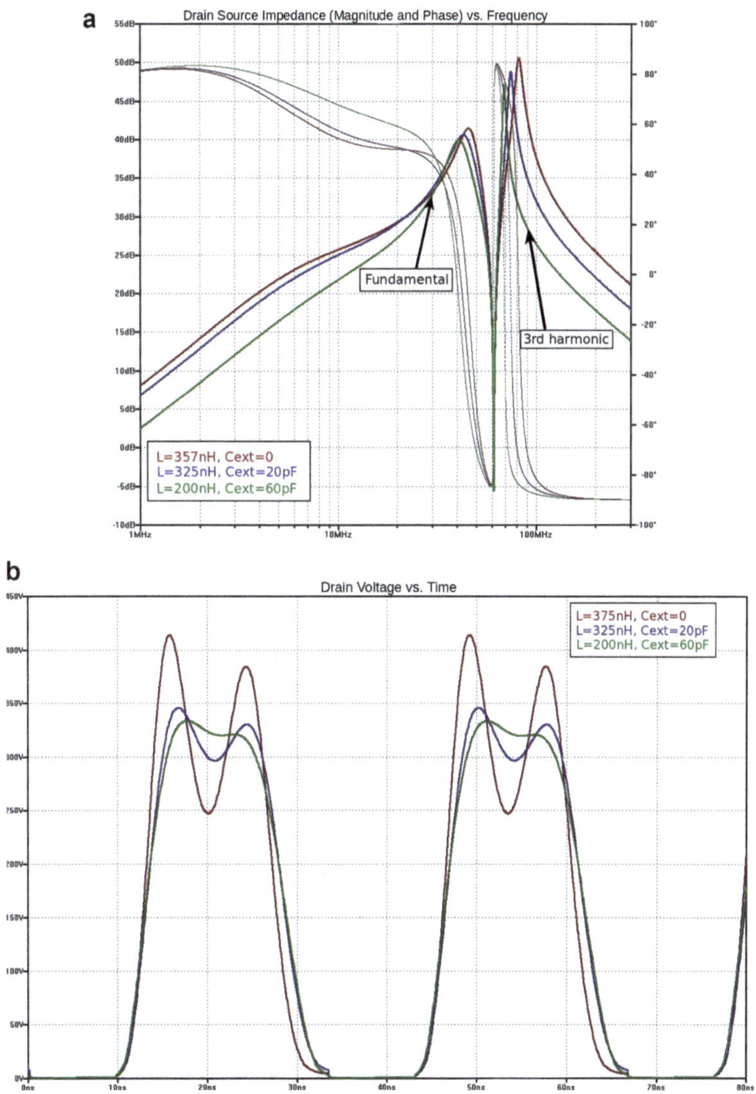

Fig. 8.13 Comparison of three simulated tuning cases. (**a**) Shows the impedance (magnitude and phase) of $Z_{ds}(f_s)$. (**b**) Shows the resulting time domain drain waveforms. These examples have the same impedance magnitude at the fundamental but different magnitude at the 3rd harmonic of the switching frequency

where f_s is the switching frequency, V_g is the gate-drive supply voltage, and Q_g is the total gate charge of the MOSFET. In the resonant PA circuits described in this chapter, ZVS operation leads to intrinsically low switching losses, and to an extent ZVS also reduces gate-drive losses due to the reduction of the *miller plateau* in the V_{gs} vs. Q_g characteristics of the device. But because the gate-drive losses in the

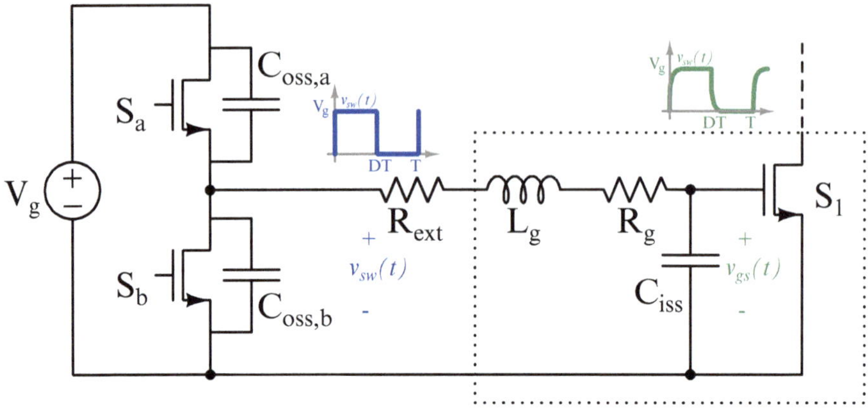

Fig. 8.14 Schematic of a simplified *"hard-switch"* gate drive. R_g, C_{iss}, and L_g represent internal gate resistance, capacitance, and parasitic inductance. R_{ext} is the external resistance added to damp unwanted gate oscillations. $C_{oss,a}$ and $C_{oss,a}$ are the device capacitance of the top and bottom switching devices of the gate-drive circuit

circuit of Fig. 8.14 are proportional to f_s, at a high frequency, $P_{g,loss}$ can become unacceptably high.

A means to reduce gate losses at high frequencies is by driving the switching devices resonantly [17–19]. Simply driving the gate with a sinusoidal voltage (i.e., Fig. 8.15), with no additional components, can significantly reduce gate losses in devices with an $R_g C_{iss}$ product that is short compared to the switching period. The losses in the driving circuit of Fig. 8.15 are $P_{g,sine} = 2\pi^2 \left(C_{iss} V_{gs,max} f_s\right)^2 R_g$. A sinusoidal gate drive naturally has relatively slow rising/falling times making it suitable for single-switch PA implementations with drain-source waveforms achieving ZDVDT at turn-on, like the Class E and E/F(Φ_2). A practical implementation of this gate-drive circuit uses a low-power Class E PA to drive C_{iss} and R_g with a sinusoidal waveform. Other driving schemes use passive components connected to the drain terminal of the FET to form a self-oscillating sinusoidal drive [11]. The effective *"on"* time of the PA FET will depend on its threshold voltage, which may be adjusted by adding a dc offset to the gate. Another consideration when selecting this type of resonant gate drive is that the slow rising gate voltage means higher conduction losses in the circuit as the FET may not be fully enhanced during a significant portion of the switching cycle [12]. The negative gate-source voltage applied to the gate of the FET can limit the applicability of this gate drive in some types of switching semiconductors.

A *trapezoidal* resonant gate drive is another interesting driving scheme for single-switch PAs that overcome some of the sinusoidal driver shortcomings. This driver has faster rise/fall times and depending on the specific implementation can apply little [20] or no negative voltage to the device gate [12, 13, 21]. When the trapezoidal waveform $V_{gs}(t)$ of Fig. 8.16 is driving a FET, the gate losses are:

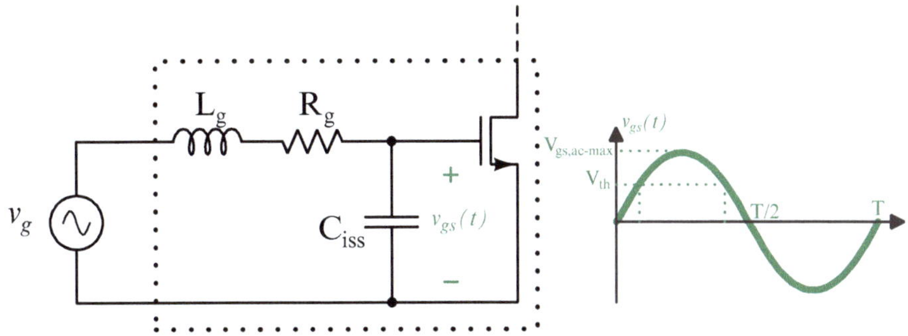

Fig. 8.15 Schematic of a simplified sinusoidal gate drive. The effective *on* time of the switch depends on the device threshold voltage. R_g, C_{iss}, and L_g represent internal gate resistance, capacitance, and parasitic inductance

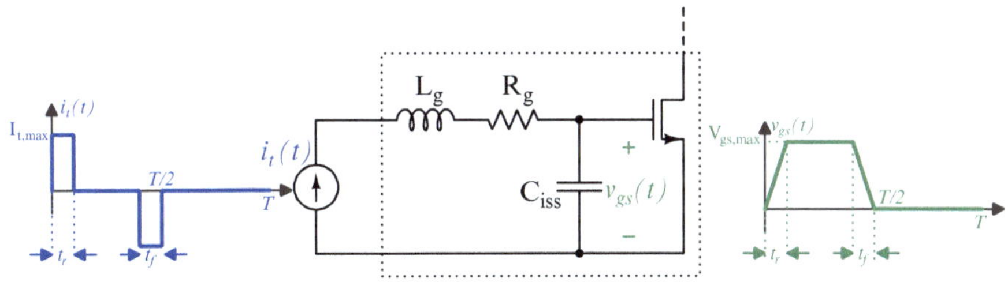

Fig. 8.16 Schematic of a simplified trapezoidal gate drive. R_g, C_{iss}, and L_g represent internal gate resistance, capacitance, and parasitic inductance

$$P_{g,trap} = C_{iss}^2 V_{gs,max}^2 R_g \left[\frac{1}{t_r} + \frac{1}{t_f} \right] f_s$$

where t_r and t_f are the rise and fall times of $vgs(t)$.

A simplified schematic of a resonant trapezoidal gate drive is shown in Fig. 8.17. The circuit is based in on the Φ_2 inverter and follows a similar tuning procedure. For this circuit, the resonant frequency formed by L_g, R_g, and C_{iss} is at a frequency significantly higher than f_s such that the impedance of looking into the gate of S_1 is capacitive. Under this condition, C_{iss} and the device capacitance of Q_{aux} take the role of C_F for tuning purposes [12].

Sinusoidal and trapezoidal gate-drive circuits can enable switched-mode PAs at frequencies reaching 100 MHz even when using silicon devices [12, 13]. The relatively low Q_g and R_g of GaN components as compared to similarly rated silicon devices makes the gate-drive design simpler and in many cases a conventional driving circuit suffices.

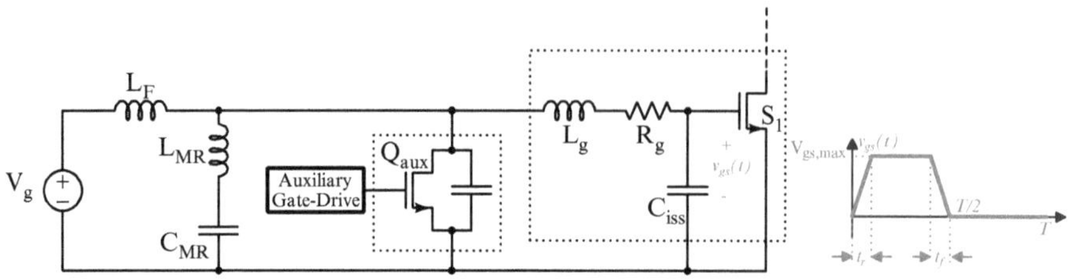

Fig. 8.17 Schematic of a resonant trapezoidal gate drive based on a Φ_2 inverter [12]. R_g, C_{iss}, and L_g represent internal gate resistance, capacitance, and parasitic inductance

8.2.3 Multi-switch Designs

Class D *"half-bridge"* power amplifiers utilize two switches operated in a complementary fashion to provide the desired output waveforms, at the expense of timing and level-shifting challenges.[2] Class E and other single-switch power amplifier designs utilize one common-referenced switch and passive elements to realize waveform shaping, eliminating both the timing and level-shifting challenges. However, these designs impose a variety of limitations, including high device stress and load and frequency range limitations due to tuning constraints for soft switching. In between these approaches are multi-switch power amplifiers that utilize a plurality of ground-referenced switches operated out of phase (typically $180°/N$ for N transistors) with power combined from the individual switch subsystems to a single output (e.g., by taking output differentially from the two subsystems in the $N = 2$ *"push-pull"* case). These designs can often leverage benefits from the interactions of the individual switch subsystems to achieve higher powers and efficiencies than are achievable with a single-switch power amplifier.

Current-Mode Class D

A widely known multi-switch amplifier is the so-called *"current-mode Class D"* amplifier and its variants (e.g., [15, 22]). Figure 8.18 illustrates the topology of the current-mode Class D inverter along with the waveforms for the variant in which the feed inductors L_f are large and carry nearly constant current. The two transistors are run at approximately 50% duty ratio 180° out of phase, and the output is taken differentially between the transistor drains. For this case, the resonant inductor Lr is tuned to be parallel resonant with the sum of C_r and C_s at the switching frequency such that a sinusoidal voltage appears at the output, while each transistor has a half-sine-wave voltage of amplitude πV_{in} and carries approximately a square-wave

[2]The split-winding transformer *"push-pull"* version of the Class D eliminates the level-shifting challenge, but transformer leakage makes this variant undesirable at high frequencies.

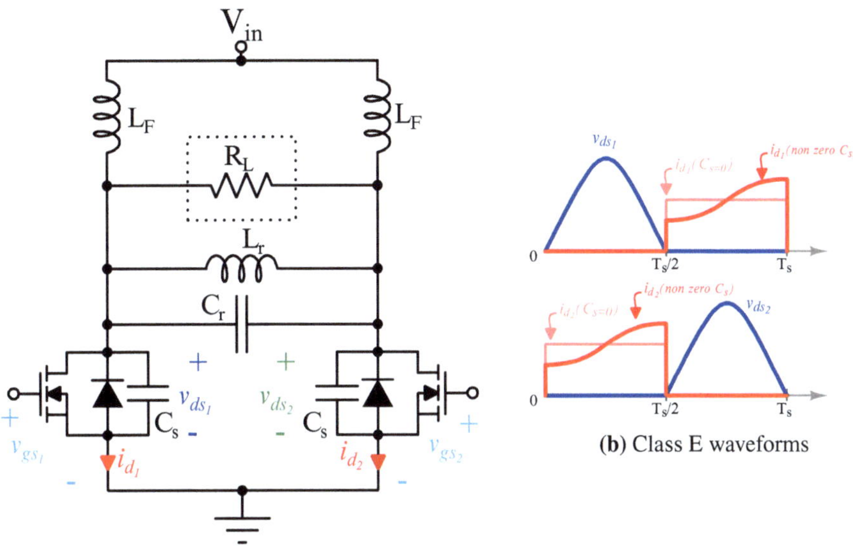

(b) Class E waveforms

(a) Current Mode Class D schematic

Fig. 8.18 (**a**) Current-mode Class D schematic. (**b**) Relevant waveforms. For the case where the feed inductors L_f are large and carry nearly constant current

current. This design naturally achieves zero-voltage switching with well-defined dv/dt at the transistors and can maintain zero-voltage switching over a wide range of load resistances, though it is sensitive to reactive load variations. It likewise naturally absorbs device capacitances (which can account for substantial portions of the resonant capacitance). When the device capacitances are small, the device current waveforms are approximately *square wave* in nature, providing low rms current stress; as the device capacitances C_s represent an increasingly large portion of the total capacitance, the device currents become lower at turn-on and higher at turnoff. If a transformer (or balun) is used to convert the differential converter output to a single-ended (ground referenced) one, the inductor L_r absorbs the transformer magnetizing inductance. Lastly, it is noted that the values of inductors L_f can be reduced and the inductors L_f can be a part of the resonant tank.

Multi-switch power amplifiers using two transistors (*"push-pull"* amplifiers) have been implemented based on a variety of tunings (e.g., Class E and Class Φ_2 operation among others [15, 23–25]), and implementations ganging four or more power amplifiers are sometimes used (e.g.,[26, 27]). Benefits of these multi-switch variants can include better waveform management (e.g., leveraging voltage and current cancellation among the phase-shifted subsystems), higher power operation as compared to single-transistor designs, and reduced component counts and sizes as compared to inphase power combining of paralleled amplifiers.

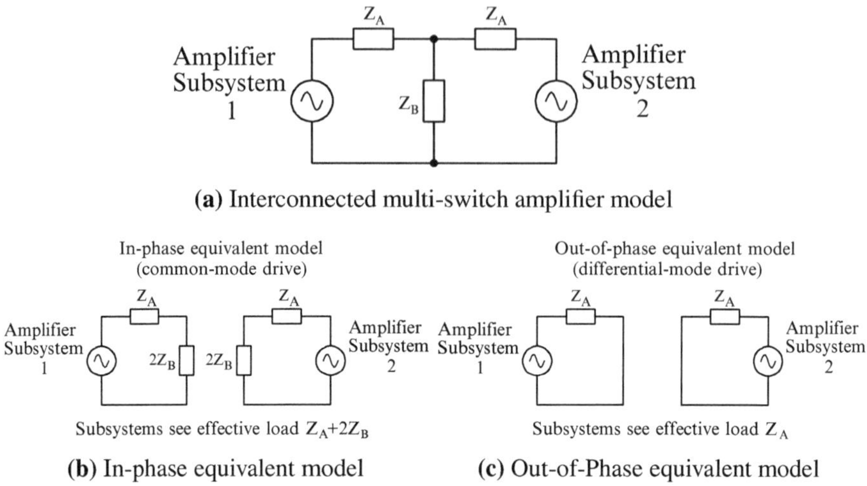

Fig. 8.19 (**a**) Multi-switch amplifier subsystems interconnected by impedance network. (**b**) Inphase circuit equivalent. (**c**) Out-of-phase circuit equivalent. The effective loading of the subsystems is different for depending upon whether the voltage components at that frequency are in phase or out of phase

A key opportunity in multi-switch amplifiers is the use of interactions among the subsystems to provide design benefits. For example, in two-switch *"push-pull"* amplifiers, the two *amplifier subsystem halves* are operated a half-cycle out of phase, such that their their even harmonic components of voltage and current are in phase, while their fundamental and odd harmonic components are out of phase. By appropriate selection of the impedances coupling between the two subsystems (Fig. 8.19), one can control the effective tuning seen in each subsystem using fewer and smaller components than is possible in a single-ended design. For example, in the current-mode Class D design of Fig. 8.18, even harmonic components of the transistor drain voltages are in phase and thus drive no currents through the differentially connected resonant tank and load, while the fundamental and odd harmonic voltage components drive current flows based on the differential impedance posed by the tank and load. Thus, filtration of the even harmonic voltage components produced by each subsystem is inherently achieved. More broadly, performance improvements of multi-switch designs over single-ended ones can be achieved through selection of the relative phases of the subsystems and the impedances interconnecting them [15, 23–27].

Multi-switch amplifiers are also sometimes used in a frequency multiplier mode wherein the fundamental switching frequency components of the subsystems cancel and one delivers a reinforced harmonic component to the output (e.g., [28, 29]). Frequency multiplier operation is usually undertaken when the required output frequency exceeds what is otherwise achievable with a given semiconductor device, as one generally pays an efficiency penalty for only delivering harmonic power to the output. Nevertheless, frequency multiplication is sometimes also used as a secondary operating mode to provide an expanded range of achievable voltages at the output (e.g., [30]).

There are various other motivations for either implementing a single power amplifier having multiple switches or multiple power amplifiers with power combining of their individual outputs. This is often done simply to achieve higher power than can be realized from one power stage. Additionally, however, one can utilize interactions among the individual power amplifiers or subsystems to compensate for load impedance variations (e.g., to preserve soft-switching operation) as proposed in [31] and/or to control the power delivered to the output (e.g., through outphasing), as described below.

8.2.4 Power Control Techniques

A key design decision in a switched-mode power amplifier system is how output power is controlled. In *"linear"* power amplifiers where the transistor is employed as a current source, increased rf input drive amplitude directly yields increased rf output. By contrast, some other means of controlling output power is needed in a switched-mode system.[3] We first consider two very widely used techniques for modulating output power in switched-mode amplifier systems – *"outphasing control"* and *"drain modulation"* – and then briefly address several additional mechanisms that are also sometimes employed. We also briefly consider how to address load variations.

Outphasing Modulation

One widely-used technique for controlling output power in switched-mode rf amplifiers is known as *"outphasing"* control [33]. This term was introduced by Henri Chireix in his classic 1935 paper [34], though it is sometimes called phase-shift control in the power electronics community. As illustrated in Fig. 8.20, an outphasing system comprises two or more switched-mode power amplifiers that are each coupled to the load by a combiner network. Treating the power amplifiers as ac voltage sources and the combining network and load as a linear network, it is apparent that the net voltage at the load is the (vector) sum of the output voltage responses to the individual power amplifier output signals. By controlling the relative phase of the two (fixed amplitude) power amplifiers, one can modulate the output voltage and hence the output power.

[3]It is possible to build a switched-mode system whose output responds to variations in rf input amplitude [32], but such systems use controls to indirectly mimic that characteristic of linear power amplifers.

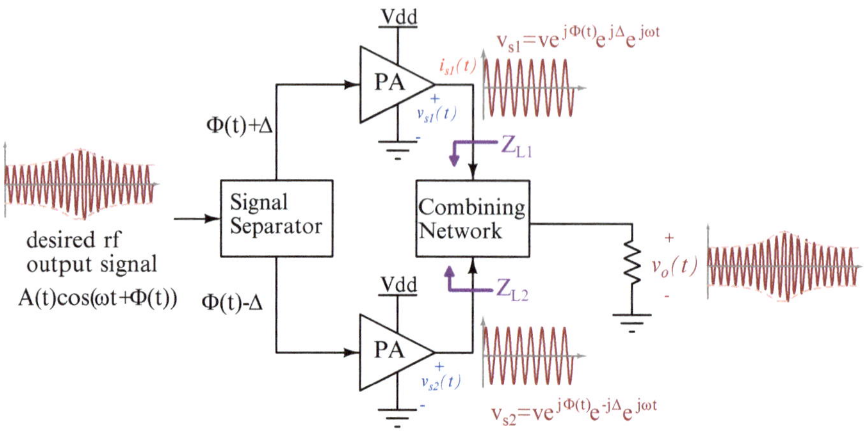

Fig. 8.20 General structure of an outphasing system using two power amplifiers and a combining network. A desired rf output is synthesized as the sum of two fixed-amplitude outputs from two switched-mode amplifiers are phase-shifted with respect to each other. By controlling the relative phase of the two switched-mode amplifiers, the amplitude of the output (and hence the output power) can be modulated

Outphasing systems can be categorized by the characteristics of the power-combining network and how the effective load impedances seen by the individual power amplifiers vary with the synthesized output amplitude. (The effective loading impedance of an amplifier is the complex ratio of voltage to current at the PA output with both amplifiers active; it is the impedance *"seen"* by the power amplifier during operation.) One approach is to utilize an *"isolating"* power combiner, in which the load impedance seen by each amplifier is constant, independent of the net output amplitude and relative phases of the power amplifiers. This approach is valuable when the load impedance is well known and the performance of the power amplifiers is very sensitive to variations in their (individual) loading. Of course, if the individual power amplifiers see constant loading (and thus provide constant output power), then the combining network must provide a place to deliver power not delivered to the load as the output is modulated. Typically, an *"isolation"* resistor is used to absorb power not delivered to the load. However, designs have been implemented in which much of this isolation power is recovered back to the input dc supply through a rectifier network (e.g., [35]).

More interesting from an efficiency point of view is the use of a lossless *"nonisolating"* power combiner. Such a combiner might comprise reactive elements and/or use transformers or transmission lines to couple the PAs to the load. In the simplest case, the load may be differentially coupled between the outputs of the power amplifiers such that the difference in the two PA output voltages appears across the load. However it is implemented, such a combiner provides some type of *"scaled vector sum"* of the power amplifier outputs to the load (and, as such combiners are typically reciprocal, the net response at the load is reflected at the power amplifier outputs). The combiner essentially allows the superposition of the responses of the individual PA outputs at the load such that phase shift between the power amplifiers provides control of the net amplitude and phase of the output waveform.

An important design consideration with a lossless power combiner is how the effective load impedances (or admittances) seen by the power amplifiers vary as the output amplitude is modulated. Consider the simple lossless combiner in Fig. 8.21. When the two amplifiers are in phase, no current flows and the effective loading admittances seen by the two amplifiers are zero and conductive. When the two amplifiers are fully out of phase, the load voltage and thus power are maximized, and the loading admittances are again conductive. As phase is swept from inphase to out-of-phase, one can continuously modulate the output voltage and power. This

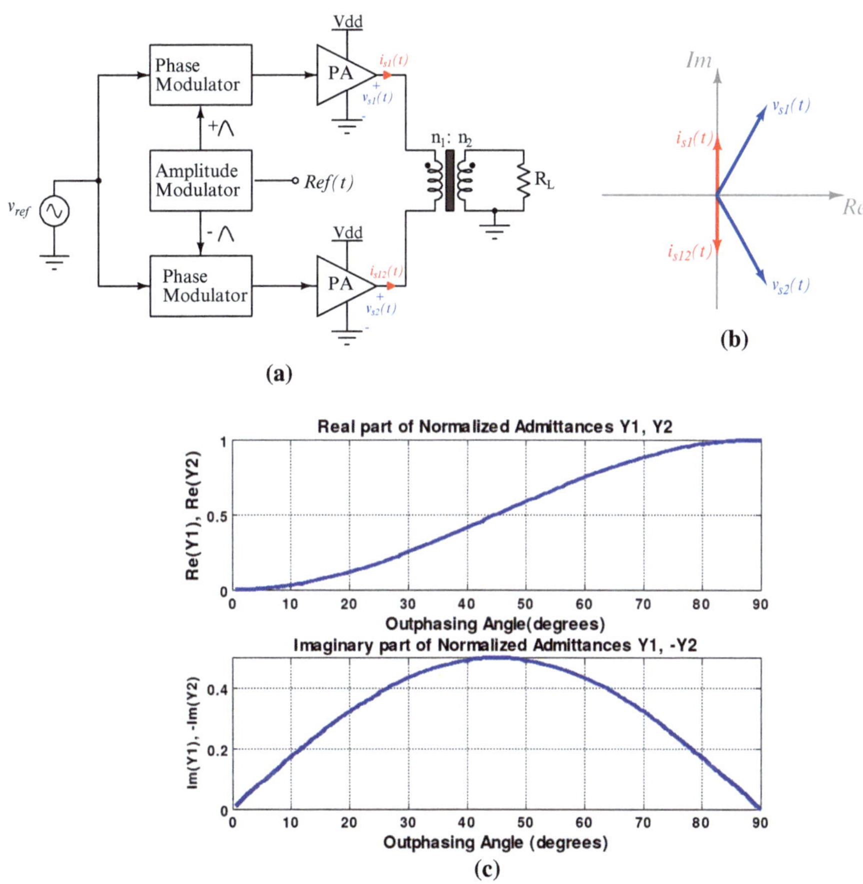

Fig. 8.21 (**a**) An outphasing system with an isolating power combiner. (**b**) Phasor plot of PA voltages and currents. (**c**) Real and imaginary components of the loading admittances seen by the power amplifiers

power variation is reflected as a variation in the effective conductances loading each amplifier. However, owing to the phase differences between the individual PA voltages and their currents, the *"leading"* power amplifier (i.e., the PA with leading phase) sees a capacitive component to its susceptance that varies with output amplitude, while the lagging power amplifier sees an inductive component to its effective loading susceptance. The loading variations seen by the power amplifiers are a key characteristic of how such an outphasing system acts to modulate output power; indeed, this variation is sometimes referred to as *"load modulation"* of the power amplifier. (Outphasing is one of many ways of realizing load modulation.)

While load modulation of the conductive (resistive) components of admittance (impedance) is central to modulating output power in such an amplifier, the output-dependent variations in susceptive (reactive) loading components can be problematic for many high-frequency switched-mode power amplifiers. Design of the combiner network to provide loading characteristics appropriate to the selected power amplifier is a key goal in outphasing systems. For example, in the classic Chireix power combiner [34], fixed complementary shunt susceptances (or series reactances) are added to provide partial cancellation of the susceptive (reactive) components of effective admittance (impedance) seen during outphasing, as illustrated in Fig. 8.22. By appropriately selecting these susceptances, one can reduce the average or worst-case susceptive loading across a given outphasing range making the effective PA loads look more nearly like variable resistors across the outphasing range. Additional susceptive/reactive loading can be provided as needed within the power amplifiers to achieve desirable operation (e.g., inductive loading to maintain ZVS soft switching.) Other outphasing combiners and approaches can provide still better loading characteristics for the power amplifiers. For example, as demonstrated in [36–39], by leveraging *"multi-way"* outphasing (using more than two amplifiers) and an appropriate lossless combiner, one can achieve nearly resistive loading of the power amplifiers over a very wide outphasing range.

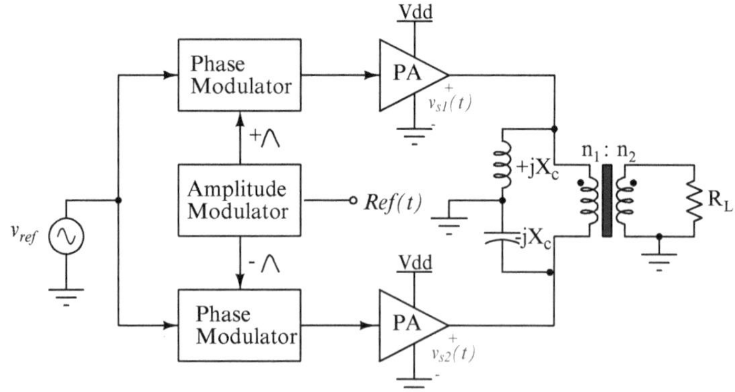

Fig. 8.22 A version of the "Chireix" power combiner. Fixed shunt susceptances are added to partially cancel the effective loading susceptance during outphasing

It may be concluded that outphasing control – when utilized with appropriate power amplifiers and combiners – can provide a very effective means to modulate output power in switched-mode power amplifier systems.

Drain Modulation

Another widely used technique for controlling output power in switched-mode rf amplifier systems is *"drain modulation"* or *"power supply modulation"*. Considering a switched-mode amplifier as a *"switched linear system"*, it is apparent that the amplitude responses of the ac voltages and currents (i.e., the system output) should be proportional to the dc power supply voltage input. We can thus modulate the output power of a switched-mode amplifier by varying the dc power supply voltage that feeds it. (This is sometimes thought of as controlling the transistor *"drain"* bias, hence the term drain modulation.) To accomplish this, one may utilize a dc-dc converter to actively control the power supply voltage feeding the PA (a *supply modulator* or *drain modulation*) and utilize the power amplifier to convert this variable-amplitude supply voltage to a variable-amplitude rf voltage (Fig. 8.23).

Considering a desired rf output signal that may have both a varying amplitude and phase (e.g., for communications), one can use drain modulation to control the output rf amplitude and the phase of the power amplifier gate-drive signal to control the output rf phase. Implementation of this approach for communications applications dates back to at least the 1950s [40] and is sometimes known as *"envelope elimination and restoration" (EER)*. Today it is much more commonly referred to as *"envelope tracking"* and may be used to provide efficiency enhancement of both switched-mode and linear-mode amplifiers (e.g., [41–45]).

A first challenge in implementing envelope tracking relates to the dynamic range over which the supply voltage can be modulated in practice. Owing principally to PA device capacitance nonlinearity, there is often only a limited range over which the supply voltage can be modulated before power amplifier performance suffers. To

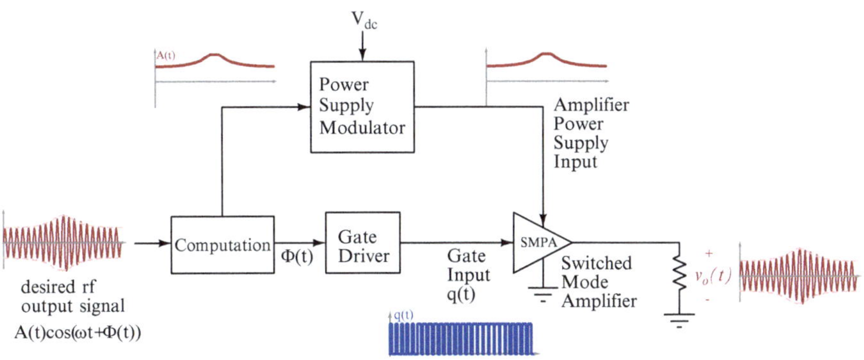

Fig. 8.23 Power supply (or "drain") modulation of a switched-mode amplifier

address this, in some systems supply voltage is only modulated over some limited range (e.g., 4 : 1) below which some secondary means may be used to modulate rf output power (e.g., outphasing, backing into linear operation of the power amplifiers, etc.). A second challenge of envelope tracking relates to bandwidth. If the output envelope need only be modulated slowly, it is straightforward to realize the supply modulator (e.g., as a buck converter). However, for applications such as communications, the required control bandwidth might be quite high (e.g., up to 20 MHz bandwidth for LTE signals), making it difficult to realize a supply modulator with the necessary degree of output bandwidth. As a consequence, for very high-bandwidth cases, the supply modulator efficiency may suffer, and in some cases, the switched-mode conversion of the supply modulator needs to be augmented with a linear amplifier to achieve the necessary control bandwidth for the envelope signal. This tends to degrade the attractiveness of envelope tracking for very high-bandwidth rf outputs.

A variant of the drain modulation approach is to use *"discrete"* drain modulation, in which the power supply modulator simply switches the PA supply among multiple discrete levels. Some secondary means of modulating the output power (e.g., outphasing) is used to *"interpolate"* among the discrete rf output powers that would otherwise arise (enabling continuous control of the output amplitude) and to mitigate any disturbance in the rf output signal arising from the abrupt supply transitions. For example, in *"asymmetric multilevel outphasing" (AMO)*, outphasing is used as a second means of control, and the individual power amplifiers may be instantaneously fed from different ones among the set of discrete supply voltages (e.g., [46, 47]).

A key benefit of discrete supply modulation is that one can efficiently adjust the supply voltage extremely rapidly (requiring only a set of dc supplies and a switching network rather than a very high-bandwidth power supply). A major challenge of discrete drain modulation is that precisely coordinated, high-speed control (via the second control means) is necessary to compensate for the abrupt supply transitions.

Other Control Approaches

There are a variety of other control approaches that can be employed in switched-mode rf applications. Frequency control, in which one incorporates a frequency-sensitive resonant tank in the design and varies frequency to adjust gain and output amplitude, can be quite effective in rf applications. However, in many applications frequency is not a free variable – it is either constrained by specifications (e.g., ISM-band limitations) or the frequency is defined as part of the desired output. Burst mode, or on/off control, in which the power amplifier is turned on and off at a frequency well below the switching frequency, is quite effective for controlling average power. It can be useful in applications such as heating and power conversion, but is unsuitable for applications in which the instantaneous rf output must be continuously maintained at a desired point. Duty ratio control can also be employed in some switching power amplifiers but is typically only effective over a very limited modulation range, though it may sometimes be effective as a secondary control means (e.g., [48]). Structural or *"parametric"* modulation of the system (e.g., realizing effectively tunable components to provide gain variation/load modulation of the amplifier) can also be used (e.g., [49]), though this approach has not received deep exploration to date. Of course, combinations of the above techniques may also be profitably employed, depending upon the application.

Load Impedance Variations

While much of the discussion above has assumed a particular load resistance, load impedance variations can be a significant issue in many applications and have a significant impact on power amplifier design. Here we briefly outline several techniques that have been employed to address this issue. One approach – which is widely used in microwave applications – is to include a nonreciprocal device such as a circulator or isolator as part of the system. This guarantees a specified load impedance for the power amplifier. However, deviations in the load impedance necessarily cause power to be delivered to some other output – commonly to an *"isolation"* resistor where it is lost, though energy recovery via a rectifier may be a possibility. The same is true of *"balanced"* amplifier configurations [50].

Other means of managing load impedance variations that do not result in lost power are certainly desirable. One possibility is to use a *"tunable impedance matching network" (TMN)* to dynamically adjust to match the varying load impedance to that desired for the power amplifier (e.g., [51]). One can also sometimes use a fixed matching network to map a load impedance range into a different range that is acceptable for the power amplifier (e.g., [5]). In applications where one controls the structure of the load, it may be possible to use resistance compression network (RCN) techniques to ensure that load variations are compressed into an acceptable range (e.g., [52–54]). Lastly, it is possible to structure a power amplifier system such that it can be controlled to dynamically adjust for load variations via load modulation; this can be done in much the way one uses load modulation to adjust output power for a fixed load impedance. Indeed, the approach proposed in [31] uses a combination of drain modulation and outphasing to adjust load variations into a range acceptable for the switched-mode power amplifiers comprising the system. Nonetheless, effectively addressing load impedance variations in switched-mode amplifier systems remains an open and potentially highly valuable area of research.

8.3 Magnetics Design for High Frequency

Magnetics for power applications in the HF to VHF frequency range pose a special challenge. Approaches used for power electronics in the hundreds of kHz can result in low Q and excessive losses, whereas traditional RF design approaches can result in large physical size and often do not prioritize efficiency. Commercial off-the-shelf parts can be used, but rarely is data available for their characteristics at high drive levels or even over the full frequency range of interest. Thus, custom-designed magnetics are often necessary.

The primary design challenges are losses in magnetic cores, eddy-current loss effects in winding conductors, and capacitance in windings. Designs that don't use a magnetic core but rely on flux through the air or other nonmagnetic materials avoid problems of core loss. Such "air-core" inductors are used in many HF and particularly VHF designs. We start this section by considering winding loss effects and apply this to air-core design. Approaches to evaluating the improvement available through adding a magnetic core are then discussed, and the potential of available magnetic materials is surveyed.

In addition to inductors, transformers are often desired for isolation and transformation. Issues in transformer design are discussed in Section 8.3.5.

8.3.1 High-Frequency Winding Losses

A changing magnetic field induces an electric field which drives current in a conductor, looping around the flux lines. At HF and VHF frequencies, such *eddy currents* can be severe, and good winding designs are based on working within the constraints they impose.

At 10s to 100s of kHz, eddy-current losses can be made negligible through the use of conductor diameters that are thin compared to the electromagnetic skin depth,

$$\delta = \sqrt{\frac{\rho}{\pi f \mu}} \tag{8.2}$$

where ρ is the conductor resistivity, f is the frequency, and μ is the conductor permeability, ordinarily equal to the permeability of free space $\mu_0 = 4 \times 10^{-7}\pi$. For example, litz wire uses many fine strands, each much smaller than a skin depth, individually insulated and twisted in configurations that limit flux linkage that would otherwise induce currents to circulate among the strands [55]. Over the HF to VHF range, the skin depth in copper ranges from about 40 µm at 3 MHz to 4 µm at 300 MHz. Although 40 µm diameter wire is readily available, the goal of making the conductor diameter much smaller than the skin depth is not feasible, and litz wire only has limited usefulness in the HF range, even with alternative strand configurations discussed in [56].

When conductors much thinner than a skin depth are not practical, one must use a conductor thicker than a skin depth and accept that ac current will flow only on the surface of the conductor, in a region approximately one skin depth deep. If the current is uniformly distributed over this surface, the resistance becomes

$$R_u = \frac{\ell \rho}{\delta c} \tag{8.3}$$

where ℓ is the path length and c is the perimeter of the conductor. However, uniform current distribution is rarely achieved in practice. An isolated cylindrical conductor would have uniform current density around its perimeter, but other shapes will have nonuniform current density, and nearby conductors or magnetic materials can also change the current distribution and introduce circulating currents. Thus, winding design for the HF and VHF range should focus on spreading current over the surface of conductors as uniformly as possible, to approach the resistance given by (8.3) as closely as possible.

Useful insight can be obtained from the fact that the tangential magnetic field H near the surface of a conductor thick compared to a skin depth is equal to the surface current density σ, so the design goal of uniform current density translates to a uniform field strength near the surface of the conductor, tangential to the surface. We can also conclude that a winding should comprise only a single layer. The current density on the surface of the winding facing the highest field region will be sufficient to support that field, and any additional winding layers only introduce additional, unnecessary losses.

If we consider a rectangular conductor, larger in the $\hat{y}$ direction than the $\hat{x}$ direction, carrying current in the $\hat{z}$ direction, we would ideally like the field H to be equal around the full perimeter, but if that is difficult, we can still approach the ideal resistance (8.3) if the two longer sides are fully utilized, as in Fig. 8.24. Thus, we

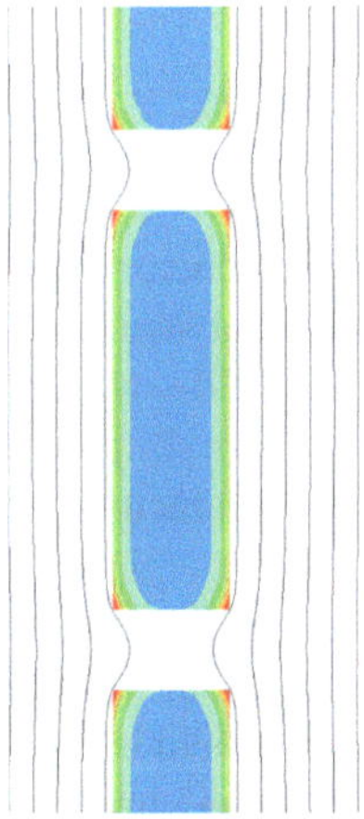

Fig. 8.24 A single-layer winding with equal field strength (indicated by field-line density) on the left and right, resulting in equal current density (colored shading) on both sides of the rectangular conductors. The resistance is 47% higher than that given by (8.3) because the top and bottom surfaces of each turn are underutilized. The use of both of the longer sides would allow the resistance to approach half of R_1 given by (8.3) with smaller spacing between turns; as shown its resistance is 55% of R_1

ideally want equal field magnitudes on both sides of the conductor. Furthermore, these fields must be in opposite directions; if they are in the same direction, the currents on opposite sides of the conductor are in opposite directions, and the net current in the conductor is zero.

Although equal field magnitude on both sides of a conductor is ideal, in practice, it is more common to have a strong field only on one side of a conductor or one side of a single-layer winding comprising several or more turns, such that the current flows primarily on one surface only. The resistance can then approach

$$R_1 = N\frac{\ell_t \rho}{\delta w/N} = N^2\frac{\ell_t \rho}{\delta w} \tag{8.4}$$

where w is the total width of a winding, N is the number of turns in it, and ℓ_t is the length of one turn. For rectangular conductors, $R_1 \approx 2R_u$ if the dimension of the conductor parallel to the field is much greater than the thickness of the conductor and we use approximately half the surface area for conducting current.

Because the current flows on the surface of the conductor, the quality and material of the surface can be important: the surface roughness should be small compared to the skin depth [57]. This is rarely an issue in the HF range but is sometimes a concern in the VHF range, and it can be worthwhile to select conductors with a smooth surface or even polish them. Silver plating is sometimes used because its conductivity is 6–8% higher than copper, depending on the purity of both. However, the skin resistance is only improved by the square root of the conductivity, i.e., an improvement of 3–4%, because the increased conductivity reduces the skin depth.

8.3.2 Air-Core Design

Common geometries for an air-core inductor include solenoids, toroids, and planar spirals. Considerations in choosing between these include ac resistance, ease of fabrication, and containment of the field—without a magnetic core to contain the field, external fields can be a concern if they cause EMI problems or induce excessive eddy currents in nearby conductors, causing power loss and reducing inductance.

A planar spiral is easy to fabricate in, for example, a printed circuit board process, but the field geometry is unfavorable both from the point of view of external field containment and for ac resistance. The field is primarily perpendicular to the surface of the winding as a whole, creating high current density on the edges of each turn and failing to utilize the larger top and bottom surfaces effectively.

In contrast, a toroid contains most of the field, nearly eliminating each of the problems associated with an external field, and is an excellent choice for HF and VHF inductors [58]. Because of the field containment, the conduction is primarily on the inner surface of the conductors, facing the interior field, which means that the resistance can approach R_1 but will be more than twice R_u. A solenoid does not contain the field as well as a toroid, but it can have a higher Q, can be simpler to fabricate, and so is also a good option to consider for many applications.

Air-Core Toroids

Toroidal air-core inductors may be constructed in several ways. They may be simply wound from magnet wire on a dielectric core. However, this leaves space wasted between turns at the outside perimeter, not used for conduction. Ideally shaped conductors, narrower at the inside diameter and wider at the outside, can be fabricated by 3-D printing [59] or depositing a conductor over the whole surface of a dielectric core and cutting slits to separate it into turns [60]. An alternative approach for easier manufacturing is to use a printed circuit board (PCB) process. Limitations of the PCB process include the limitations imposed by the process on the via geometries for vertical connections between layers and the necessity of a rectangular cross section for the flux path. These limitations are minor – tightly placed vias can provide low resistance for the vertical path, and although rounded corners for the cross section of the flux path allow the current path to be slightly shortened, it is shown in [60] that even if this profile is optimized, it provides only about an 8% improvement. High-performance PCB toroidal inductors have been widely applied [61, 62].

The inductance of an air-core toroid with many turns is dominated by the inductance associated with the toroidal field contained by the winding. However, there is also a *poloidal* field outside of the winding (Fig. 8.25), because of the single-turn loop that the current takes around the toroid. The total inductance is accurately estimated by including both of those components, resulting in

$$L = \frac{N^2 h \mu_0}{2\pi} \ln\left(\frac{d_o}{d_i}\right) + \frac{d_i + d_o}{4} \mu_0 \left[\ln\left(8\frac{d_o + d_i}{d_o - d_i}\right) - 2\right] \tag{8.5}$$

in terms of the inner diameter d_i, outer diameter d_o, height h, and number of turns N [63].

A first estimate of the resistance can be obtained assuming that the current paths are perfectly vertical at the vias and radial on the top and bottom surfaces.

$$R_t = \frac{N^2 \rho}{2\pi \delta} \left(\frac{2h}{d_o} + \frac{2h}{d_i} + \ln\frac{d_o}{d_i}\right) \tag{8.6}$$

A more accurate estimate is obtained in [64] by considering the effect of the cut between turns, the diagonal skew of the turns on the top and bottom surfaces, and the current on the outside surface associated with the external, poloidal field.

Of these effects modifying (8.6), the biggest factor is, for small numbers of turns, the diagonal skew factor which can, for example, increase the resistance by a factor of three for five turns [64]. Fortunately, this severe impact can easily be mitigated. If a toroidal winding requires a small number of turns, it can be implemented with a set

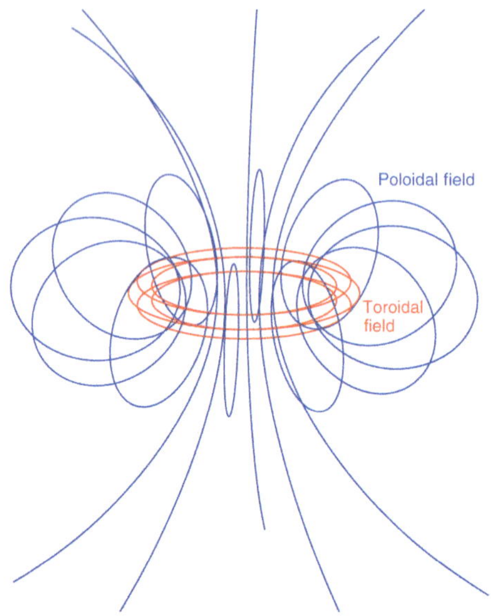

Fig. 8.25 Toroidal and poloidal fields. The poloidal field in a toroidal inductor is produced by the single-turn current around the toroid and is normally much smaller than the toroidal current

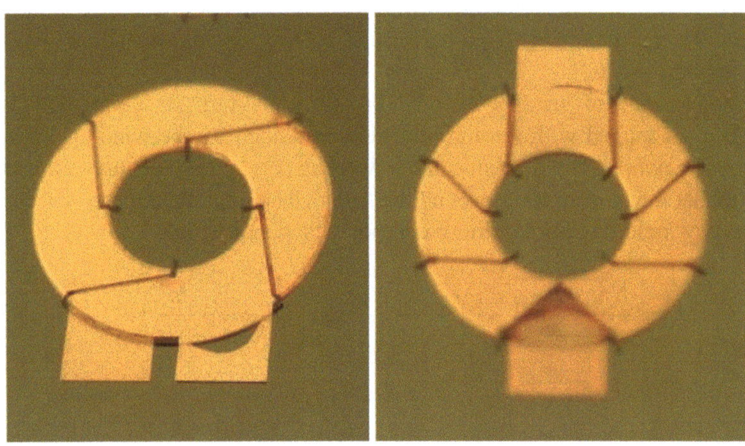

Fig. 8.26 A four-turn toroidal inductor implemented two ways: left, a conventional winding; right, two parallel four-turn windings

of parallel windings, each with the same number of turns, as shown in Fig. 8.26. This greatly reduced the skew angle and the resulting impact on resistance. For example, a four-turn inductor implemented with two windings in parallel, each of which has four turns, has the skew of an eight-turn winding, resulting in a 40% reduction in winding loss compared to a simple five-turn winding [65]. This configuration also reduces capacitance by keeping opposite ends of the winding apart from each other and reduces the external poloidal field.

Note that both the resistance and the inductance are approximately proportional to the number of turns squared. Thus, the ratio between inductance and resistance and thus the quality factor Q are approximately constant, independent of the number of turns, as long as a winding requiring a small number of turns is implemented using enough parallel windings to keep the top and bottom conductor directions approximately radial.

Air-Core Solenoids

An air-core solenoid wound from magnet wire can be easily made or purchased. At high power levels, copper tubing is used rather than solid wire to avoid wasting copper and sometimes to allow for liquid cooling. A solenoid can also be made in a printed circuit board process, with the flux path parallel to the board. A long solenoid has a much stronger field inside than out, and the field is tangent to the conductors near the center of its length. This is a similar scenario to an air-core toroid, and the performance achievable is similar. The solenoid can achieve higher Q at the expense of having a larger external field. A conductive shield can be considered to contain that field [66]. If it is thick compared to a skin depth, the ac field will be effectively blocked, but if there is not insufficient space allotted for the field, the inductance will be reduced and the losses increased, so it is typically more practical to use a toroid if a low external field is required.

In the case of a solenoid long enough that the field inside is uniform, the energy storage is proportional to the volume, and the loss is proportional to the surface area of the winding. This indicates that a shorter, fatter solenoid will have a higher Q. Changes in that direction get us out of the realm of the applicability of the long solenoid approximation, and good designs are those that balance these two effects [67–70].

The comparative performance of air-core toroids and solenoids depends on the specific constraints under which they are optimized, but, for example, if PCB designs for both are optimized under the same constraints of height and area, the solenoid can have about 50% higher Q if the area allotted is not very large compared to the square of the board thickness t^2. As the area allotted becomes much larger than t^2, the achievable Q becomes no better than that of a toroid. A simple solenoid has much lower self-capacitance than a comparable simple toroid, where the capacitance is dominated by the capacitance between the first and last turns. However, the use of a parallel-winding toroid as in Fig. 8.26 reduces this difference.

8.3.3 Magnetic Core Materials and Performance

An ideal lossless magnetic material with a relative permeability μ_r would increase the inductance, and thus the quality factor, of a toroidal inductor by a factor of μ_r. By scaling the number of turns down to return to the same inductance value, one can use the core to reduce the resistance by a factor of μ_r for the same inductance, if the losses in the core are negligible. Given that magnetic materials HF frequency range typically have relative permeability in the range of 10 to 100, the potential for dramatic performance improvements is evident. However, magnetic materials also introduce their own power loss, and so careful examination of magnetic material losses is essential.

A useful way to evaluate the capability of a magnetic material in a power application is the "performance factor," defined as the product of peak flux density and frequency at a specified loss density. The performance factor is proportional to the VA product that the component can handle for a given winding NI product. Thus, it can be considered to be an assessment of the magnetic material's contribution to the power handling capability. The performance factor was measured for a wide range of materials in [71, 72]. Figure 8.27 shows the envelope of the performance-factor data in [71, 72] along with the envelope for commercial materials in a 2013 databook [73] and estimates for recently introduced or improved materials from Fair-Rite (80 and 67 MnZn and NiZn ferrites) and Ferroxcube (3F46 MnZn ferrite). Although there is a general trend of higher performance factor at higher frequencies, these gains do not continue indefinitely: from the perspective of magnetic core capability, there is little advantage to frequencies over 10 to 20 MHz. However, magnetic material manufacturers have only just begun to focus on power applications of materials in the HF range, and the improvements already realized suggest that further improvements may be possible.

If performance factor data is used to help select an operating frequency, it can be useful to incorporate the variation in winding performance as a function of frequency to create a *modified performance factor*. Although the appropriate

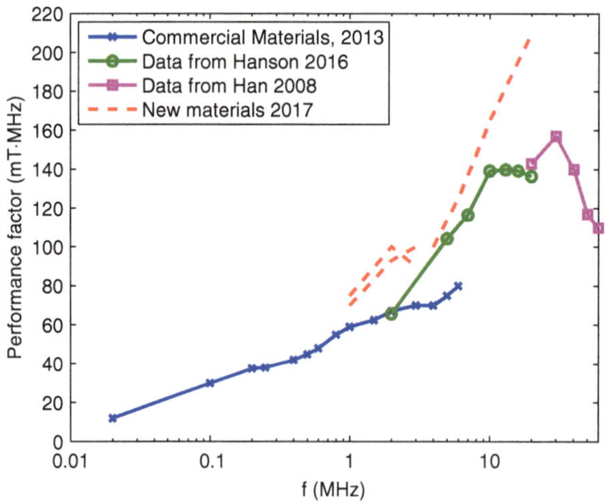

Fig. 8.27 Performance factor $B \cdot f$ for commercial magnetic materials, based on a loss density of 500 mW/cm^3

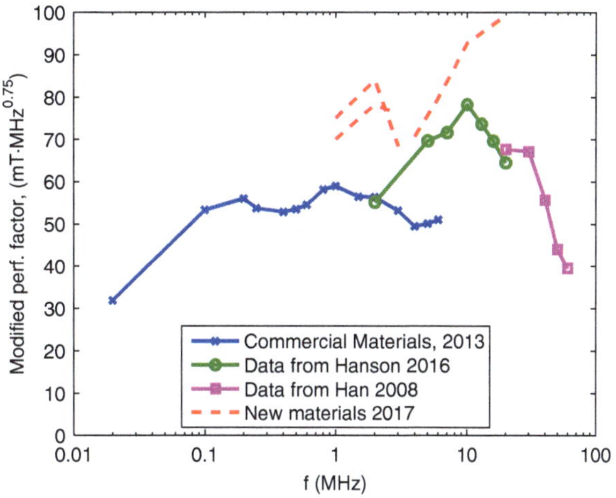

Fig. 8.28 Performance factor $B \cdot f^{0.75}$. The modification accounts for skin-effect losses in the winding

modification depends on the details of the assumptions about the winding design constraints, considering skin-effect limited winding loss in a single-layer winding as in (8.4) leads to a modified performance factor $B \cdot f^{0.75}$ as plotted in Fig. 8.28 [71]. From this perspective, magnetic component performance does not depend as strongly on frequency, and the frequency can be chosen based on other considerations. However, in the VHF range, magnetic material losses are high enough that an air-core design is usually preferred. With an air core, increased frequency only improves the available performance: quality factor is proportional to the square root of frequency [74].

8.3.4 Design with Magnetic Cores

With a magnetic core, a key design tradeoff is the balance between core losses and winding losses. In a transformer, the choice of the number of turns directly affects both winding loss and core loss. In an inductor, changing the number of turns while maintaining the required inductance value entails the use of an air gap, with the length set differently according to the number of turns. This assumes that the inductance would be too high if the ideal number of turns was used without a gap, which is almost always the case [71]. The total loss is minimized when core loss and winding loss are similar, but usually the optimum is to have slightly more winding loss than core loss. This can be demonstrated using the Steinmetz model for core loss in which the core loss per unit volume is given in terms of peak ac flux density $\hat{B}$ as $P_v = k\hat{B}^{\beta}$, where β is an experimentally determined exponent, usually between 2 and 3 [75]. Thus, core loss $P_c \propto N^{-\beta}$, and, with winding loss $P_w \propto N^2$, the optimum distribution between winding loss and core loss is $P_w = \frac{2}{\beta} P_c$.

However, the implementation of a gap affects the shape of the field, as does the use of a core. And, as discussed in Sect. 8.3.2, the shape of the field in the winding region has an important effect on winding loss. To spread the current over the surface of the winding and thus obtain low resistance, the field must be parallel to the winding surface, and its strength must be uniform. In contrast, an air gap leads to a concentration of field strength in one region and to a curved field that has components perpendicular to the conductor. Thus the introduction of a gapped core has the potential to exacerbate winding loss. However, a carefully implemented core also has the potential to improve winding loss by shaping the field to be closer to the ideal.

Consider an inductor on a high-permeability magnetic core with a rectangular winding window and a gap in one leg of the core. Any conductor in the region of the concentrated magnetic field near the gap will incur high losses. One way to reduce this problem is to replace the gapped core leg with an ungapped bar of lower permeability magnetic material, to achieve the same reluctance and inductance that would have been achieved with the gap and the high-permeability core material. This approach, sometimes called a "distributed gap," avoids the field concentration and resulting winding losses of a gap. With a rectangular winding window, it results in field lines parallel to the low-permeability core leg, and current can be spread evenly over the surface of a winding facing it.

However, if one chooses a magnetic material based on what has the lowest loss, it is unlikely to have the right permeability to avoid the need for a gap. A low-permeability material can be approximated by the use of multiple smaller gaps, spaced evenly along a core leg. This approach, called a "quasi-distributed gap," can approach the performance of a distributed gap if the distance between gaps (the gap pitch) is adequately small or if the winding is spaced far enough away from the gaps: the spacing should be no closer than about one quarter the gap pitch [76].

Although one generally thinks of a gapped core as making winding loss worse than without one, a well-designed core configuration can actually reduce winding loss. A good example of this is a solenoid. A standard air-core solenoid has a much stronger field on the inside surface of the winding than the outside and primarily conducts on that surface. With a high-permeability rod core, the field inside the winding is shorted out by the low reluctance of the rod. The field is then stronger on the outside surface, and the winding conducts primarily on the outside. Tuning the permeability to an intermediate value can balance the field strength on the inside and outside surfaces of the winding, allowing both to be used for conduction. In [77], this is done with a quasi-distributed gap, achieving $Q = 620$ at 3 MHz.

Although most winding designs for the HF and VHF range use conductors that are thick compared to a skin depth, foil is economically feasible to produce in much thinner than the thinnest economically feasible litz strands. Effective use of thin-foil conductors in the HF to VHF range requires overcoming a number of challenges as reviewed in [78]. One of these is ensuring that field lines are parallel to the foil layers. A magnetic core can be used to shape field lines parallel to the foil as part of the strategy for such a design [79–81].

8.3.5 Transformers and Impedance Transformation

At typical power electronics frequencies, transformer design is in many ways similar to, but easier than, inductor design. There's usually no need for an air gap, and so there are no issues with field shapes near the gap, and interleaving can be used to further reduce proximity-effect loss. However, in the HF and VHF range, capacitance between windings becomes a bigger issue, particularly if interleaving is used and if at least one of the ports is used for high voltage or high impedance. Moreover, depending on the application, leakage inductance may become an important consideration. Interleaving and close spacing of windings can reduce leakage inductance but may result in intolerably high capacitance.

Although one can perform HF and VHF transformer design as being similar to lower-frequency design, but with more difficult constraints and tradeoffs, a different conceptual approach, that of "transmission-line transformers" [82–84], can help direct the designer toward designs that avoid detrimental capacitive effects. The concept is to construct windings from paired conductors that act as transmission lines with well-defined characteristic impedance. When they are used as part of a matched-impedance system, the capacitance and inductance do not limit bandwdith or degrade signal integrity. The impedance transformation can be designed by configuring transmission lines in series at one end and in parallel at the other. Doing so requires both conductors of some lines to float at potentials different from ground. Winding them on a core allows this by providing a common-mode choke effect.

Although the conceptual approach of transmission-line transformers is very different from that traditionally taken in power electronics, the resulting physical implementation can typically also be understood as an autotransformer that is very well configured to minimize detrimental capacitance and leakage inductance effects. The use of opposing current in adjacent conductors also tends to minimize proximity effect losses. So the concept can be considered as a way to obtain a good autotransformer design, rather than as an alternative to a good autotransformer design.

A truly distinct approach to impedance and voltage transformation is to use an LC-resonant matching network. This can be a simple and efficient approach. The bandwidth is inherently limited, which can be a disadvantage, but can also be useful if reduced gain away from resonance is used for converter control. Although basic design approaches for matching networks have been well established for many decades, only recently have design approaches been developed considering losses and maximizing efficiency, starting with [85], which assumed a maximum inductor quality factor Q and chose the number of stages to maximize the efficiency under that constraint. A similar analysis in [86] constrains the volume of the magnetics rather than Q and accounts for the fact that one large inductor could have higher Q than multiple smaller inductors. But both of these analyses constrain the impedance between stages to real values. Higher efficiency is possible if this constraint is removed [87]. Near optimal designs could be obtained by selecting the number of stages based on [86] and then, if a multi-stage network is chosen, performing the detailed design based on [87].

References

1. B. Molnar, Basic limitations on waveforms achievable in single-ended switching-mode tuned (Class E) power amplifiers. IEEE J. Solid-State Circuits **19**(1), 144–146 (1984)
2. E.L. Owen, History [origin of the inverter]. IEEE Ind. Appl. Mag. **2**(1), 64–66 (1996)
3. D. Prince, The inverter. Gen. Elect. Rev. **28**, 676–681 (1925)
4. I. T. U. (ITU), Radio regulations articles. Booklet (2016). ISM bands outlined in Article 5.150
5. D.C. Hamill, Impedance plane analysis of Class DE amplifier. Electron. Lett. **30**(23), 1905–1906 (1994)
6. S.-A. El-Hamamsy, Design of high-efficiency RF Class-D power amplifier. IEEE Trans. Power Electron. **9**(3), 297–308 (1994)
7. N. Sokal, A. Sokal, Class E—a new class of high-efficiency tuned single-ended switching power amplifiers. IEEE J. Solid-State Circuits **SC-10**(3), 168–176 (1975)
8. M.K. Kazimierczuk, D. Czarkowski, *Resonant power converters* (Wiley, New York, 2012)
9. M. Kazimierczuk, K. Puczko, Exact analysis of Class-E tuned power amplifier at any Q and switch duty cycle. IEEE Trans. Circuits Syst. **34**, 149–159 (1987)
10. N.O. Sokal, Class-E RF power amplifiers. *American Radio Relay League* (ARRL), QEX, pp. 9–20, Jan/Feb 2001
11. J.M. Rivas, R.S. Wahby, J.S. Shafran, D.J. Perreault, New architectures for radio-frequency dc-dc power conversion. IEEE Trans. Power Electron. **21**(2), 380–393 (2006)
12. J.M. Rivas, D. Jackson, O. Leitermann, A.D. Sagneri, Y. Han, D.J. Perreault, Design considerations for very high frequency dc-dc converters, in *37th IEEE Power Electronics Specialists Conference*, 2006, pp. 18–22
13. R.C. Pilawa-Podgurski, A.D. Sagneri, J.M. Rivas, D.I. Anderson, D.J. Perreault, Very-high-frequency resonant boost converters. IEEE Trans. Power Electron. **24**(6), 1654–1665 (2009)
14. L. Roslaniec, A.S. Jurkov, A.A. Bastami, D.J. Perreault, Design of single-switch inverters for variable resistance/load modulation operation. IEEE Trans. Power Electron. **30**, 3200–3214 (2015)
15. S.D. Kee, I. Aoki, A. Hajimiri, D. Rutledge, The Class-E/F family of ZVS switching amplifiers. IEEE Trans. Microwave Theory Tech. **51**, 1677–1690 (2003)
16. J.M. Rivas, Y. Han, O. Leitermann, A. Sagneri, D.J. Perreault, A high-frequency resonant inverter topology with low voltage stress, in *IEEE Power Electronics Specialists Conference (PESC)* (IEEE, 2007), pp. 2705–2717
17. R.L. Steigerwald, Lossless gate driver circuit for a high frequency converter, U.S. Patent 5,010,261, 23 Apr 1991
18. D. Maksimovic, A MOS gate drive with resonant transitions, in *IEEE 22nd Annual Power Electronics Specialists Conference (PESC '91 Record)*, June 1991, pp. 527–532
19. P. Dwane, D.O. Sullivan, M.G. Egan, An assessment of resonant gate drive techniques for use in modern low power dc-dc converters, in *Twentieth Annual IEEE Applied Power Electronics Conference and Exposition (APEC)*, vol. 3, Mar 2005, pp. 1572–1580
20. L. Gu, W. Liang, J. Rivas-Davila, A multi-resonant gate driver for very-high-frequency (VHF) resonant converters, in *IEEE 18th Workshop on Control and Modeling for Power Electronics (COMPEL)*, July 2017, pp. 1–7
21. F. Hattori, H. Umegami, M. Yamamoto, Multi-resonant gate drive circuit of isolating-gate GaN HEMTs for tens of MHz. IET Circuits, Devices Syst. **11**(3), 261–266 (2017)
22. H. Kobayashi, J.M. Hinrichs, P.M. Asbeck, Current-mode Class-D power amplifiers for high-efficiency rf applications. IEEE Trans. Microwave Theory Tech. **49**, 2480–2485 (2001)

23. F. Raab, Idealized operation of the Class E tuned power amplifier. IEEE Trans. Circuits Syst. **24**, 725–735 (1977)
24. Z. Kaczmarczyk, W. Jurczak, A Push-Pull Class-E inverter with improved efficiency. IEEE Trans. Ind. Electron. **55**, 1871–1874 (2008)
25. J.S. Glaser, J.M. Rivas, A 500 W push-pull dc-dc power converter with a 30 MHz switching frequency, in *2010 Twenty-Fifth Annual IEEE Applied Power Electronics Conference and Exposition (APEC)*, Feb 2010, pp. 654–661
26. I. Aoki, S.D. Kee, D.B. Rutledge, A. Hajimiri, Distributed active transformer: a new power-combining and impedance-transformation technique. IEEE Trans. Microwave Theory Tech. **50**, 316–331 (2002)
27. S. Jeon, D.B. Rutledge, A 2.7-kW, 29-MHz Class-E/F_{odd} amplifier with a distributed active transformer, in *IEEE MTT-S International Microwave Symposium Digest*, June 2005, p. 4
28. K. Shinoda, T. Suetsugu, M. Matsuo, S. Mori, Idealized operation of Class DE amplifier and frequency multipliers. IEEE Trans. Circuits Syst. I: Fundam. Theory Appl. **45**, 34–40 (1998)
29. R. Zulinski, J. Steadman, Idealized operation of Class E frequency multipliers. IEEE Trans. Circuits Syst. **33**, 1209–1218 (1986)
30. W. Inam, K.K. Afridi, D.J. Perreault, Variable frequency multiplier technique for high-efficiency conversion over a wide operating range. IEEE J. Emerging Sel. Top. Power Electron. **4**, 335–343 (2016)
31. D.J. Perreault, A new architecture for high-frequency variable-load inverters, in *2016 IEEE 17th Workshop on Control and Modeling for Power Electronics (COMPEL)*, June 2016, pp. 1–8
32. T.W. Barton, D.J. Perreault, Theory and implementation of rf-input outphasing power amplification. IEEE Trans. Microwave Theory Tech. **63**, 4273–4283 (2015)
33. T. Barton, Not just a phase: outphasing power amplifiers. IEEE Microwave Mag. **17**, 18–31 (2016)
34. H. Chireix, High power outphasing modulation. Proc. Inst. Radio Eng. **23**, 1370–1392 (1935)
35. P.A. Godoy, D.J. Perreault, J.L. Dawson, Outphasing energy recovery amplifier with resistance compression for improved efficiency. IEEE Trans. Microwave Theory Tech. **57**, 2895–2906 (2009)
36. D.J. Perreault, A new power combining and outphasing modulation system for high-efficiency power amplification. IEEE Trans. Circuits Syst. I: Regul. Pap. **58**, 1713–1726 (2011)
37. T.W. Barton, D.J. Perreault, Four-way microstrip-based power combining for microwave outphasing power amplifiers. IEEE Trans. Circuits Syst. I: Regul. Pap. **61**, 2987–2998 (2014)
38. A.S. Jurkov, L. Roslaniec, D.J. Perreault, Lossless multi-way power combining and outphasing for high-frequency resonant inverters, in *Proceedings of The 7th International Power Electronics and Motion Control Conference*, vol. 2, June 2012, pp. 910–917
39. T.W. Barton, A.S. Jurkov, P.H. Pednekar, D.J. Perreault, Multi-way lossless outphasing system based on an all-transmission-line combiner. IEEE Trans. Microwave Theory Tech. **64**, 1313–1326 (2016)
40. L.R. Kahn, Single-sideband transmission by envelope elimination and restoration. Proc. IRE **40**, 803–806 (1952)
41. J.J. Yan, C. Hsia, D.F. Kimball, P.M. Asbeck, Design of a 4-W envelope tracking power amplifier with more than one octave carrier bandwidth. IEEE J. Solid-State Circuits **47**, 2298–2308 (2012)
42. T.M. Aitto-oja, High efficiency envelope tracking supply voltage modulator for high power base station amplifier applications, in *IEEE MTT-S International Microwave Symposium*, May 2010, pp. 1–1
43. M. Rodríguez, Y. Zhang, D. Maksimović, High-frequency PWM buck converters using GaN-on-SiC HEMTs. IEEE Trans. Power Electron. **29**, 2462–2473 (2014)
44. M. Vasic, O. Garcia, J.A. Oliver, P. Alou, D. Diaz, J.A. Cobos, Multilevel power supply for high-efficiency rf amplifiers. IEEE Trans. Power Electron. **25**, 1078–1089 (2010)
45. V. Yousefzadeh, E. Alarcon, D. Maksimovic, Three-level buck converter for envelope tracking applications. IEEE Trans. Power Electron. **21**, 549–552 (2006)

46. P.A. Godoy, S. Chung, T.W. Barton, D.J. Perreault, J.L. Dawson, A 2.4-GHz, 27-dBm asymmetric multilevel outphasing power amplifier in 65-nm CMOS. IEEE J. Solid-State Circuits **47**, 2372–2384 (2012)
47. P.A. Godoy, S. Chung, T.W. Barton, D.J. Perreault, J.L. Dawson, A highly efficient 1.95-GHz, 18-W asymmetric multilevel outphasing transmitter for wideband applications, in *2011 IEEE MTT-S International Microwave Symposium*, June 2011, pp. 1–1
48. S. Chung, P.A. Godoy, T.W. Barton, D.J. Perreault, J.L. Dawson, Asymmetric multilevel outphasing transmitter using Class-E PAs with discrete pulse width modulation, in *2010 IEEE MTT-S International Microwave Symposium*, May 2010, pp. 264–267
49. H.M. Nemati, C. Fager, U. Gustavsson, R. Jos, H. Zirath, Design of varactor-based tunable matching networks for dynamic load modulation of high power amplifiers. IEEE Trans. Microwave Theory Tech. **57**, 1110–1118 (2009)
50. K. Eisele, R. Engelbrecht, K. Kurokawa, Balanced transistor amplifiers for precise wideband microwave applications, in *1965 IEEE International Solid-State Circuits Conference on Digest of Technical Papers*, vol. VIII, Feb 1965, pp. 18–19
51. A.S. Jurkov, A. Radomski, D.J. Perreault, Tunable impedance matching networks based on phase-switched impedance modulation, in *2017 IEEE Energy Conversion Congress and Exposition (ECCE)*, Oct 2017, pp. 947–954
52. A.A. Bastami, A. Jurkov, P. Gould, M. Hsing, M. Schmidt, J.I. Ha, D.J. Perreault, Dynamic matching system for radio-frequency plasma generation. IEEE Trans. Power Electron. **33**, 1940–1951 (2018)
53. Y. Han, O. Leitermann, D.A. Jackson, J.M. Rivas, D.J. Perreault, Resistance compression networks for radio-frequency power conversion. IEEE Trans. Power Electron. **22**, 41–53 (2007)
54. T.W. Barton, J.M. Gordonson, D.J. Perreault, Transmission line resistance compression networks and applications to wireless power transfer. IEEE J. Emerging Sel. Top. Power Electron. **3**, 252–260 (2015)
55. C. Sullivan, R. Zhang, Analytical model for effects of twisting on Litz-Wire losses, in *IEEE Workshop on Control and Modeling for Power Electronics (COMPEL)*, June 2014
56. B.A. Reese, C.R. Sullivan, Litz wire in the MHz range: Modeling and improved designs, in *IEEE 18th Workshop on Control and Modeling for Power Electronics (COMPEL)*, 2017, pp. 1–8
57. S. Hinaga, M. Koledintseva, P. Anmula, J. Drewniak, Effect of conductor surface roughness upon measured loss and extracted values of PCB laminate material dissipation factor, in *Proceedings of the Technical Conference on IPC Expo/APEX*, 2009, pp. S20–S22
58. C.R. Sullivan, D. Harburg, J. Qiu, C.G. Levey, D. Yao, Integrating magnetics for on-chip power: a perspective. IEEE Trans. Power Electron. **28**(9), 4342–4353 (2013)
59. W. Liang, L. Raymond, J. Rivas, 3-D-printed air-core inductors for high-frequency power converters. IEEE Trans. Power Electron. **31**(1), 52–64 (2016)
60. C.R. Sullivan, W. Li, S. Prabhakaran, S. Lu, Design and fabrication of low-loss toroidal air-core inductors, in *IEEE Power Electronics Specialists Conference (PESC)* (IEEE, 2007), pp. 1754–1759
61. G. Zulauf, W. Liang, J. Rivas-Davila, A unified model for high-power, air-core toroidal PCB inductors, in *IEEE 18th Workshop on Control and Modeling for Power Electronics (COMPEL)* (IEEE, 2017), pp. 1–8
62. S. Orlandi, B.A. Allongue, G. Blanchot, S. Buso, F. Faccio, C.A. Fuentes, M. Kayal, S. Michelis, G. Spiazzi, Optimization of shielded PCB air-core toroids for high-efficiency dc-dc converters. IEEE Trans. Power Electron. **26**(7), 1837–1846 (2011)
63. S. Ramo, J. Whinnery, T.V. Duzer, *Fields and Waves in Communication Electronics*, 3rd edn. (Wiley, New York, 1994)
64. J. Qiu, D.V. Harburg, C.R. Sullivan, A toroidal power inductor using radial-anisotropy thin-film magnetic material based on a hybrid fabrication process, in *IEEE Applied Power Electronics Conference and Exposition*, Mar 2013

65. J. Qiu, A.J. Hanson, C.R. Sullivan, Design of toroidal inductors with multiple parallel foil windings, in *IEEE Workshop on Control and Modeling for Power Electronics (COMPEL)* (IEEE, 2013), pp. 1–6
66. T. Simpson, Effect of a conducting shield on the inductance of an aircore solenoid. IEEE Trans. Magn. **35**, 508—515 (1999)
67. R. Medhurst, HF resistance and self-capacitance of single-layer solenoids. Wirel. Eng. Part 1, **24**(2), pp. 35–43, Part 2 **24**(3), pp. 80–92, Feb/mar (1947)
68. R. Medhurst, Q of solenoid coils. Wirel. Eng. **24**(9), p. 281 Sept. (1947)
69. M. Callendar, Q of solenoid coils. Wirel. Eng. (Correspondence) **24**(6), p. 185 (1946)
70. T. Lee, *Planar Microwave Engineering*, chap. 6 (Cambridge University Press, Cambridge 2004)
71. S.L. C.S. A.J. Hanson, J.A. Belk and D. Perreault, Measurements and performance factor comparisons of magnetic materials at high frequency. IEEE Trans. Power Electron. **31**(11), 7909–7925 (2016)
72. Y. Han, G. Cheung, A. Li, C.R. Sullivan, D.J. Perreault, Evaluation of magnetic materials for very high frequency applications. IEEE Trans. Power Electron. **27**(1), 425–435 (2008)
73. Soft Ferrites and Accessories Data Handbook, Ferroxcube International Holding B.V. Poland 2013
74. D.J. Perreault, J. Hu, J.M. Rivas, Y. Han, O. Leitermann, R.C. Pilawa-Podgurski, A. Sagneri, C.R. Sullivan, Opportunities and challenges in very high frequency power conversion, in *Twenty-Fourth Annual Applied Power Electronics Conference and Exposition (APEC)* (IEEE, 2009), pp. 1–14
75. C.P. Steinmetz, On the law of hysteresis (reprint of article from 1892). Proc. IEEE **72**, 197–221 (1984)
76. J. Hu, C.R. Sullivan, AC resistance of planar power inductors and the quasi-distributed gap technique. IEEE Trans. Power Electron. **16**(4), 558–567 (2001)
77. R. Yang, A. Hanson, D. Perreault, C. Sullivan, A low-loss inductor structure and design guidelines for high-frequency applications, in *IEEE Applied Power Electronics Conference*, Feb 2018
78. C.R. Sullivan, Layered foil as an alternative to Litz Wire: multiple methods for equal current sharing among layers, in *IEEE 19th Workshop on Control and Modeling for Power Electronics*, 2014
79. A.L.F. Stein, P.A. Kyaw, C.R. Sullivan, High-Q self-resonant structure for wireless power transfer, in *IEEE Applied Power Electronics Conference and Exposition (APEC)*, Mar 2017, pp. 3723–3729
80. P.A. Kyaw, A.L.F. Stein, C.R. Sullivan, High-Q resonator with integrated capacitance for resonant power conversion, in *IEEE Applied Power Electronics Conference and Exposition (APEC)*, Mar 2017, pp. 2519–2526
81. P.A. Kyaw, A.L.F. Stein, C.R. Sullivan, Fundamental examination of multiple potential passive component technologies for future power electronics. IEEE Trans. Power Electron. (2017). http://dx.doi.org/10.1109/TPEL.2017.2776609
82. C. Trask, Transmission line transformers: theory design and applications. Part 1, in *High Frequency Electronics*, 2005, pp. 46–53
83. C. Trask, Transmission line transformers: Theory design and applications. Part 2, in *High Frequency Electronics*, 2006, pp. 26–32
84. R. Mack, J. Sevick, *Sevick's Transmission Line Transformers: Theory and Practice*, 5th edn. *Scitech Publishing imprint of the IET* (IET, 2014)
85. Y. Han, D.J. Perreault, Analysis and design of high efficiency matching networks. IEEE Trans. Power Electron. **21**(5), 1484–1491 (2006)
86. P.A. Kyaw, A.L. Stein, C.R. Sullivan, Analysis of high efficiency multistage matching networks with volume constraint, in *IEEE 18th Workshop on Control and Modeling for Power Electronics (COMPEL)* (IEEE, 2017), pp. 1–8
87. A. Kumar, S. Sinha, A. Sepahvand, K.K. Afridi, Improved design optimization approach for high efficiency matching networks, in *2016 IEEE Energy Conversion Congress and Exposition (ECCE)* (IEEE, 2016), pp. 1–7

Index

G. Meneghesso et al. (eds.), *Gallium Nitride-enabled High Frequency and High Efficiency Power Conversion*, Integrated Circuits and Systems,
https://doi.org/10.1007/978-3-319-77994-2

MIX
Papier aus verantwortungsvollen Quellen
Paper from responsible sources
FSC® C105338

If you have any concerns about our products,
you can contact us on
ProductSafety@springernature.com

In case Publisher is established outside the EU,
the EU authorized representative is:
Springer Nature Customer Service Center GmbH
Europaplatz 3, 69115 Heidelberg, Germany

Printed by Libri Plureos GmbH
in Hamburg, Germany